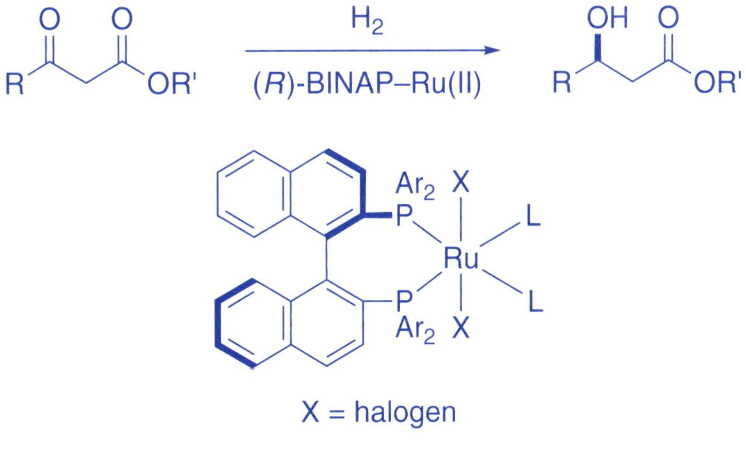

X = halogen

LIVES
IN
CHEMISTRY

LIVES IN CHEMISTRY — LEBENSWERKE IN DER CHEMIE

Major goals in chemical research often take a long time to achieve. In chemistry, the research process — from the hypothesis to experiments and interpretation — often consists of many rapid steps that proceed more quickly than those in neighboring disciplines. The works in the present series show how such a process can translate into major scientific achievements.

The series is steered by an Advisory Board that has been appointed by the Executive Committee of the History of Chemistry Division of the German Chemical Society (Gesellschaft Deutscher Chemiker, GDCh). The project is sponsored by Dr. Karl Reuter (RCA, Freiburg, Germany). In the first volume of "Lives in Chemistry", Günther Maier describes how the series was born. Like chemistry, the series is international, and chemistry is broadly defined, thus incorporating many neighboring disciplines.

In science, life and work are often closely interwoven. The series consists of important, contemporary documents that are stimulating, instructive, and entertaining — and as diverse as people can be.

Thanks go to all those who contributed to the realization of this idea.

Berlin, Spring 2025
Carsten Reinhardt,
Chairman Advisory Board

PUBLISHED TITLES IN THIS SERIES

GÜNTHER MAIER
DAS WAR'S – ERINNERUNGEN EINES DOKTORVATERS
(DONE—MEMORIES OF A PHD SUPERVISOR, 2021)

GERHARD ERTL
MEIN LEBEN MIT DER WISSENSCHAFT (2021)

HENRI BRUNNER
BILD UND SPIEGELBILD: KLEINER UNTERSCHIED – GROSSE AUSWIRKUNGEN
(LOOKING GLASS CHEMISTRY: SMALL DIFFERENCES—GIGANTIC EFFECTS, 2021)

DIETER OESTERHELT AND MATHIAS GROTE
LEBEN MIT LICHT UND FARBE – EIN BIOCHEMISCHES GESPRÄCH
(LIFE WITH LIGHT AND COLOR—A BIOCHEMICAL CONVERSATION, 2022)

STEPHEN B. H. KENT
INVENTING SYNTHETIC METHODS TO DISCOVER HOW ENZYMES WORK (2022)

FRANZ EFFENBERGER
VON AROMATEN UND HETEROCYCLEN ZUR BIO- UND NANOTECHNOLOGIE
(FROM ARENES AND HETEROCYCLES TO BIO- AND NANOTECHNOLOGY, 2023)

GERHARD ERTL
MY LIFE WITH SCIENCE (2023)

HORST KESSLER
NMR: MEIN KOMPASS IN DER ORGANISCHEN UND MEDIZINISCHEN CHEMIE
(NMR WAS MY COMPASS IN ORGANIC AND MEDICINAL CHEMISTRY, 2023)

LARRY E. OVERMAN
DESIGNING SYNTHETIC METHODS AND NATURAL PRODUCTS SYNTHESIS (2024)

HUBERT SCHMIDBAUR
FROM CHEMICAL CRAFTSMANSHIP TO THE ART OF GILDING ATOMS (2024)

KATHARINA KOHSE-HÖINGHAUS
BURNING FOR SCIENCE—A WOMAN IN A TECHNICAL FIELD (2025)

SIGRID PEYERIMHOFF
AB INITIO – EIN LEBEN FÜR DIE QUANTENCHEMIE
(AB INITIO—A LIFE IN QUANTUM CHEMISTRY, 2025)

ALBERT ESCHENMOSER
PRIZES ARE GOOD BUT DISCOVERIES ARE BETTER (2025)

RYOJI NOYORI
RESEARCH SHOULD BE FRESH, SIMPLE, AND CLEAR (2025)

THE INTERNATIONAL LIVES IN CHEMISTRY SERIES
PAYS TRIBUTE TO OUTSTANDING SCIENTISTS,
DESCRIBES THEIR ROUTE TO SUCCESS,
IS AN INSPIRATION FOR THE FUTURE.

THE LIVES IN CHEMISTRY BOOK SERIES HAS BEEN RECOGNIZED
AS ONE OF "THE BEST GERMAN BOOK DESIGNS" IN DESIGN,
CONCEPTION, AND WORKMANSHIP.

L-I-C.ORG

RYOJI

NOYORI

RESEARCH

SHOULD BE

FRESH,

SIMPLE,

AND CLEAR

BIBLIOGRAPHIC INFORMATION OF THE GERMAN NATIONAL LIBRARY

THE GERMAN NATIONAL LIBRARY LISTS THIS PUBLICATION IN THE DEUTSCHE NATIONALBIBLIOGRAFIE; DETAILED BIBLIOGRAPHIC DATA ARE AVAILABLE IN THE INTERNET AT HTTP://DNB.DNB.DE.

LIVES IN CHEMISTRY – LEBENSWERKE IN DER CHEMIE

EDITED BY THE ADVISORY BOARD OF THE HISTORY OF CHEMISTRY DIVISION OF THE GERMAN CHEMICAL SOCIETY (GESELLSCHAFT DEUTSCHER CHEMIKER, GDCh)
ISSN 2747-8696
HTTPS://DOI.ORG/10.47261/LIC

L-I-C.ORG

SUPPORTING MATERIAL
FOR ANY AVAILABLE SUPPORTING MATERIAL, PLEASE VISIT L-I-C.ORG/1135

COPY EDITOR
ELIZABETH HUGHES

GRAPHIC DESIGN AND TYPESETTING
ANDREAS TÖPFER, BERLIN (WWW.ANDREASTOEPFER.DE)

PRINT AND BOOOKBINDING PROCESSING
LIVONIA PRINT, SIA, JŪRKALNES IELA 15/25, 1046 RIGA (WWW.LIVONIAPRINT.LV)

FIRST REPRINT 2025

ISBN 978-3-86225-135-3 (HARDCOVER)
ISBN 978-3-86225-564-1 (E-BOOK/PDF)
ISBN 978-3-86225-579-5 (E-BOOK/EPUB)
HTTPS://DOI.ORG/10.47261/1564

GRAPHIC DESIGN—LOOK & FEEL

LIKE CHEMICAL RESEARCH, THE DESIGN OF THE LIVES IN CHEMISTRY SERIES (LIC) IS INNOVATIVE AND DRIVEN BY EXPERIMENTATION. READING A LIC BOOK IS A SENSORY EXPERIENCE: MAKE ANNOTATIONS, ROTATE THE BOOK TO REVEAL LANDSCAPE FORMATS, FEEL THE DIFFERENT TEXTURES, ENJOY THE SCREEN-PRINTED COVER, AND MAKE USE OF THE VERY SPECIAL SLIPCASE. INSPIRED BY THE CONCEPT OF A LABORATORY NOTEBOOK, IT INVITES TO ENGAGE ACTIVELY—MUCH LIKE YOU WOULD IN THE RESEARCH LAB.

LIC'S UNIQUE LOOK & FEEL HAVE BEEN DEVELOPED BY ANDREAS TÖPFER, AN AWARD-WINNING GRAPHIC DESIGNER, ILLUSTRATOR, AND ARTIST.

COVER PICTURE

THE BINAP-RUTHENIUM-CATALYZED ASYMMETRIC HYDROGENATION OF β-KETO ESTERS TO β-HYDROXY ESTERS IS HIGHLY ENANTIOSELECTIVE (>99% *EE*), EXTREMELY EFFICIENT, AND CAN BE RUN EVEN AT > 1,000 KG SCALE—A HALLMARK REACTION OF THE NOYORI CHEMISTRY.

SPECIAL COLOR

FOR EACH BOOK, THE PUBLISHER, GRAPHIC DESIGNER, AND AUTHOR CHOOSE A SPECIAL COLOR AND THUS SET FURTHER PERSONAL ACCENTS, PANTONE 286U HERE. THIS MAKES LOOKING AT AND READING THE BOOK AN EXPERIENCE, FROM THE SLIPCASE, THE COVER, THE TEXTS AND IMAGES, TO THE APPENDICES.

PAPER

MUNKEN PRINT WHITE 15 PAPER REPRESENTS SUSTAINABLE FORESTRY MANAGEMENT AND COMBINES THE BEST PROPERTIES OF WOOD-FREE AND WOOD-CONTAINING PAPER (EU ECOLABEL, FSC-C020637).

TYPOGRAPHY

THE AKZIDENZ GROTESK NEXT FONT IS USED IN THIS BOOK. THIS FONT WAS CREATED IN 1880 BY THE ROYAL PRUSSIAN PUNCH CUTTER FERDINAND THEINHARDT AND WAS ORIGINALLY CALLED "ROYAL GROTESK". IT WAS USED IN CONTEMPORARY PUBLICATIONS OF THE PREUSSISCHE AKADEMIE DER WISSENSCHAFTEN (PRUSSIAN ACADEMY OF SCIENCES). SANS SERIF FONTS WERE OFTEN EMPLOYED IN SCIENTIFIC PUBLICATIONS BEFORE THE BEGINNING OF THE TWENTIETH CENTURY. IN CONTRAST TO THIS, THE SERIF FONT CENTURY SCHOOLBOOK IS USED FOR GENERAL PARTS OF THE TEXT.

Table of Contents

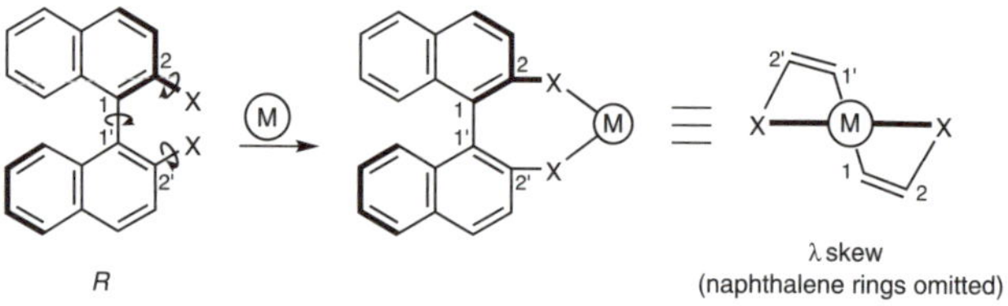

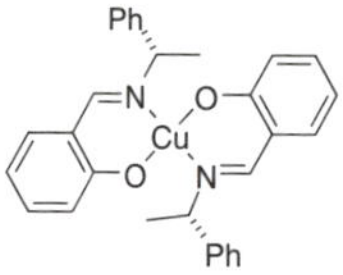

chiral organometallic molecular catalyst

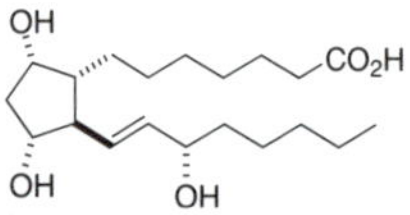

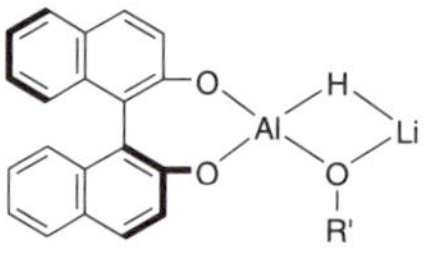

(S)-BINAL–H

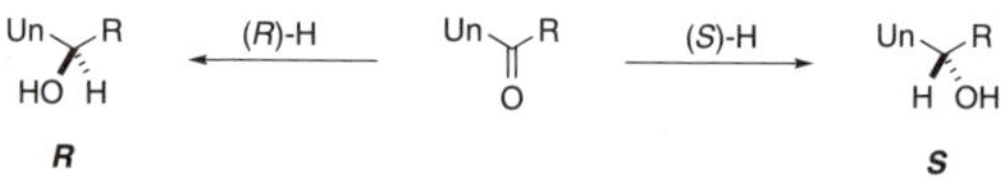

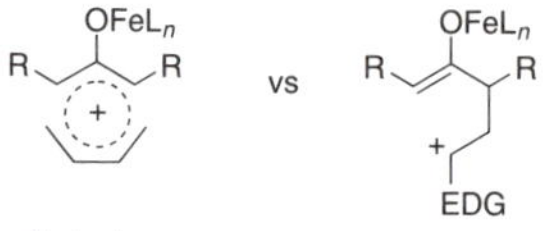

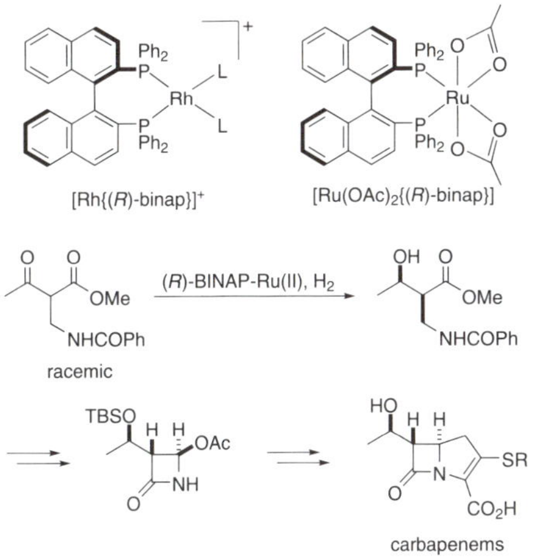

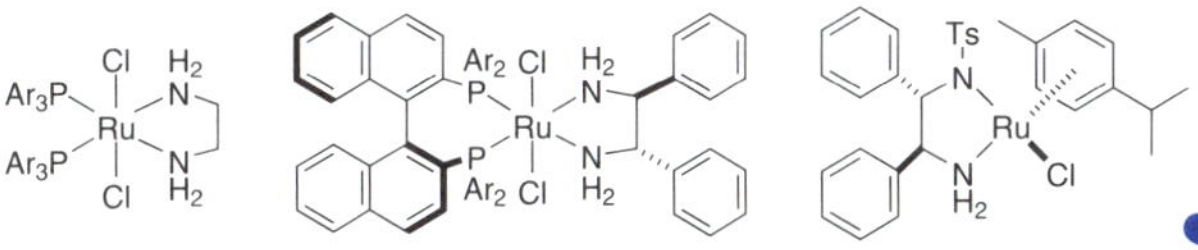

"NH" effect　AH catalyst　ATH catalyst

ESSENTIALS

“

“We must venture to consistently create new knowledge to open new worlds of the unknown.”

“[...] it was and still remains my belief that, to a large extent, the discovery of efficient catalytic reactions relies on serendipity and originates from sound, comprehensive scientific knowledge.”

“I believe that Japanese culture permeates every aspect of our research in science and technology. I fully agree with the statement of Louis Pasteur, ‘Science has no borders, but scientists have their fatherland.’ There are ethnic and regional characteristics in approaches to research.”

“I have been contemplating the origin of unique creativity in scientific research, and I now strongly believe that ethnic culture is important in nurturing scientific originality.”

"Science is objective, but scientific research is a human activity, and a scientific endeavor is a work of art."

"Both the scientist and artist seek through their work to elevate themselves, and both enrich our lives spiritually and materially. Science and art originate from the same source, and the only difference is in how they express themselves."

"Conduct research in a fresh and straightforward manner, has been a motto of mine, and I prefer simplicity over complexity."

"There are limits to any person's ability to think. It is the multiplication of different ways of thinking and the accumulation of skills that create new power. I believe that a 'good mix' of people with different abilities and roles is the best organization for academic research."

“While careful planning is important in research, one should not overlook the good fortune that comes our way by chance. I have always believed that unexpected discoveries bring about scientific advancement. For this reason, the significance of students is immeasurable. They have a fresh way of thinking and young energy that professional researchers, indoctrinated with a variety of information and knowledge, do not possess. What is most gratifying is that they sometimes follow absurd ideas and make unimaginable ‘mistakes.’”

“I am not a fan of compromise and prefer to argue for what is right.”

“Basically, science pursues the truth. Scientists must always be aware of the Socratic ‘knowledge of ignorance.’ They must have the ability to recognize that we know nothing. However, the accumulated scientific discoveries of our forebears have transformed all things of heaven and earth from myth to reality.”

"Research is about consideration and action, and is conducted by people. I am somewhat uncomfortable with the modern research trend of 'having a computer think and a robot work to come up with the answers,' simply to search for results that can be presented. This may be effective, but it does not bring the real pleasure of research. In academia, which is driven by intrinsic motivation, I would like to see distinctive ideas rather than efficiency. I have a habit of trying to convince myself with my own logical system and different perspectives. I am pleased when I find consistency, but what is more important for me are the contradictions I encounter and to be troubled by them. I believe that individual subjectivity is essential, and especially, the power of intuition based on experience is decisive."

"I was very fortunate to have collaborators with a rich sense of serendipity who transformed my rather foolish idea (pre-catalyst) into a new concept or a powerful principle or method (true catalyst) through their keen experimental observation skills."

PROLOGUE

PROLOGUE

Scientific research is a never-ending journey of knowledge. There is more meaning in experiencing various encounters and making a good journey itself rather than reaching the destination. Excellent research nurtures talented youths and contributes to the society. The passage of my life has been decided by innumerable coincidences and a very few inevitabilities. I followed the guideposts on the road as I went, but as I encountered the unexpected, I stopped and ended up wandering. This was all so fascinating. Looking back on the path I have taken, I cannot be sure I made the right decisions at the crossroads I came across. It was not easy to accomplish what I intended to do, but sometimes, things eventually went the way I had wished. My life has not been straightforward.

My childhood started in Kobe, Japan, and I moved to Kyoto, Nagoya, and then to Tokyo. As my environment changed, I took on each new challenge with sincerity, one after another, and before I knew it, I had arrived at where I am today. Originally, my intention was not to devote myself to academia. Because of my family's background, I was raised to be an industrial engineer, and I also received education to become one. However, the opportunities I had in graduate school and the guidance of my mentor led me to a long career as a university academic. As a consequence, I have gained many acquaintances not only in Japan but also in other countries as well as in industry and have lived a happy life as a researcher for half a century.

Science teaches us the principles of the natural world. It provides us with the most powerful way of thinking to understand all things in the universe objectively and to deal with society in a rational manner. I am not sure how well I have fulfilled my role as an educator, as I did not have the broad knowledge that an academic or scholar is expected to have. However, having lived my life with a strong passion for chemistry, I believe that as a scientific researcher, I was able to play a certain role in the progress of science by challenging the unknown and the impossible.

Furthermore, science-based technology is the only way for Japan, a country with scarce natural resources, to survive the international competition of this century, and it is also a pillar of international cooperation to solve various problems for the survival of humanity. From this perspective, I feel that I was able to make a small contribution to the development of chemical technology. Although I am oriented toward self-realization based on my own ideas and intentions, I have avoided living a self-centered life. Therefore, at the request of the government, I have also participated in the administration of science, technology, and education, and have been involved in the management and administration of a national research institute. During this time, I have shared my values and beliefs in many places, and I have argued that our planet, with its 8 billion people, faces many serious sustainability issues, including climate change, resource and energy depletion, biodiversity loss, and the spread of infectious diseases. Together with the younger generation, I would like to devote the remainder of my life with sincerity to the creation of a new society.

In this autobiography, I am given the opportunity to reflect on my young days and my

60 years of experience as a researcher, educator, and administrator. As a chemist, I have published research papers in journals and presented my scientific findings at international conferences. However, it never occurred to me to write about my personal life and journey. Many things have remained closed until now.

The most important thing to remember is that human life is unique to each person and no two journeys are the same. Looking back, I should say that I have been fortunate enough to have lived a fulfilling life, both publicly and privately. Some coincidences certainly worked positively, but I may have been unaware of many negative coincidences. The hardships of my impoverished youth have transformed to good memories in retrospect. Time moves in one direction and never goes backwards. My family environment has also changed dramatically. In this book, there are events that my long-time wife, Hiroko, and my sons, Eiji and Koji, do not know about. Under the tutelage of predecessors from around the world and in many other fields, together with the help of seniors and friends, I was able to navigate ups and downs, and overcome difficulties. I must once again express my gratitude to them, many of whom have already passed away, and whom I truly miss nowadays.

I would be grateful if readers who shared the same era with me could sympathize with the fact that we have seen similar landscapes. Those who will live in the future must now decide what they should be and where they should stand. For the younger generations, who are full of sensitivity and have different cultural backgrounds, I would be more than happy if the footsteps of a Japanese chemist could serve as a small guide.

WHERE DID I COME FROM?

1

 DOI.ORG/10.47261/1564-1

1.1 Memories of my childhood

On a night of heavy wind and rain, I was born on September 3, 1938, in a suburb of Kobe (now known as Ashiya) as the first son of Kaneki and Suzuko Noyori. Our family moved to Kobe soon after I was born. I grew up with two younger brothers, Tatsuhiko and Masaharu, and a younger sister, Yoshiko, in the pleasant district of Rokko, blessed with beautiful natural surroundings. The port city, which prospered through trade and shipping, had a rich international flavor; there were Indians, Italians, Russians, and people of many other backgrounds. After World War II, a residence for the families of the United States Occupation Forces called Rokko Heights was established, adding to the diversity of the city.

My pre-war childhood was spent playing with wooden horses and paper balloons, going to the zoo, and visiting my grandparents. I went to a day care once a week, and it was while drawing pictures, singing songs, and playing in the schoolyard that I learned the rules of living in a group. Sporadic memories of the war still occupy my mind. I recall the brass window guards being taken and used for the war effort. I also remember wearing an air-raid hood and carrying a sling bag containing dry bread and minimal daily necessities, as we evacuated to the air-raid shelter of a neighboring community. It was hard for a young child to understand why this world had been shaped in this way.

In April of 1945, I was to enter an elementary school affiliated to Kobe University, however, just half a year before I was meant to start, US Air Force B29s heavily bombed Kobe, reducing the central city to ashes. My mother, two younger brothers, and I evacuated to the nearby countryside, 80 km west of Kobe until the end of the year. My father, who had to work, remained in Kobe and was able to visit us only on the weekends.

Later in 1945, I entered an elementary school in the countryside. Although my grades were not bad, I quickly saw that the children from the *satoyama* (countryside) had skills that I did not. Not only did I lack the muscle strength to run around the rice paddies and wheat fields, but I also lacked the wisdom to live in nature. I was intimidated by the huge bodies of cows and horses and was also afraid of snakes. I was not good at fishing and could not even climb persimmon (*kaki*) trees to pick the fruit. At Kobe, children caught butterflies, dragonflies, and cicadas, but here in the countryside, they made sparrow traps, stilts, and wove sandals out of rice straw.

The countryside was a safe and natural environment, but a completely different world than what my family was used to. Light (from kerosene), heat (from wood and coal), and water (from ponds and wells) all had to be obtained on our own. People lived together in villages of a few dozen households. Instead of going to the market, we grew vegetables in the fields and went into the mountains to pick bracken, royal fern, and other wild vegetables, as well as to collect chestnuts. Sugar was

Fig. 1.1.
Ryoji Noyori, 200 days old.

scarce, and sweets were only eaten on special occasions, while wild silverberries (*gumi*), wild strawberries, jujube fruits, and taro buds were eaten as snacks. Potatoes were stored under the floor for the winter. The adults would go to the river and catch large quantities of sweetfish (*ayu*), which they would then dry. For a child who grew up eating freshly caught fish from Akashi (a place west of Kobe famous for good fish), this preserved fish was not very tasty. We children learned how to survive from watching the adults, and living in the countryside had a great influence on me.

The end of the war in Japan was announced on August 15, 1945. My mother made me sit in the living room and listen to the Emperor's radio broadcast, but I hardly understood what it meant. I don't think my parents were ever in favor of the war. Even after the war ended, our country was terribly short of food and supplies. We returned to Kobe, our house fortunately had not been burned down, but our family of six still had to live frugally for years. The misery and poverty during and following the war can never be forgotten. From these experiences, I felt strongly from an early age that war should be avoided under any circumstances.

1.2 A mischievous city kid

In the New Year, I was finally able to start at Mikage Elementary School (now known as Kobe University Elementary School), which I was originally supposed to attend.

The school was located on a hill with a panoramic view of the Hanshin area and of the harbor. It was a two-story wooden building with a large playground and surrounded by green trees and streams. There were plenty of examples for natural science and research materials, and no shortage of places to play in the mud. We delighted in playing in the woods next to the school and the sewage trench that we called Devil's Cave, all the more so because the adults forbade it. The school and our parents were certainly aware of our misbehavior, but they often overlooked it. They may have thought it necessary to develop the ability to recognize potential dangers with some degree of self-responsibility.

The miserable aftereffects of the war were felt at school too. The US military sprayed us children with large amounts of DDT (dichlorodiphenyltrichloroethane). They told us it was for hygiene purposes, but they simply assumed we were infested with fleas and lice. At school the quality of stationery was poor, and we used brittle pencils, sandy erasers, and writing pads that looked like yellowed dust paper. Eventually, schools began to provide lunch, but the menu of cobbled bread and skimmed milk powder was not well received by any of us. The children wore clothes made by their mothers on manual sewing machines or with needle and thread. Rubber-made shoes were not easy to come by, so I wore straw sandals to school for a few months until the teacher spoke to my mother about it.

Having just returned from the evacuation site, I was not used to how the other pupils behaved in mathematics class. Some of my classmates proactively asked questions to the teacher, which was not usual in

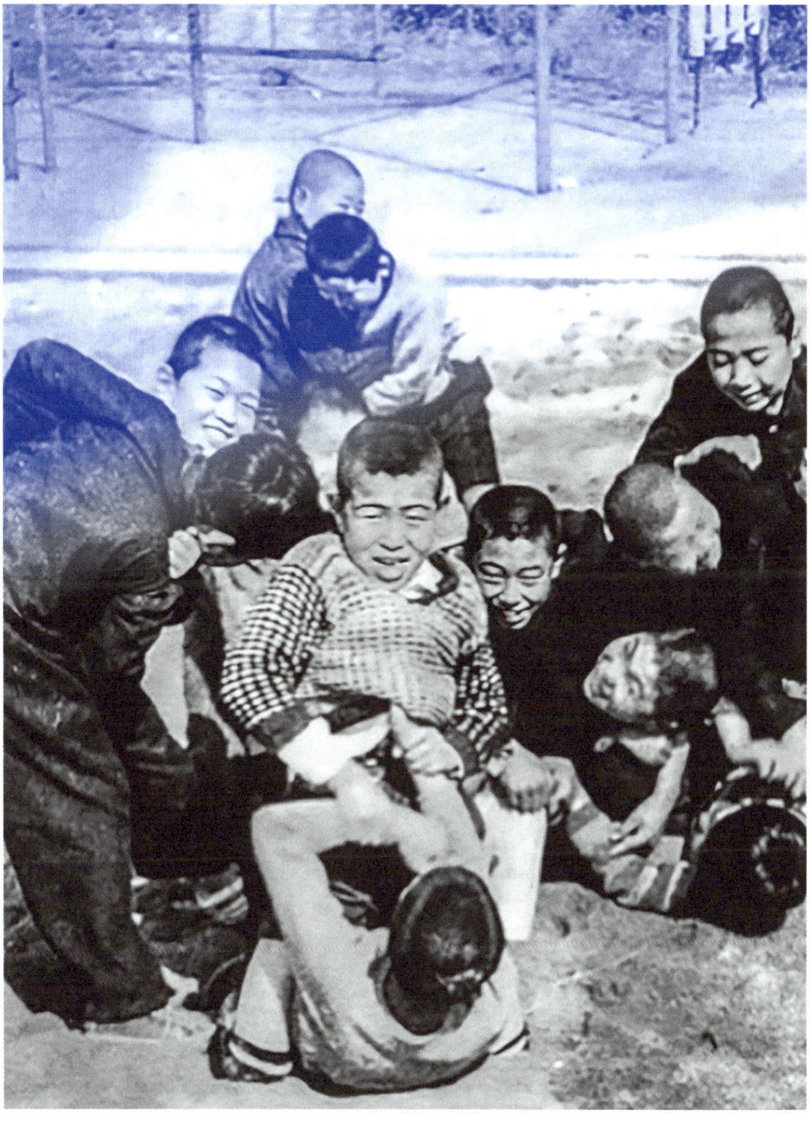

Fig. 1.2. On the playground in fifth grade (author in the center).

the countryside, and for a while I was rather reserved and had a feeling of inferiority. Nevertheless, after a year or two, I gained enough confidence to make mathematics a subject that I became good at. Looking back, it appears to me that the school provided guidance that allowed each student to affirm his or her own existence and gain confidence. I am grateful to my parents for allowing me to go to this "model school."

When people asked my mother about me at that time, she did not hesitate to answer that I was a mischievous child. Both of my parents were passionate about education, and they must have felt that education was the only thing they could leave behind for their children in a society that had lost everything by the brutal defeat in the war. However, they valued friendships and love for learning that come from school over grades and homework. My mother tried to make me learn to play the piano and to speak English at church, but I did not fit in at all, and the only thing that lasted for a while was painting lessons at Sunday school. I was more interested in playing outdoors, and neither of my parents discouraged me from doing so.

Until I was in the sixth grade (1950), I was not given any pocket money, and my mother decided which books I should read. I felt somewhat inconvenienced; perhaps this was the educational policy of many middle-class families at the time. I was forced to read many masterpieces written for children, such as *Treasure Island, Heart*, and *Little Lord Fauntleroy*, as well as historical and scientific texts, including Jean-Henri Fabre's *Insect Life*. I repeatedly read Plutarch's *Parallel Lives*, a biography that vividly depicts the tumultuous lives of ancient Greek and Roman heroes, and I came to admire Alexander the Great and Caesar for their grand visions along with their brave and courageous ways of life.

1.3 Helping in the family

Together with my two younger brothers, I often helped with the household chores. We grew vegetables and flowers, tended the chickens, cooked, heated the bath, burned garbage, cleaned, and did carpentry work. Plastic was not available yet, so we made good use of wood and crafted with planers, saws, chisels, hammers, nails, and screws. This required a working knowledge of mechanics and geometry. We were impressed by the skills of the carpenters, who visited our home from time to time, and tried to learn their techniques, but their tools were different, and it was not easy to imitate them. Looking back, these experiences were a treasure trove of tacit knowledge—real science practice—which could not be found in the subjects of modern private tuition schools.

Japan's post-war reconstruction had just begun, and all kinds of goods were in short supply. In the downtown district of Sannomiya, the black market under the elevated railway tracks was swarming with people looking for daily necessities and PX (post exchange) supplies released by the US military. The lack of food was especially difficult to bear. I remember my mother visiting a farmer in the countryside and exchanging food for her kimono that she had as her trousseau. Fortunately, fish was fresh, and on holidays, I had looked forward to watching the fishmongers with their carts at the kitchen door. For some reason, beef was always readily available in Kobe and was often served on the dinner table. I was often sent on errands to a nearby store, where alcohol, tempura oil, and soy sauce were sold by weight, and I never bought a full bottle.

1.4 My parents

We all must open our own path in life. However, along with meeting respectable teachers and friends, I believe that family environment has the greatest influence on nurturing a child's aspirations. I would like to mention a little bit about my late parents' upbringing and their way of life.

My father, Kaneki Noyori (1911–1982), was born in Nagoya, the third son of Sanshiro and Chinatsu Shiokawa. My grandfather Sanshiro grew up in Tokyo and was Branch Manager of the Bank of Japan in Nagoya, and later involved in the creation of several major regional banks.

Figs. 1.3.–1.4. Author's parents, Kaneki and Suzuko Noyori.

Noyori is the family name on my mother's side and can be traced back to a region in the eastern part of Fukuoka Prefecture (in the Kyushu area located in southern Japan). The family was a great landowner who made its living as sake brewer. I remember visiting their large residence when I was a child, but all their land was lost during their absence in the postwar agricultural land reform, and I have never returned to the original domicile of Noyori.

My mother, Suzuko (1914–2005), was the eldest daughter of Jiro and Ishi Noyori and was born in Nagoya, where her father worked for Ministry of Railways. Jiro's elder brother, Tatsuji Noyori, worked at the general trading company Mitsui & Co. and was the first president of Mitsui Insurance Company (now known as Taiju Life Insurance Company). Because he and his wife, Nobu, had no heirs, they adopted my father, Kaneki, to inherit the Noyori family name and marry Suzuko.

Growing up in a family of bankers, my young father was encouraged to become an engineer by Kouki Kudoh, the husband of my mother's paternal aunt Kinu. Kudoh was a hard-working engineer who had been the right-hand man of Shitagau Noguchi, the founder of Japan's chemical industry and founder of the Nicchitsu Konzern Group. Noguchi's love for chemical technology was so strong that he named his three sons Ryuji, Bosshi, and Hisha after the chemists that he most admired, Luigi Casale, Carl Bosch, and Franz Fischer, respectively.

Educated at the University of Tokyo, my father joined Kanebo, a large spinning enterprise, and then moved to Kaneka Corporation, which was established by combining its non-textile and chemical business sectors. As a very capable engineer with a doctorate degree, he long directed the research and development division, resulting in the development of numerous useful technologies, and later served on the Board of Directors. At that time, the chemical industry in Japan was dominated by imported industrial technologies, but he believed that the development of the nation's own industrial technology was the key to Japan's economic recovery.

My father who did not like to socialize at work always came home at 5:30 p.m. every day and we all had dinner together. On days when *sukiyaki* was served, the house was exceptionally lively. My father would talk about industrial technology, and we children would report on the daily events at school. Although my mother was busy preparing meals, she was a good listener, and she had a great deal of knowledge about science and industry, which was unusual for a housewife in those days. After dinner, the children would study independently, but at 8:30 p.m., we would all gather to have tea together with cakes and cookies.

My mother was raised mainly in Tokyo, although her father's job also took her family to Nagoya, Kobe, and Sendai. She studied English literature at the Women's College, the predecessor of the University of the Sacred Heart. She had a strong desire for knowledge in all fields and learned various arts such as *waka* poetry, calligraphy, and Japanese painting. She also believed that it was important to serve society, and for more than fifty years she worked on a manual typewriter to translate books in many subjects into Braille. For her work, she received a letter of appreciation from the Pope, which she took pride in for the rest of her life. In her later years, she participated in public lectures at Kobe University and other institutions as the eldest auditing student and enjoyed socializing at cultural centers. However, she was still a woman born in the early Taisho era. She placed the highest priority on being her husband's companion as well as the education of her children, and throughout her life she avoided all luxury and respected simplicity.

1.5 My admiration for Professor Yukawa

In 1939, on the eve of war between Britain and Germany, Japanese residents gathered at the port of Hamburg, where the last repatriation ship, the *Yasukuni Maru*, was waiting. Among the passengers were two researchers from RIKEN. One was Shinichiro Tomonaga, who had studied under Heisenberg and received the Nobel Prize in Physics in 1965. The other was Hideki Yukawa, who was the first Japanese scientist to be invited to give a lecture at the Solvay Conference in Brussels. Unfortunately, the international conference was cancelled, delaying the international debut of his meson theory, so instead Yukawa traveled on to the US to meet Einstein, Oppenheimer, Fermi, and others. My parents were also on this boat.

On the recommendation of Shingo Tsuda, President of Kanebo, my parents traveled to Europe and the US so my father could learn about chemical technology internationally in order to help the Japanese industry. My mother and father happened to travel with Professor Yukawa from Kobe to Naples as well as across the US on the *Yasukuni Maru*, and back to Yokohama via San Francisco on the *Kamakura Maru*. They said that during the long voyage, Professor Yukawa gave an understandable lecture on elementary particles in a soft Kyoto dialect. Ten years after the voyage, in 1949 Professor Yukawa became the first Japanese Nobel laureate, receiving the prize in physics. Japan had been ruined, but our scientific intellect remained unaffected. I was then eleven years old and was especially delighted with this great event, because the laureate was an acquaintance of my parents. My parents were equally thrilled, and any news was frequently brought up at our family dinners – even that Professor Yukawa had tripped on the stairs at the memorial Nobel Prize dinner.

Even as children, our father had guided my two brothers and myself to study engineering. At a time when chemical, mechanical, and electrical were considered the top three fields in engineering, each of us chose a different path. My first younger brother, Tatsuhiko, born in 1940, studied mechanical engineering at the Tokyo Institute of Technology and joined Kobe Steel Ltd. He later served as an executive of Nabtesco Corporation. Masaharu, my second brother, was born in 1942 and majored in electronic engineering at Osaka University before joining Matsushita Electric Industrial Co. (now known as Panasonic). My younger sister, Yoshiko, who was born in 1947, studied piano at the Department of Music at Kobe College. She married Shunichi Murahashi (Professor Emeritus of Osaka University), who is also a close friend of mine. I am exceptionally grateful to my parents for bringing us up in good health.

1.6 Feeling the power of chemistry

After hearing the news of Professor Yukawa's Nobel Prize in 1949, when I was eleven years old and had just entered Nada Junior High School, my father took me to a conference presentation on a new fiber called

Fig. 1.5. Professor Yukawa (second from the left) and my parents (on the right).

Nylon. I was the only child in the audience. The lecturer was the president of the Toray Industries, Inc., which had acquired a technological license from DuPont in the US. He proudly explained, "This new fiber can be synthesized from coal, water, and air, and it is thinner than a spider's thread, yet stronger than steel wire." I was highly impressed: here was a new material created by chemistry from almost nothing. I was so moved that I went home and told my younger brothers about it. From this moment, I dreamed of becoming a chemist, hopefully in industry, who could invent all kinds of materials that would benefit the society and help Japan's economic recovery from the devastation of the war.

Fig. 1.6. The Noyori family. From the left: Masaharu, Ryoji (in 6th grade), Suzuko, Yoshiko, Tatsuhiko, Kaneki in the back.

1.7 The importance of physical strength *and* academic capability

I attended Nada Junior and Senior High School, a prestigious six-year private boys' school in Kobe, and my favorite subjects were mathematics and science. Actually, my entrance to Nada Junior High School was not so easy. I was so overwhelmed at the entrance examination that I misunderstood the meaning of the question in mathematics, which was a subject I was supposed to be good at. I panicked and couldn't get around to the other questions, so I thought I had failed the exam. I was so discouraged that I crawled under the bed covers in tears and could not eat for nearly three days. A few days later, led by my mother, I reluctantly went on the train to go and see the exam results. Contrary to my expectations, I found my exam number on the announcement board and went wild with joy. I had never been so surprised and happy in my life.

My school was founded by the same family as the Kodokan Judo School (Jigoro Kano developed judo from *jujutsu*), so the sport was very popular with the pupils. During the six years of junior and senior high school, I practiced judo every day except on Sundays, and I eventually achieved a blackbelt. I learned much discipline through this traditional activity. It starts and ends with a bow (*rei*). Soft and fair goes far. In addition to training the body and mind, Japanese martial arts cultivate a sense of responsibility through civility and adherence to norms. They have qualities that are not found in sport competitions in which people simply compete to win or lose. The friendly rivalry with both senior and junior students naturally improved my motivation for schoolwork as well as my concentration. However, I still remember the freezing cold tatami mats during practice sessions held in the early mornings of winter before school classes. And I will never forget the warmth of the potato soup that I had after the practices.

But academic ability alone is not enough to have a presence amongst the students. At Nada Junior and Senior High School, nerd students were looked upon in scorn. Rather, overall competence and trustworthiness were required. The role of the homeroom teacher was important in a class, but students were also autonomously involved in keeping order within the classroom. Although some of the students were a bit flirty, they had to be strong in fights and wrestling. In addition to arm strength and athletic ability, students were required to have a certain degree of education along with a variety of knowledge, the ability to cooperate and assemble with classmates, the skills to plan and carry out jokes and tricks, and finally a good personality with a sense of humor. I had a good academic record, but I think my social position was established through my athleticism and friendship with my classmates.

My English teacher had a passion for film, especially for US and European films, and I often pleaded to hear his critiques on movies with Greta Garbo, Marlene Dietrich and Ingrid Bergman. The Western history

Fig. 1.7. Author (17 years old) in judo wear at Nada Senior High School.

teacher sparked in interest in the subject through the intimate relationships of heroes and emperors, starting with Alexander the Great, Caesar, Cleopatra, Marie Antoinette, and Josephine. With these side stories, the contents of the lessons naturally entered our minds. My first chemistry class in my first year of junior high school was given by Dr. Kazuo Nakamoto, who had studied at Osaka University and afterwards become an inorganic chemistry professor at the Illinois Institute of Technology and later at Marquette University in the US. "This is how it will be taught in universities", he would say, and would usually end the class about 10 minutes ahead of time, which made me feel that university teachers were different. Decades later, I met Dr. Nakamoto again at Nagoya University and in Chicago and enjoyed talking about the past. At the Senior High School, Junzaburo Ouchi was a highly experienced teacher, who taught chemistry. He was very knowledgeable about facts on academia and industry outside of textbooks. In the final year, we quickly finished going through the usual textbook and learned the introductory part of organic chemistry from an English textbook. My appetite for chemistry was further whetted by the guidance of many enthusiastic teachers.

Fig. 1.8. Author in the classroom at Nada Senior High School.

During high school, I came to admire Ichiro Sakurada, the inventor of Vinylon, the first Japan-made synthetic fiber. He was a professor of chemistry at Kyoto University. At the time, chemistry departments in the faculty of engineering were attracting the best high school students due to the rapid development of the petrochemical industry. Together with my passion for nylon that I had when I entered junior high school, my desire to study at Kyoto University's faculty of engineering became concrete.

2

ENTERING KYOTO UNIVERSITY IN THE SPUTNIK YEAR

DOI.ORG/10.47261/1564-2

2.1 Moving from Kobe to Kyoto in 1957

In 1957, I entered Kyoto University at the age of eighteen, the year the Soviet Union launched its satellite Sputnik into space. This event clearly demonstrated the power of science-based technology and astonished young science students in Japan. The world was changing.

Japan's economic circumstances urged the government to rapidly revive industrial productivity, and thus there was a growing momentum for the enhancement of science and engineering at universities. At that time, the ratio of humanities to science students in Japan was about three to one, but plans were in place to increase the number of science and engineering students by 8,000, and by the year 1961, 20,000 additional students were to be enrolled. I entered as the ninth graduating class under the new university system, and it was still a time when less than 10 percent of the population of my generation went to study at university. The competition was much harsher compared to nowadays, when more than half of the students go to university and apply to several. Then, it was common to take the entrance exam for only one university. I applied to Kyoto University because two of my role models were professors there – Hideki Yukawa in the School of Science and Ichiro Sakurada in the School of Engineering. I decided to pursue my interest in nylon and studied in chemistry in the School (or Faculty) of Engineering, though I later decided to major in organic chemistry instead of polymer chemistry.

In the first two years of university, students spent their time studying liberal arts, but after that, we entered the Department of Industrial Chemistry as undergraduates, which greatly supported Japan's postwar reconstruction. The size of the chemistry department in the School of Engineering was much larger than that in the School of Science, and was probably the largest of its kind in Japan.

Being released from family constraints in Kobe, I felt a sense of freedom for the first time in my life. At the suggestion of my mother's best friend, I stayed in Shimogamo, a northern district of Kyoto. My lodgings were conveniently located near the main campus of Kyoto University, but first year students first took liberal arts courses at the Uji campus, which was about an hour south by city bus and train. I spent my free time socializing with students from all over Japan, and enjoyed having a good time. My friends and I drank cheap *sake* (Japanese rice wine) and talked about the future of industry and of Japan. When we missed the last train back to Kyoto from Uji, we went straight into an all-night mahjong game. As this was a game that I had played since I was a child, I taught the beginners around me how to play. When I returned to Kyoto in the evening, I was often tempted by Kawaramachi, a main district of the downtown, where I saw many movies, including most of the popular

Fig. 2.1. Author in the second year of university with his grandfather Sanshiro Shiokawa, 1958.

French, Hollywood, and Akira Kurosawa films. I also joined the university's rugby football club and learned the spirit of "one for all, all for one." I only stayed in the club for one year; the rigorous daily training was too much for me. However, this hard but precious experience was unforgettable throughout my entire life.

I think all of the 600 or so students in my School were male as were the faculty members. The class of first year students was a mixture of students from chemistry as well as from electrical, mechanical, civil engineering, metallurgy, among other departments, all with various aspirations. Professor Hideki Yukawa once came to the Uji campus to give a talk on philosophy titled "Chaos". The hall was packed, and the lecture content was beyond the comprehension of many students, including myself, but we were thrilled to see this science hero in person for the first time.

It was not that I was not good at or disliked learning. It was just that I was fascinated by something fresher and more stimulating than the lectures in the liberal arts courses. When I became a third year undergraduate student and began to commute to the traditional red brick building in the main campus, I started focusing on my studies. In the beginning, I found it hard to read English and German in the textbooks due to my lack of study in liberal arts, but I led my classmates in experimental subjects. In analytical chemistry experiments, we had to wait for long hours, not only for operating the balances and laboratory tables, but also for the magnetic crucible firing. While we waited, we would usually go out to the courtyard corridor. My classmates were all brilliant, and many of them had high aspirations after their graduation. Although they were able to study well, they were not necessarily familiar with the industrial world. I had a little more information about industry because of my family environment. Thus, in our free time, I would pick up a copy of the Japan Company Handbook (quarterly corporate report) and explain about the company and its executive board members including their backgrounds. In addition, I was also the manager of the class baseball team and recruited players for our team, so I got to know many people in my class.

2.2 Joining Keiiti Sisido's laboratory of organic chemistry in 1960

In the third year of the undergraduate course, we studied organic chemistry using Louis Fieser's textbook, *Introduction to Organic Chemistry*, which shifted my interest from practical polymer chemistry to fundamental organic chemistry. Although this field did not appear so systematic to me at the time, the molecular structures were highly diverse, and I felt that the applications of them might benefit the society in many ways. Much later, I learned the view of our farsighted predecessor, Justus von

Liebig (1803–1873), who stated, "without a knowledge of chemistry, the statesman must remain a stranger to real life in the state and its organic development and perfection. Without it, his gaze will not be sharpened nor his spirit awakened to what is truly useful or harmful to the country and human society". In the following year, in 1960, I joined Professor Keiiti Sisido's laboratory in the Department of Industrial Chemistry to develop further practical training. This was my starting point, where Associate Professor Hitosi Nozaki warmly guided me in many ways.

Geschichte des Deutschen Museums. Akteure, Artefakte, Ausstellungen (eds.: Wilhelm Füßl, Helmuth Trischler). Prestel, München **2003**.

The laboratory environment was very hospitable. Unfortunately, although Japan's economy had been improving, the academic environment was still not robust enough to promote high-level education and research in science. In organic chemistry laboratories, the structures of organic compounds were determined mainly by combustion analysis. The hand-operated Beckman ultraviolet (UV)-visible spectrometer was the sole reliable spectroscopic tool. Kyoto University was equipped with only one infrared (IR) spectrometer for the whole university. No nuclear magnetic resonance (NMR) instrument was installed at this leading national university, and only a national nuclear energy research institution, GENKEN, and one private company, Kurehabo Co., owned primitive 40 MHz NMR machines. In later years, I would frequently take our synthetic samples to the laboratory of Shionogi Pharmaceutical Company in Osaka for NMR measurements on a Varian A60 instrument.

Since neither silica gel nor active alumina were available for column chromatography, recrystallization or derivatization to crystalline compounds were crucial for obtaining analytically pure substances. Oily compounds were purified by large-scale distillation or sometimes steam distillation. Because only very limited types of solvents and reagents were commercially available, we had to synthesize the necessary materials according to the standard procedures written up in *Organic Syntheses*. For example, we synthesized benzophenone, triphenylphosphine, diborane, and even dimethoxyethane to use as a solvent. Reactions were conducted on a rather large scale using 0.5 to 2 L vessels.

We spent long hours and studied very hard in the departmental library. *Chemical Abstracts* were indispensable because access to the original literature was limited. In a time before the widespread use of photocopiers, we had to make an abstract or take full notes after reading the article from the beginning to the end. Looking back, the progress of research was very slow, but what I remember is the enthusiasm of the teachers and students for learning. We had not yet developed competitive spirit of the Western world, but we were fascinated by chemistry and desired to contribute to its advancement.

The undergraduate course of Japanese universities bestows research education in the fourth year through apprenticeship in independent laboratories. In the 1960s, the major purpose of the School of

Fig. 2.2. Graduation photo of author (back row, 3rd from left) with classmate, Kazuhide Tani, later involved in menthol synthesis (back row, 2nd from left), Sisido (front row, center) and Nozaki (front row, left), 1961.

Engineering was to nurture the workforce for industry. More than half of the students went to industry after completing their bachelor's degree, and some students continued their training under the same mentor in graduate school, which consists of a two-year master and a three-year doctor course. The curriculum of science and engineering remained much less systematic compared to the US and Europe. Most faculty members had no long-term experience abroad. Fortunately, Professor Sisido had stayed at the University of Chicago as a visiting scientist and Professor Nozaki had spent a postdoctoral year at Cornell with Jerrold Meinwald, and both of my mentors appreciated critical thinking.

I finished my undergraduate degree in March 1961 and proceeded to the master's course as planned. I continued to study under Sisido and increasingly spent all night in the lab working on my experiments that I lost a lot of weight and my cheeks became sunken. Sometimes I'd run into old friends, who would look at me surprised, realizing that my focus had completely shifted to science.

Our research still relied largely on our own imagination and inspiration because we had little aid by the instruments we had. It was perhaps in 1967, when the Nozaki laboratory became equipped with a 60 MHz NMR machine. As such, Japan's young researchers at that time started to conduct scientific research in their own style. Fortunately, the research environment including internationalization improved in the mid-1960s through the various efforts of the government and universities.

2.3 My first research topic: *o*-xylylene chemistry (1960–1966)

Numbers in square brackets refer to publications in the appendix, letters to literature notes at the end of a chapter.

In the early 1960s, physical organic chemistry flourished, particularly focusing on structures and reactivities of organic molecules, sparking the full imagination of researchers. My first research topic given to me by Nozaki was the chemistry of *o*-xylylene (*o*-quinodimethane) 1 (**Fig. 2.3.**). He was interested in the conformation of the bridged hydrocarbon 4 [1], which was synthesized by thermolysis at 300 °C of cyclic sulfone 2 or benzocyclobutene 3 in the presence of anthracene. The reaction generates first the reactive intermediate 1 proven by its trapping with dienophiles such as acrylonitrile or methyl acrylate to afford 6. Knowing that the reaction of benzyl halides with Fe powder in a water suspension generates a benzyl radical, Nozaki suggested I prepare 4 more readily by Fe reduction of dibromoxylene 5 in the presence of anthracene that possesses high affinity to radicals at the *meso* positions. However, the trial was unsuccessful, although the generation of 1 was evident by its trapping with dienophiles, giving Diels-Alder adducts 6. We were puzzled to see reactive species 1 generated by thermolysis of 2 or 3 undergoing both [4 + 2] and [4 + 4] cycloadditions, while the same intermediate 1 formed from dibromide 5 gives only [4 + 2] adducts. This phenomenon happened prior to the Woodward-Hoffmann proposal in 1965, which was based on the molecular orbital symmetry conservation principle. Then, we considered that: (1) *o*-xylylene 1 exists in an equilibrium of its singlet 1S and triplet state 1T; (2) under rather mild conditions (< 100 °C), it reacts in a singlet state 1S to give the [4 + 2] adducts; (3) at high temperature, however, triplet 1T undergoes the thermally forbidden [4 + 4] reaction via the radical mechanism. Although any detailed theoretical scrutiny lacked half a century ago, this view was consistent with the finding that high-temperature thermolysis of 2 in the absence of any acceptors, forms its dimer, [2,2]-*ortho*cyclophane **[2]**.

[2] H. Nozaki, R. Noyori, and K. Sisido: Chemistry of Xylylenes and Related Compounds. *Nippon Kagaku Zassi* **1966**, *87*, 641–651.

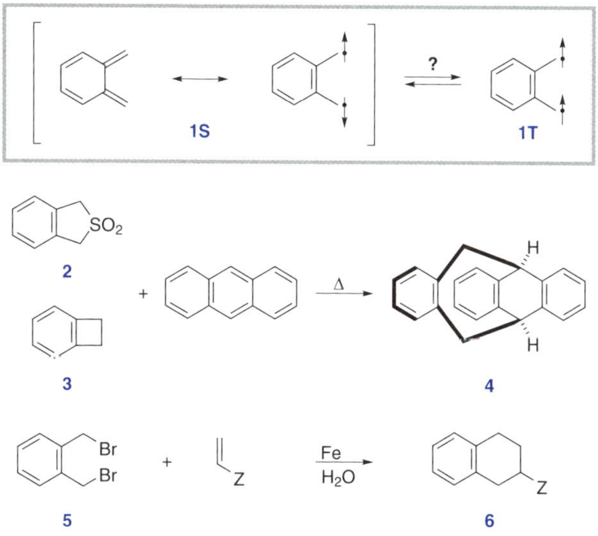

Fig. 2.3. Diverse reactivity of *o*-xylylene. Z = CN, CO_2Me.

2.4 The "immortal Noyori" survives an explosion

In 1963, my daily mentor Professor Nozaki was appointed to chair a new organic chemistry laboratory. I admit that when I first received the offer to become an instructor, I was grateful for the recognition of my abilities and the trust placed in me, but since I was just finishing my master's degree, I felt it was an inappropriate appointment and initially declined the offer. However, I was encouraged by Professor Nozaki, who said that I could obtain my PhD after becoming an instructor, and I decided to take up the offer, although I still dreamed of becoming a leader in industry. Up until then, I had been an equal to and befriended with the students. After becoming an instructor, I used the leadership skills I learned through judo to set an example for the students and support the laboratory as a staff member.

Unlike today's highly professional laboratories, this new lab attracting the best and proudest students was an educational and research organization in which everyone was equal, a group of amateurs with no experience but who trained each other in friendly rivalry to become the leaders of tomorrow. Fully enjoying this free atmosphere, I had to explore unique science in my sub-research group with undergraduate and master course students. "Experiments are serious business. If you fail, do not think about starting over. If you have to start over, try something new instead." "The way you set up the experiment shows whether it will work or not. Are you sure you want to use this awkward setup? You should not do this." Eventually, I became unwillingly known as the "demon sergeant" in the Department of Industrial Chemistry. However, the students understood my way and our relationship of trust was that of close seniors and juniors, and we could say whatever we wanted to in our discussions.

"Baaang!" The flask I had peeked into suddenly exploded. A graduate student, Masaaki Yamabe, who had been carrying out his experiment on the other side of the table, fell out of his chair with a loud thud. I had a big cut across my right cheek and chin and a lot of blood was

Fig. 2.4. Injured author at his parents' home in Kobe, 1963.

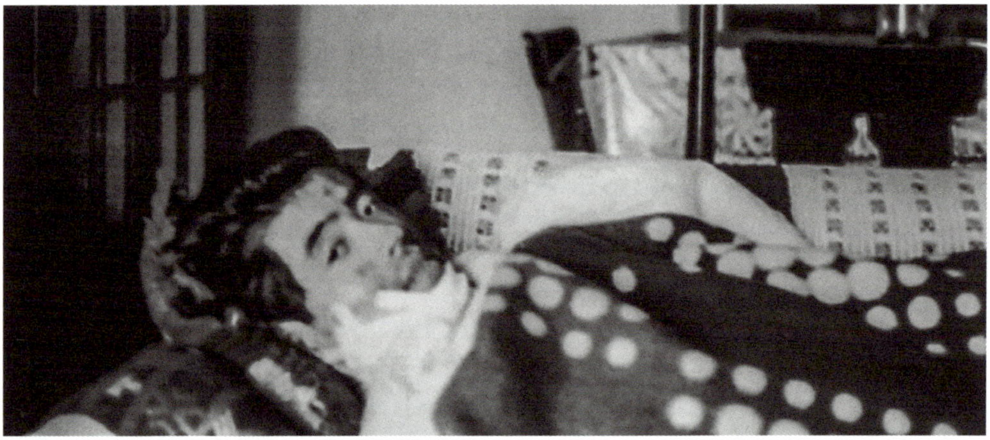

Fig. 2.5. Nozaki group of Kyoto University in 1964. From the left, students: R. Urakabe, M. Yamabe and N. Kozaki, Professor: H. Nozaki, and Instructors: author and K. Kondo.

spurting out. I immediately ran to the sink and tried to wash off the blood, but Yamabe shouted, "We have to go to the hospital!" We ran down the stairs and climbed over the fence around the premises. Yamabe quickly caught a taxi; I was bleeding so much that the driver asked me if I was a *yakuza* (Japanese gangster).

I had only been an instructor for a little over a month and had been working on the amino-functionalization of polypropylene, and was heating the powdered polymer with ethyl azidoformate in an oil bath. Because azide compounds are explosive, we had calculated the theoretical amount of gas to be generated and as a precaution used an open system for the container, but it was a recklessly large-scale experiment. Rather than a simple explosion, the result was more like a detonation and there was no trace of the container I had used. I wrote "EXPLOSION" and in large letters "regretful disappointment" in my lab notebook.

After the anesthesia from the surgery wore off, I spent one night at Nozaki's home in Osaka and, on Yamabe's insistence, another at my parents' home in Kobe. The next day, I reassured my worried parents that I was fine, and I returned to the laboratory at Kyoto University with the bandages still on me. I was greeted with both surprise and disappointment by my students, who had expected me to be absent for a month or so, their faces saying, "if only you could take a little more time off ..." But I did not want to leave chemistry for even a moment. In addition to "demon sergeant," I became known as "the immortal Noyori."

I was later told that documentation had been filed stating that I had conducted the experiment under the order of the professor, thus I was given worker's compensation insurance of more than 50,000 Japanese yen (several times my monthly salary), which allowed me to purchase a typewriter for my research.

2.5 Reflecting on Japan's organic chemistry in the early 1960s

Japan already had a long tradition in natural products chemistry, which was further strengthened by the 1964 IUPAC conference, which brought some of the fields brightest stars to Kyoto (**Fig. 2.6.**). Its great success owes much to Koji Nakanishi then at Nagoya University, who played a key role as a chemistry ambassador with his extensive academic network. We were encouraged by structural elucidation of tetrodotoxin by the groups of Yoshimasa Hirata, Toshio Goto (Nagoya University), and Kyosuke Tsuda (University of Tokyo) in competition with Woodward.

Furthermore, polymer chemistry, which is closely related to the chemical industry, was quickly regaining its prominence in Japan's chemical world, especially at Kyoto University. In this regard, organometallic catalysis chemistry received much attention by researchers at Osaka including Shunsuke Murahashi, Nobue Hagihara, Sei Otsuka, Hiroshi Yamazaki, and Akira Nakamura, among others. The standards of basic organic chemistry were generally still not very high, but an exception was Tetsuo Nozoe's school at Tohoku University, which specialized in the non-benzenoid aromatics chemistry and was a unique strength in Japan. Organic synthesis then was mostly achieved by raw stoichiometric methods. Wataru Nagata at Shionogi Research Laboratory, an inventor of organoaluminium-aided hydrocyanation of enones [A], was well recognized, while the young Teruaki Mukaiyama of Tokyo Institute of Technology led synthetic organic chemistry as a spirited leader. I deeply acknowledge his unfailing efforts.

Under the influence of American and European chemistry, our generation was attracted by such major topics as aromaticity, short-lived

Fig. 2.6. IUPAC Natural Product Chemistry Symposium in Kyoto, 1964. From the left: Shigehiko Sugasawa, George Büchi, Derek H. R. Barton, Robert B. Woodward, Mikhail M. Shemyakin, Vladimir Prelog, Gilbert Stork, Theodor Wieland, Carl Djerassi, John M. Robertson.

reactive species, classical versus non-classical carbocations, Hammett rule, strained hydrocarbons. In the early 1960s, organic chemistry in Japan was still an empirical art. There were few reliable theoretical predictions. The situation changed drastically in 1963, when Charles C. Price of the University of Pennsylvania (later President of the American Chemical Society), a Quaker with a strong religious faith and a passion for education, visited Japan. He gave a series of lectures to graduate students at both Kyoto University and Osaka University, explaining the basics of physical organic chemistry in a logical and quantitative manner. I had just become an Instructor, and this was my first lecture that I heard in English, but I was able to understand quite well since I had "chemical ears." We were recommended to read *Mechanism and Structure in Organic Chemistry*, a textbook written by Edwin S. Gould. As such, most of my research curiosity has arisen from the physical organic or mechanistic thinking nurtured since then.

Edwin S. Gould: *Mechanism and Structure in Organic Chemistry*. Holt, Rinehart and Winston, New York, **1964**.

From this time onwards, although rare, prominent foreign academics began to visit Japan, and those of us who were craving for new knowledge and ideas began to travel from Kyoto to Osaka to attend the lectures by these visitors, not wanting to miss the opportunity. In addition, young researchers began to go to universities in the US as postdoctoral fellows to be trained and gain overseas experience. Looking back on those days, I was interested in a wide and shallow range of topics.

It was in 1965 that Woodward and Hoffmann published a series of papers on molecular orbital (MO) symmetry conservation rules in the *Journal of the American Chemical Society* to the excitement of the worldwide chemistry community **[B]**. Marshall Gates, the editor of the journal, quickly handled these manuscripts through decisive action himself, taking only 50 days and 35 days, for their first and third papers respectively, from their manuscript submission to publication in print. These sensational theories were somewhat difficult to understand for me as an experimentalist, but they were so orderly and beautiful that, in my cold Kyoto lodgings, I repeatedly read and tried to understand the first paper that appeared in January. It was one of the most memorable papers in my life, and it was truly mind-blowing.

[B] R. B. Woodward and R. Hoffmann: Stereochemistry of Electrocyclic Reactions. *J. Am. Chem. Soc.* **1965**, *87*, 395–397.

[C] K. Fukui, T. Yonezawa, and H. Shingu: A Molecular Orbital Theory of Reactivity in Aromatic Hydrocarbons. *J. Chem. Phys.* **1952**, *20*, 722–725.

This research shared the same conclusion as Kenichi Fukui's frontier molecular orbital theory published in 1952 **[C]**. Learning its novelty from neighboring experts, I attended his course lectures in the mid-1960s, but my background was insufficient to understand the significance of his theory. It was in 1976, when Fukui published a 300-page monograph (from Maruzen Co., in Japanese) on the whole story in a more understandable manner, whereby we all became convinced of the true generality of this basic theory. Although Woodward passed away in 1979, Fukui shared a 1981 Nobel Prize in Chemistry with Hoffmann. Later, I had the opportunity to frequently interact with both Fukui and Hoffmann.

Fig. 2.7. Kenichi Fukui (1918–1998), Nobel Prize for Chemistry 1981.

2.6 Implications of molecular catalysis

[D] J. Halpern: Coordination Chemistry & Homogeneous Catalysis. *Chem. Eng. News* **1966**, *44*, 68–75.

Creativity often emerges at the interface of different disciplines. In the mid-1960s, most organic chemists focused on purely organic structures and reaction mechanisms, and were totally indifferent to inorganic coordination chemistry. However, I was most impressed by a brief account given by Jack Halpern at Chicago in *Chemical & Engineering News* in 1966 **[D]**, in which he stated that tetracoordinated d^8 Rh(I) or Ir(I) complexes behave similarly to structurally simple 6e carbene species. That is, Rh(I) species, like singlet CH_2, reacts with a C=C linkage to form a π complex or metallacyclopropane, and inserts into an H–H, H–X (X = halide), or C–X bond. Similarly, pentacoordinated d^6 Ru(II) and d^7 Co(II) complexes act like 6e carbocations and 7e free radicals, respectively. This simple analogy [E] convinced me of the bright future of organometallic reagents and catalysts. I reconfirmed this belief by listening to Halpern's brilliant lecture in Kyoto. Afterwards, I met very frequently with this highly influential scientist and am grateful to his guidance throughout my long career.

This scientific encounter encouraged me to gradually depart from classical organic chemistry and shift towards the direction of organometallic chemistry for organic synthesis, now known as OMCOS (Organometallic Chemistry Directed Toward Organic Synthesis). Modern organic synthesis no longer exists without organometallic molecular catalysis. In the 1960s, however, the concept of "molecular catalysis" had not emerged. Catalysis continued to refer to industrial processes, converting coal- and petroleum-based or other inorganic raw materials to chemically useful products. Efficient catalysts had been discovered largely by coincidence or experience, and classified merely by their solubility in reaction media, i. e. heterogeneous versus homogeneous. The Fe / Al / K-based catalyst of the Haber-Bosch ammonia synthesis or the Ziegler-Natta olefin polymerization catalysts were heterogeneous, while metal carbonyls for the Reppe reaction or for hydroformylation (oxo synthesis) were a typical example of the latter. In sharp contrast, the notion of molecular catalysis utilizes structural and electronic characteristics of coordination complexes to the full extent, providing a logical way to achieve efficient chemical reactions. By definition, a molecule has a fixed elemental composition, a definite atom connectivity, a stereo-defined configuration though sometimes dynamic in nature, and some conformations, and any molecule can be designed and synthesized at will. This in turn could expand synthetic targets to various useful commodity chemicals and even high value-added fine chemicals possessing a precise functionalized structure. But this fundamental notion was not fully verified during the mid-1960s.

2.7 Asymmetric catalysis with chiral organometallic molecular compounds

In the 20th century, asymmetric catalysis was among the most significant topics, in both science and technology, because it allows the production of a wide range of natural and unnatural substances with high enantiomeric purity. In principle, asymmetric catalysis is achievable by a molecular catalyst, consisting of a metallic element M and chiral organic ligand(s) L* (Fig. 2.8.). Such a compact organometallic catalyst L*–M with a molecular weight of less than 1,000 (or 20 Å in length) defines the stereochemical outcome of a reaction in an absolute sense. Ideally designed artificial catalysts can kinetically discriminate with $\Delta\Delta G^{\ddagger}$ of ca. 10 kJ/mol between enantiotopic atoms, groups, or faces in prochiral molecules. This chemistry thus provides an ideal means for *multiplication* of molecular chirality; a single man-made chiral molecule can produce a huge number of left- or right-handed molecules. The periodic table lists many catalytic metal elements, while the permutation of chiral organic ligands is unlimited. Precise chemical synthesis needs a wide variety of asymmetric reactions, and only the concept of molecular catalysis can cope with this requirement. Molecular architecture capable of creating chiral catalysts has converted the chemist's dream into reality. How was this chemistry born then?

Fig. 2.8. Principle of multiplication of chirality. (Top) Ordinary metal catalysts M form enantiomeric products in equal amounts. (Below) A single, well-designed chiral molecular catalyst, (*S*)- or (*R*)-L*–M, converts a large number of prochiral molecules at will to either left- or right-handed products. M = catalytic metal element. L* = chiral organic ligand.

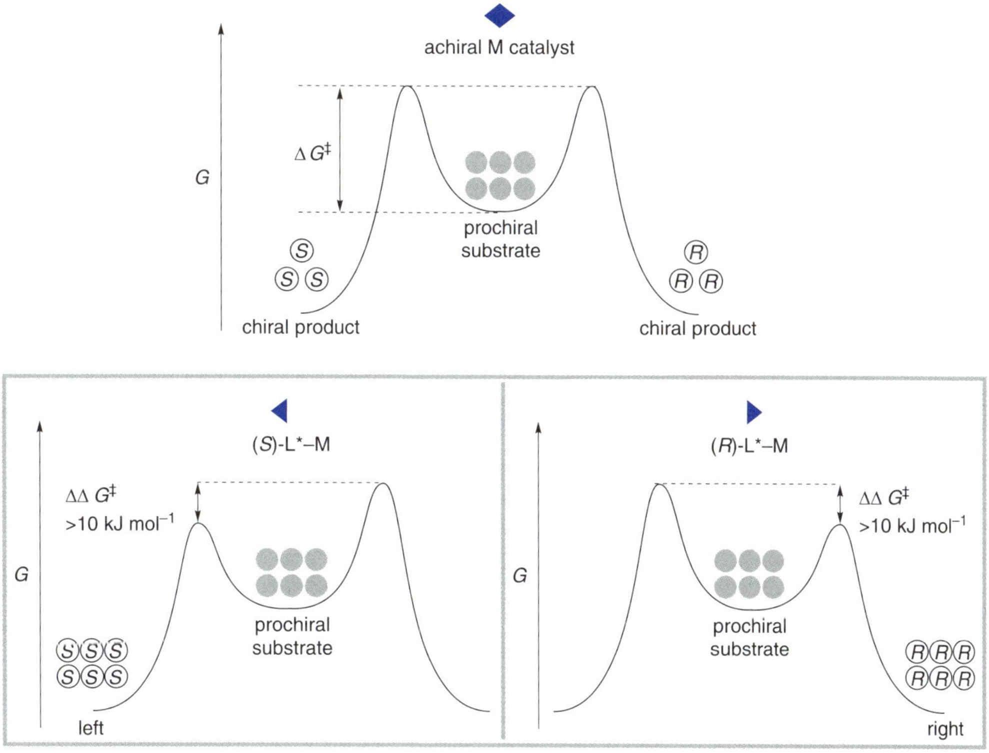

2.8 Birth of asymmetric catalysis with chiral organometallic compounds (1966 – 1968)

The progress of science is totally unpredictable [F]. This general asymmetric method was born by chance in 1966, when I was a twenty-seven year old instructor in Professor Nozaki's group in Kyoto. As illustrated in **Fig. 2.9.**, when styrene was reacted with ethyl diazoacetate at 58 – 60 °C in the presence of a 0.01 molar equivalent of chiral Sacconi's Schiff-base-Cu(II) complex 7, the non-racemic cyclopropane products were produced with 6 – 10 % optical yield (enantiomeric excess, *ee*) and in 72 % chemical yield [3 – 5] [G]. The *cis* and *trans* isomers are diastereomers.

Similarly, the reaction of simple diazomethane and *trans*-1-phenylpropene catalyzed by (*S*)-7 affords (1*R*,2*R*)-*trans*-1-methyl-2-phenylcyclopropane with 8 % *ee*. Moreover, the Cu-catalyzed decomposition of allyl diazoacetate and similar olefinic diazoketones undergo intramolecular cyclopropanation to give the bicyclic products in a partially resolved form. To my knowledge, these cyclopropanations, albeit with low enantioselectivity, were the first examples of asymmetric catalysis affected by a chiral organometallic molecular complex. In place of soluble chelate complex 7, solid Cu(II) tartrate could also be used as a chiral catalyst but less effectively.

Structure determination of the *cis*- and *trans*-cyclopropanecarboxylic esters was not easy, because routine NMR instruments remained unavailable. We first obtained the analytically pure carboxylic acids and compared the IR spectra with those of separately synthesized authentic racemic samples. Determination of the enantiomeric composition was even more laborious, because a polarimeter with a 5 cm quartz cell was the only instrument for this purpose. The pure esters had to be collected in ca. 100 mg quantities by repeated preparative GLC (gas liquid chromatography) separation on a 2 m Apiezon L 30 % column, which required tough work over a few nights. We were delighted to know that the analytically pure esters thus obtained were definitely non-racemic **[3]**. Finally, the optical purities and absolute configurations were determined after conversion to carboxylic acids and by comparison of the signs and optical rotation values with the reported ones [H; the absolute configuration of the *cis* product was still undecided].

[3] H. Nozaki, S. Moriuti, H. Takaya, and R. Noyori: Asymmetric Induction in Carbenoid Reaction by Means of a Dissymmetric Copper Chelate. *Tetrahedron Lett.* **1966**, *7*, 5239–5244.

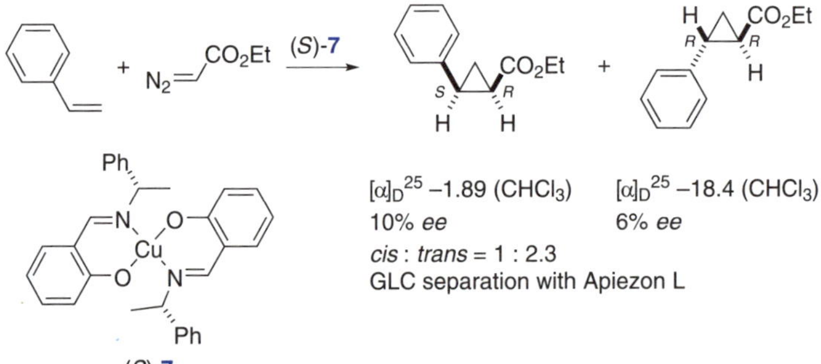

(*S*)-7

$[\alpha]_D^{25}$ –1.89 ($CHCl_3$)
10% *ee*

$[\alpha]_D^{25}$ –18.4 ($CHCl_3$)
6% *ee*

cis : *trans* = 1 : 2.3
GLC separation with Apiezon L

Fig. 2.9. The first asymmetric catalysis using a chiral complex.

2.9 The asymmetric carbene reaction

I have to admit that this is an already fossilized scientific fact. It may be totally meaningless for the current young generation with plentiful knowledge and resources. However, for some curious minds, I hope this might provide an otherwise unnoticed hint from this archeological study.

Since **Fig. 2.9.** itself tells nothing about the motivation of the then vibrant young researchers, one would ask what was the original purpose of this study? The research culture in the 1960s was totally different from now; the chemistry of reactive intermediates such as carbocations, free radicals, carbanions, and others was in full bloom. Back then, we were intrigued by the behavior of carbene intermediates generated from diazoalkanes by photolysis or thermolysis with and without transition metals. Since Herzberg's postulate in 1959 **[I]**, the relation between the structure, spin multiplicity, and reaction modes of carbenes was considered a most fascinating subject, in which many eminent chemists participated. Singlet CH_2 with a bent σ^2 structure (first excited state) and triplet CH_2 with a less bent, σ^1p^1 structure (ground state) are interconvertible by intersystem crossing (**Fig. 2.10.**), as we speculated earlier on in the *o*-xylylene chemistry (**Fig. 2.3.**). Singlet carbenes are known to undergo stereospecific *cis*-cycloaddition to alkenes with retention of configuration but also randomly insert into C–H linkages, while the triplet species undergo cycloaddition to olefinic linkages in a stepwise, and non-stereospecific manner, and react with C–H bonds by a radical mechanism. Thus, both singlet and triplet behave non-selectively, though in different fashions. Notably, transition metals, particularly Cu powder, oxides, or salts, efficiently catalyze the decomposition of diazoalkanes, resulting in highly selective carbene reactions, i. e. stereospecific cyclopropanation without C–H insertion. The role of the transition metals was yet unclear at this point.

[I] G. Herzberg and J. Shoosmith: Spectrum and Structure of the Free Methylene Radical. *Nature* **1959**, *183*, 1801–1802.

Earlier in 1952, Peter Yates noted that the reaction of α-diazoketones and alcohols in the presence of Cu powder gave only α-alkoxy-substituted ketone products resulting from carbene O–H insertion, instead of esters via the Wolff rearrangement, and proposed the formation of an α-keto-carbene intermediate bound to the Cu surface with the valence electrons completing the octet of the carbenic carbon [J]. This sensible explanation was accepted in many publications claiming the reactive species to be a carbene bound to Cu. [For examples, see references in [K].] George Hammond, however, postulated the possibility of the intervention

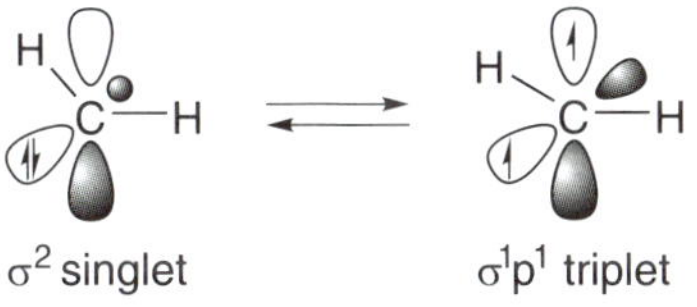

Fig. 2.10. Structures of carbene.

of a charge-transfer complex between triplet CH_2 and Cu powder or soluble Fe(III) tris(dipivaloylmethide) [L]. He considered that the delivery of the CH_2 fragment to an olefin occurs via a triplet 1,3-biradical, as inferred from the rapid quenching of anthracene phosphorescence with Fe salt, an effect found by George Porter (1967 Nobel Laureate in chemistry) **[M]**. Its spin conversion to the singlet state could be instantaneous in the field of the paramagnetic heavy metals, resulting in the stereospecific cyclopropanation without C–H insertion. This interpretation was also reasonable and roused my curiosity in terms of molecularity.

[M] G. Porter, M. R. Wright: Modes of Energy Transfer from Excited and Unstable Ionized States. Intramolecular and Intermolecular Energy Conversion Involving Change of Multiplicity. *Disc. Faraday Soc.* **1959**, *27*, 18–27.

Since a "chiral molecular cause" must kinetically reflect on the "chiral consequence," we wanted to chemically confirm the existence of the carbene species covalently bound to the Cu atom. Thus, we hit upon the idea that an optically active Cu molecular catalyst serves as a diagnostic tool for the hypothesis. Asymmetric induction, if observed in the olefin cyclopropanation with chiral catalyst (*S*)-7 (**Fig. 2.9.**), would provide definite evidence for the intervention of a carbene-Cu intermediate. This was our simple scientific logic, and we were excited to observe the asymmetric induction.

In 1966, I was lucky enough to first encounter the metal-catalyzed asymmetric reaction. We were thrilled with the findings, but outside the team, no one appreciated its significance at the time because this reaction was so small. I recall that: (1) the enantioselectivity was up to only 55 : 45 or 53 : 47, slightly favoring one enantiomer over the other, and (2) the three-membered ring-forming reaction was slightly peculiar, but not generally useful. The *Journal of the American Chemical Society* rejected our manuscript and instead was published in *Tetrahedron Letters* [3]. After three decades, I was pleased that this finding was eventually recognized as one of the reasons that I was selected to receive the 2001 Wolf Prize in Chemistry shared with Henri B. Kagan and K. Barry Sharpless. I consider, in general, that more acknowledgements need to be made of what initially seem to be radical, or even stupid ideas of young researchers.

In fact, this discovery resulted not from a particular synthetic purpose. Rather, it came by accident from our mechanistic speculation. At the time, since designed "asymmetric catalysis" via intermolecular chirality transfer remained too ambitious, we used the now uncommon term "asymmetric induction." I must confess that we were unable to assert its potential utility. Two years later, in 1968, when I moved from Kyoto to Nagoya University, I no longer worked on this interesting but seemingly synthetically insignificant subject. Nothing comes from nothing. As so often happens, however, this "science in action" totally differed from the story of the resulting technological invention and innovation. This serendipitous encounter was later developed to useful technologies or even innovations unforeseen by its discoverers; they gained answers to questions they were not even asking (**CHAPTER 9**).

2.10 Challenging a famous statement of Louis Pasteur

Fig. 2.11. Louis Pasteur (1822–1895).

Young researchers face a multitude of options when choosing their scientific topic. This was also the case with me. Clearly, chirality is an intrinsic feature in science, technology, and even society. [For example, see references in [N].] Molecular chirality is among the unimaginable sorts of exciting subjects and eventually became my major and life long focus being triggered by the encounter with the asymmetric carbene reaction.

I recall now that my academic career has long challenged the work of Louis Pasteur, who stated in 1851, "Dissymmetry is the only and distinct boundary between biological and nonbiological chemistry. Symmetrical physical or chemical force cannot generate molecular dissymmetry." Rigorously speaking this was not true, but was accepted as valid from a practical point of view. As such, practical access to optically active organic compounds had long been nature's monopoly.

Many organic compounds have a right and left form, known as enantiomers. This difference in forms can have a large effect in biology and living organisms. The phenomenon of chirality occurs because the receptors and enzymes in our bodies are mostly proteins made up of only *S*-configured amino acids. Certain small organic molecules are bound reversibly to the binding sites of enzymes via various weak interactions, thereby prohibiting or promoting their biological actions. In order to form such a precise molecular complex, the structures of the small and large molecules must be complementary both electronically and spatially. Also, such biologically active compounds must have an appropriate surface potential that is generated largely by oxygen or nitrogen functionalities in the molecules. Since the presence of a defined *R* or *S* stereogenic center(s) makes the chiral molecules single-handed, we say from chemists' point of view that "life depends on molecular chirality."

There are numerous examples of enantiomer effects, which are frequently dramatic, as exemplified in **Fig. 2.12.** Enantiomers often smell and taste differently. The structural difference between enantiomers can be serious with respect to the actions of synthetic drugs. Chiral receptors in the human body interact only with drug molecules having the appropriate absolute configuration, resulting in marked differences of pharmacological activities of enantiomers. A compelling example was provided by the tragic administration of racemic thalidomide to pregnant women in the 1960s. (*R*)-Thalidomide has desirable sedative properties, while its *S* enantiomer is teratogenic and induces fetal malformations [O]. Such problems arising from inappropriate molecular recognition should be avoided at all costs. Notably, thalidomide is now approved to treat a series of cancers including multiple myeloma and other diseases. Thus, for preparative purposes, generally

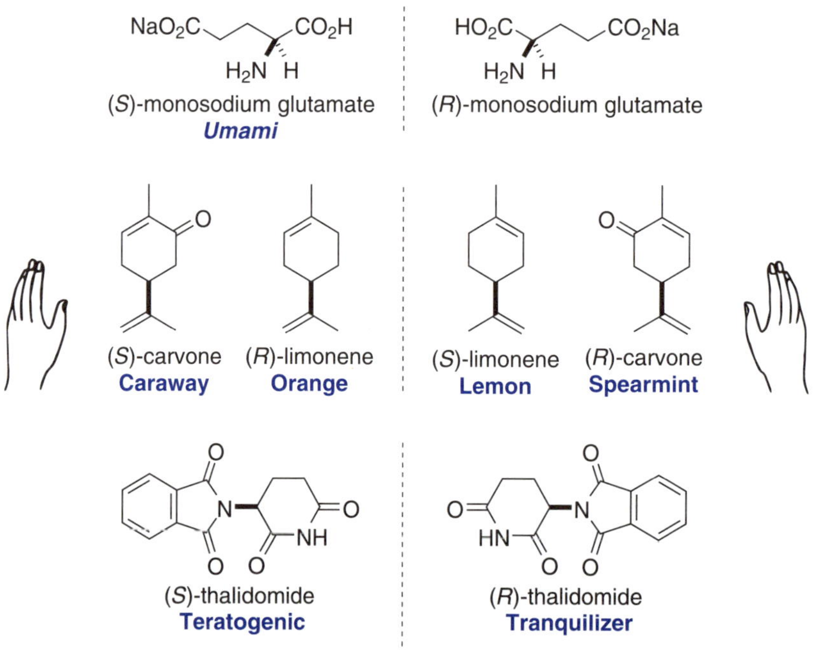

Fig. 2.12. Enantiomers and their biological activities.

reliable methods are required for the synthesis of enantiomeric organic compounds, however practical means were not available until half a century ago, which was in the 1970s.

Despite the recent rapid progress of biomolecular therapy, involving macromolecules such as bioactive proteins and monoclonal antibodies, chiral molecular medicines continue to serve as a critical part of this important field [P]. The significance of three-dimensionally well-defined, unnatural man-made compounds with molecular weights in the hundreds will not disappear. Instead, the promising antibody-conjugate drugs will further expand the utility of such chiral small molecules.

2.11 Primitive asymmetric organometallic alkylation (1968 – 1971)

In those immature days, I was reckless enough to challenge many unsolved problems without any sophisticated, feasible research design. These trials were joyful, but their consequences were totally unacceptable in the scientific community. For example, upon considering the resonance, RM ↔ R••M ↔ R^-M^+, I was interested in controlling the property of main group organometallics. Organolithiums and Grignard reagents (commercially unavailable at that time) would serve not only as simple carbanion sources, but also behave somewhat more functional by modifying M with an appropriate auxiliary. First, I intended to observe an asymmetric induction caused by chiral ligands.

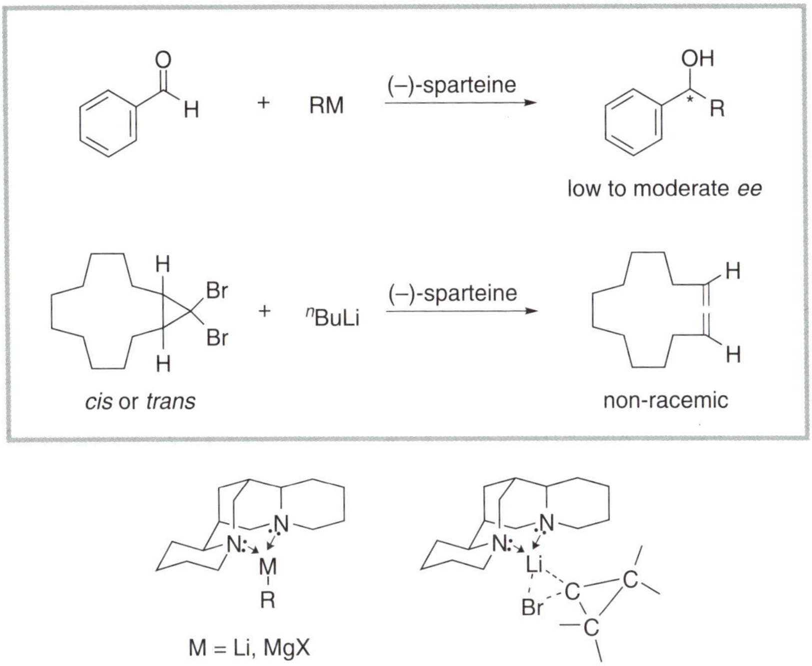

Fig. 2.13. Asymmetric induction in organometallic reactions in the presence of (–)-sparteine.

Since the reactivity of *n*-butyllithium was known to be enhanced by the chelating tertiary diamine, *N,N,N′,N′*-tetramethylethylenediamine (TMEDA), we expected that a chiral bidentate diamine (–)-sparteine might induce an asymmetric bias in the reaction of carbonyl compounds (**Fig. 2.13.**). In fact, the reaction with benzaldehyde gave the *R* addition product with 6 % *ee*. Reaction with ethylmagnesium bromide resulted in the chiral addition product with 22 % *ee* **[6]**. This was among the earliest examples of asymmetric alkylation of this sort, although a feeble asymmetric induction (5 %) was reported earlier [Q] in a Grignard addition in the presence of (2*R*,3*R*)-dimethoxybutane. Moreover, we also noticed the occurrence of asymmetric induction in the butyllithium-aided ring-opening of dibromocyclopropanes to allene products, excluding the involvement of a free cyclopropylidene intermediate [6]. In this respect, we had to wait more than ten years for this primitive finding to fully blossom by exploring organozinc chemistry in 1985 at Nagoya (**CHAPTER 5**).

[6] H. Nozaki, T. Aratani, T. Toraya, and R. Noyori: Asymmetric Syntheses by Means of (–)-Sparteine Modified Organometallic Reagents. *Tetrahedron* **1971**, *27*, 905–913.

2.12 Photochemistry of cyclic enones (1965 – 1975)

I could not focus my research attention on one topic. The experience in *o*-xylylene and carbene chemistry stimulated my curiosity in spin chemistry and photo-activated molecular states. We studied the photochemical behavior of various olefinic and ketonic substrates, particularly, medium-sized cyclic enones in polar media. Although irradiation

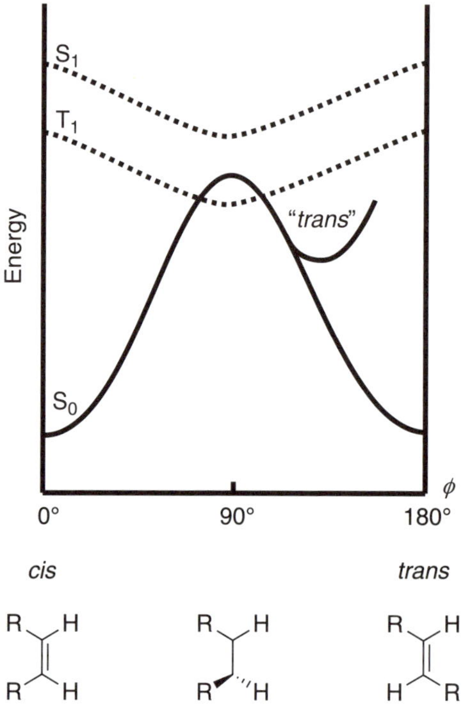

Fig. 2.14. Photoisomerization of medium-sized cycloalkenes. ϕ = C(sp^2)–C(sp^2).

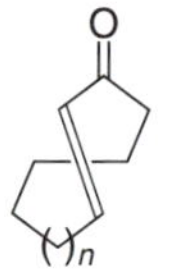

n = 0, 1

Fig. 2.15. Strained "*trans*"-2-cycloalkenones.

of 2-cyclohexenone in methanol gave the [2 + 2] cyclodimer as a major product accompanied by a small amount of 3-methoxycyclohexanone, 2-cycloheptenone and 2-cyclooctenone react readily with alcohols, acetic acid, water, and diethylamine, resulting in Michael-type adducts [7–9]. Such unique polar reactivity is rather different from the extensively studied chemistry of six-membered unsaturated ketones and was interpreted in terms of a highly strained but ground-state "*trans* isomer" (**Fig. 2.14.**) [10], instead of the commonly presumed photoexcited polar states. This was already beautifully shown in 1965 by E. J. Corey **[R]** and Philip E. Eaton **[S]** for 2-cycloheptenone. We found that 2-cyclooctenone in 2-methyltetrahydrofuran is converted to the stable *trans* enone isomer at –78 °C, which gives the alcohol adducts at room temperature in the dark [8]. Later, 1-acetyl- and 1-nitrocyclooctene were found to behave in a unique manner [11–13].

Thus, I was interested in many unordered topics but finished almost nothing. One of the unique findings, however, was exemplified by the photoreactions of acridine, quinoline, and isoquinoline with an alkanoic acid, causing the reductive alkylation under decarboxylation. Use of pivalic acid gives *tert*-butylated benzopyridines [14, 15].

[R] E. J. Corey, M. Tada, R. LaMahieu, and L. Libit: *trans*-2-Cycloheptenone. *J. Am. Chem. Soc.* **1965**, *87*, 2051–2052.

[S] P. E. Eaton and K. Lin: *trans*-2-Cycloheptenone. *J. Am. Chem. Soc.* **1965**, *87*, 2052–2054.

2.13 Receiving my PhD (Dr. Eng.) degree

Ever since hearing the Toray Nylon presentation and partly due to my father's influence, I had long planned to become an industrial engineer after graduating from Kyoto University and join Toray after obtaining my PhD (**Chapter 1**). When I finished my two-year master's course in 1963, my mentor Hitosi Nozaki was promoted to Full Professor in the same department and appointed me as an instructor to lead a subgroup in his newly founded laboratory.

At the time, unlike the standard Western university systems in which all professors were independent, Japanese universities took an old-fashioned *koza* (chair) system where a laboratory group consisted typically of one Full Professor, one Assistant Professor, and several lecture-free Instructors or Research Assistants pursuing mostly the Full Professor's research topics. Japanese industries then also provided a much better research environment to achieve ambitious research goals, and my father was not necessarily in favor of my pursuing a scholarly life. Professor Nozaki, however, encouraged me to pursue my own curiosity within his group. The university allowed us duty-intensive, non-doctoral Instructors, to submit our doctoral theses at least one year later than normal graduate students did. During my time as an instructor, I gradually started to convince myself to pursue an academic career. I strongly hoped to broaden my scientific expertise by studying abroad under the supervision of an eminent chemist. Fortunately, Professor E. J. Corey at Harvard University kindly agreed to accommodate me as a postdoctoral fellow after receiving a doctorate degree. This plan, however, was postponed for a particular reason, which is mentioned in the next chapter. Thus, without going through the formal doctoral program, I received my PhD in September 1967.

2.14 Kyoto's Hanamachi district as a cultural classroom

I sometimes wonder how in the world I had any kind of energy at the time to pursue other things outside university. While I was passionate about chemistry, I was also eager to learn more about the culture of Kyoto at night. As a poorly paid university instructor, there was only so much I could do in the Hanamachi (geisha quarter) district of Pontocho and Gion. However, even though it was a far cry from the old days when spiritual culture was in full swing, the bars had famous proprietresses and talented madams who gathered like-minded people and nurtured the future leaders of society. In addition to a wide range of famous scholars in science and humanities, there were monks from famous temples, active film directors, scriptwriters, actors, conductors, journalists, painters, potters, antique dealers, and Nishijin-ori brocade masters. I listened to the conversations of these top-notch people in their casual wear. Naturally, each person was different, but while we talked about trivial things, these were people who had had various experiences during and after the war in their youth. The active people in their fields were broad-minded, and they were generous enough to encourage ignorant young people like myself.

I went to the laboratory even on Sundays and holidays. On my few days off, I continued to go and see the movies, and very occasionally went to see *kabuki* (traditional Japanese performance art) plays.

Each person has his or her own path to self-creation, and this 11-year period of time in Kyoto occupies a large place in my heart. To begin with, I deeply acknowledge the superb scholarly guidance of my mentors and also the warm friendship with honorable students, particularly my lab mates at the Sisido / Nozaki research groups. However, scientific studies are not everything. Various non-academic, experienced people in this cultural town taught me how things work in the real world. Such informal lessons at evening parties were extremely valuable to me when I was socially immature. I felt that the basic culture of life there, had significantly affected the sensibility in my subsequent professional career.

This stimulating life came to an end when I moved to Nagoya at the age of twenty-nine, and after that life became completely work-centered. Although I had a few favorite bars in Kyoto and Nagoya, I became an inflexible chemistry addict who put work first before anything else. I imagine that the *okami* (madams) who used to talk to me must have been very bored with my conversations. One has to keep learning throughout one's life. I hope the readers will excuse my coarse behavior, which is due to the fact that I have been absorbed in my work since the age of thirty and have not studied much about life in general.

References for chapter 2

[A] W. Nagata and M. Yoshioka: Hydrocyanation of Conjugated Carbonyl Compounds. *Org. React.* **1977**, *25*, 255 – 476.

[B] R. B. Woodward and R. Hoffmann: Stereochemistry of Electrocyclic Reactions. *J. Am. Chem. Soc.* **1965**, *87*, 395 – 397.

[C] K. Fukui, T. Yonezawa, and H. Shingu: A Molecular Orbital Theory of Reactivity in Aromatic Hydrocarbons. *J. Chem. Phys.* **1952**, *20*, 722 – 725.

[D] J. Halpern: Coordination Chemistry & Homogeneous Catalysis. *Chem. Eng. News* **1966**, *44*, 68 – 75.

[E] R. Hoffmann: Building Bridges Between Inorganic and Organic Chemistry (Nobel Lecture). *Angew. Chem., Int. Ed.* **1982**, *21*, 711 – 724.

[F] R. M. Roberts: *Serendipity: Additional Discoveries in Science*. John Wiley & Sons, New York, **1989**.

[G] H. Takaya, Doctoral Thesis, Kyoto University, **1970**.

[H] Y. Inoue, T. Sugita, and H. M. Walborsky: Cyclopropanes – XVII: The Absolute Configurations of *trans*-1,2-Cyclopropanedicarboxylic Acid and *trans*-2-Phenylcyclopropanecarboxylic Acid. *Tetrahedron* **1964**, *20*, 1695 – 1699.

[I] G. Herzberg and J. Shoosmith: Spectrum and Structure of the Free Methylene Radical. *Nature* **1959**, *183*, 1801 – 1802.

[J] (a) P. Yates: The Copper-catalyzed Decomposition of Diazoketones. *J. Am. Chem. Soc.* **1952**, *74*, 5376 – 5381. See also: (b) R. Casanova and T. Reichstein: Methoxyketone aus Diazoketonen. Steroide, 5. Mitteilung. *Helv. Chim. Acta* **1950**, *33*, 417 – 422; (c) M. S. Newman and P. F. Beal III: A New Synthesis of α-Alkoxy Ketones. *J. Am. Chem. Soc.* **1950**, *72*, 5161 – 5163.

[K] (a) P. S. Skell and R. M. Etter: Electrophilic Character of Carbethoxycarbene. *Chem. Ind.* **1958**, 624; (b) C. E. H. Bawn, A. Ledwith, and J. Whittleston: Mechanismus der Polymerisation aliphatischer Diazo-Verbindungen, katalysiert durch Kupfer und Seine Salze. *Angew. Chem.* **1960**, *72*, 115; (c) G. Wittig and K. Schwarzenbach: Über Methylenierte Metallhalogenide, I. *Justus Liebigs Ann. Chem.* **1961**, *650*, 1 – 20; (d) R. Huisgen and G. Juppe: Benzo-norcaradien-carbonsäure und Benzo-cycloheptatrien-carbonsäure. *Chem. Ber.* **1961**, *94*, 2332 – 2349; (e) E. Müller and H. Fricke: Katalysierte Homologisierung von Benzol mit Diazomethan zum Cycloheptatrien. *Justus Liebigs Ann. Chem.* **1963**, *661*, 38 – 43; (f) W. Kirmse: *Carbene Chemistry*. Academic Press, New York, **1964**, 13.

[L] K. R. Kopecky, G. S. Hammond, and P. A. Leermakers: The Triplet State of Methylene in Solution. *J. Am. Chem. Soc.* **1962**, *84*, 1015 – 1019.

[M] G. Porter, M. R. Wright: Modes of Energy Transfer from Excited and Unstable Ionized States. Intramolecular and Intermolecular Energy Conversion Involving Change of Multiplicity. *Disc. Faraday Soc.* **1959**, *27*, 18 – 27.

[N] (a) M. Gardner: *The New Ambidextrous Universe.* 3rd rev. ed. W. H. Freeman & Co., New York, **1990**; (b) E. Heilbronner and J. D. Dunitz: *Reflections on Symmetry*. VHCA-VCH, Basel, Weinheim, **1993**; (c) R. Hoffmann: *The Same and Not the Same*. Columbia University Press, New York, **1995**; (d) I. Hargittai and M. Hargittai:

In Our Own Image, Personal Symmetry in Discovery. Kluwer Academic and Plenum Publishers, New York, **2000**; (e) D. Schattschneider: *M. C. Escher – Visions of Symmetry*. Thamems & Hudson, London, **2004**. Youths learn a lot from Henri Brunner's illustrative book: H. Brunner: *Rechts oder links: In der Natur und anderswo*. Wiley-VCH, Weinheim, **1999**.

[O] G. Blaschke, H. P. Kraft, K. Fickentscher, and F. Köhler: Chromatographic Separation of Racemic Thalidomide and Teratogenic Activity of its Enantiomers. *Arzneim.-Forsch.* ***1979***, *29*, 1640 – 1642.

[P] E. J. Corey, B. Czakó, and L. Kürti: *Molecules and Medicine*. John Wiley & Sons, Hoboken, **2007**.

[Q] H. L. Cohen, G. F. Wright: Reactions of Grignard Reagents in Optically Active Solvents. *J. Org. Chem.* **1953**, *18*, 432 – 446.

[R] E. J. Corey, M. Tada, R. LaMahieu, and L. Libit: *trans*-2-Cycloheptenone. *J. Am. Chem. Soc.* **1965**, *87*, 2051 – 2052.

[S] P. E. Eaton and K. Lin: *trans*-2-Cycloheptenone. *J. Am. Chem. Soc.* **1965**, *87*, 2052 – 2054.

STARTING AN INDEPENDENT CAREER AT NAGOYA UNIVERSITY

3

 DOI.ORG/10.47261/1564-3

3.1 An invitation from Nagoya University

Just after I earned my PhD in the fall of 1967, I received an unexpected offer from the Faculty of Science at Nagoya University to lead a newly created organic chemistry laboratory. This invitation perplexed me. I had not been prepared at all for such an honorable opportunity; I was a mere 29-year-old instructor at Kyoto enjoying daily research with students on the reactions of short-lived organic species. Nonetheless, an interview was immediately set up with some faculty members of Nagoya University to discuss the perspectives of the chemistry department. Being too young and inexperienced to be appointed Full Professor, I was offered a position as an independent Associate Professor of chemistry, which was unusual in the Japanese *koza* system. I was deeply honored by the offer from the reputable chemistry department and accepted this appointment.

Before making this life-changing decision, I consulted with my father, Kaneki. Like myself, he had a Dr. Eng. degree, but was a research supervisor at a chemical company and was very proud of his successful career. He knew many people at major universities, so he said to me: "In a company, if the research is really impressive, they will invest a lot of money and a lot of people. That's why they can do great work. There are a lot of excellent professors at universities, but I feel sorry for them because they are not working in a good research environment. I cannot strongly agree with your decision to go to Nagoya." However, perhaps sensing my devotion to the university as a workplace and the situation I was in, my father gave me a word of advice: "An academic must be first in class to be successful, so be prepared for that." After that, he never expressed his opinion again.

3.2 Meeting a giant in natural products organic chemistry

I arrived at Nagoya in February 1968. It was Professor Yoshimasa Hirata (1915 – 2000), who took the initiative for this unusual appointment. Since I had never met him personally before, I had much to learn from this world-renowned chemist who achieved outstanding accomplishments in natural products chemistry. He elucidated the structures of numerous biologically significant substances including potent toxins such as tetrodotoxin (pufferfish poison) and palytoxin. At the time, I was still inexperienced and unfamiliar with the culture of Nagoya's organic chemistry and the rigors of the scientific world at large. Hirata was a sincere chemist, truly devoted to his research. This natural products research giant would drive his own van, sometimes with students, to search and collect plants from Rebun Island in the north and from Kyushu and Ishigaki Island in

Fig. 3.1. Yoshimasa Hirata at Nagoya University in the 1970s.

the south, occasionally with the help of local people. When he returned to his laboratory, still dressed in boots and work clothes, he would spread the leaves and fruits on the floor to dry and extract biologically significant and often toxic compounds. Hirata's research was more oriented on natural fields instead of the more ordinary laboratory-focused approach. Unlike most contemporary scientists, Hirata always began his projects from scratch, so his research originality was unparalleled. I was told that Hirata's well-known research project from 1955 to 1964 on tetrodotoxin advanced largely thanks to the dedication of his wife, Hisako; she processed the organs of large quantities of pufferfish from the market at home and delivered them to the university laboratory. Research in Hirata's lab was conducted even on Sundays and holidays. His one and only hobby was growing oriental orchids.

There was neither a secretary nor a special office for the professor. The central table in the meeting room, which also served as the library, was always piled with teacakes and oranges, and no matter how important the visitors were, they always had their meetings there. Both staff and students would discuss things over tea or lunch together at this table. According to Hirata, this was the place which immediately revealed the abilities and personalities of the students.

Hirata's research philosophy was: "Never do research that you know the answer to, and of course never do research that others can do." His laboratory was not so well equipped and was somewhat improvised. Hirata himself had even been developing special resin materials for chromatographic separation of water-soluble substances such as palytoxin, since conventional silica or alumina could not be used. His lab's research topics stood out distinctly from others; they were not doing chemistry for chemistry's sake, but had a grand and unique style of research that addressed chemical challenges in the world of life sciences. He was a person with foresight.

As a newcomer, I first met with Kiyoyuki Yamada and Shosuke Yamamura, who had returned from Stanford University and MIT, respectively, and I learned the Nagoya University style from them. They were very warm to me from the beginning. Hirata was a patient and generous mentor, and he was very close to his lab members, including students and staff. I felt the atmosphere was totally different from what I experienced in Kyoto. Compared to other universities in Japan, Nagoya University was much more internationalized, because of the strong ties with Harvard University forged through the personal connections of Hirata himself and his school; Koji Nakanishi and Toshio Goto both worked with Louis F. Fieser, and Yoshito Kishi worked with Woodward. I had no experience abroad at this point. I learned about the culture of US universities and the latest trends in research there through visitors. To my surprise, there was an American postdoctoral researcher funded by the National

Institutes of Health (NIH) working at the Hirata lab. I sometimes wondered whether he was able to comprehend the culture of Nagoya's organic chemistry during his stay.

Soon after I arrived in my new position at Nagoya, Hirata gently told me, "I will do natural products organic chemistry, and you can do the rest. I would like you to make organic chemistry in Nagoya better and more visible worldwide." This was the only thing he ever asked of me. I simply promised to try my best toward achieving this goal. I was thus allowed to do whatever I wanted, but this was not an easy task, because our department had only two organic chemistry laboratories, Hirata's and mine. This generous, fatherly advisor consistently helped me in many aspects during his time at Nagoya University and beyond. For instance, the Hirata and Noyori laboratories shared the instruments for infrared (IR), ultraviolet (UV), and nuclear magnetic resonance (NMR) spectroscopy, and for mass spectrometry (MS). Mimicking the format of the Woodward Seminar at Harvard (**CHAPTER 4**), both groups held joint evening seminars, where young researchers with different expertise presented current topics.

Inevitably, the research and education environment in Japan and worldwide has changed. Thus, although my academic career was accompanied with some violent fluctuations, I had never forgotten the request from Hirata. It was only in 2021, when I finally felt that a great weight was lifted from my shoulders through this occasion (**CHAPTER 9**).

3.3 My memories with Satoru Masamune

I was thus destined to lead organic chemistry at Nagoya University since the late 1960s, when I was still not well recognized by the international community. There were probably many people who felt uneasy about my seemingly reckless appointment. Since I feel that it is about time that I can mention this, I would like to share a story of how I came to be offered the professorship. It was actually Satoru Masamune (1928 – 2003) that Nagoya University initially wanted to appoint as a Professor, but they were unsuccessful in doing so. Masamune was an organic chemist who studied at the laboratory of his father-in-law, Professor Tetsuo Nozoe (1902 – 1996), at Tohoku University in Japan, and then moved to the UC Berkeley as a graduate student, where he obtained his PhD with Henry Rapoport in 1957. In the mid-1960s, Masamune was a rising chemist recognized around the world and was promoted to a special position at the University of Alberta in Canada. When I visited Edmonton in May 1977, I heard this story from Masamune himself.

Soon after my visit, Masamune became a Professor at MIT and continued his research overseas for the rest of his life. Masamune was an extremely strict researcher with an intense work ethic who valued only absolute first-class research and ignored even highly respected researchers if they failed to meet his standards. For some reason, perhaps as I was a substitute for him, he and I shared a common bond, and he seemed to keep an eye on my research over the years.

In November 1981, I was invited by Masamune to stay at MIT as a Karl Pfister Visiting Professor. This visiting professorship was established in 1964 with the support of the Merck Research Lab. Karl Pfister (1919 – 2011), who graduated from MIT, was Director of R & D at Merck. In fact, there was a plan to welcome me as a professor at MIT. The environment was of course much more attractive than Nagoya University, however, I could not accept. At that time, my research on prostaglandins synthesis and BINAP chemistry was going well, and I wanted to make ensure the future of my research group in Nagoya. I also had to consider what was best for my two sons, who were then eight and three years old. If I had gone to the US at that time, I wonder whether I would have been able to develop more. I have yet to decide whether I was right or wrong about this decision, but I still cannot shake off the feeling of disgust at my own passivity. I was 43 years old at that time, and this was when I started to be aware of my role in the Japanese scientific community.

In the fall of 2001, when I was awarded the Nobel Prize, Masamune called me from Boston and congratulated me by saying, "50 years have passed since my father, Nozoe, paved the way for an internationalization of Japanese chemistry and for attaining international standards, and your achievements have demonstrated the success of this development."

3.4 Launching the Noyori Laboratory (1968)

In April 1968, I started my own research lab together with Hidemasa Takaya (1940 – 1995) and Masao Katô, who both accompanied me from Kyoto University, as well as some undergraduate students at Nagoya. At this stage, our expertise was limited to physical organic chemistry of short-lived reactive species including carbenes, nitrenes, free radicals, and photo-excited molecules. However, I had long sought to explore organic synthesis utilizing organometallic chemistry. In the 1960s, however, organometallic chemistry was largely comprised as a branch of inorganic chemistry, so not many researchers were aware of its high potential in organic synthesis.

Fig. 3.2. Launch of the Noyori Laboratory at Nagoya University in 1968. Front row middle: author. Front row right: H. Takaya. Second row right: M. Katô.

3.5 Photo-induced Nazarov chemistry (1968 – 1975)

Upon starting research in my lab at Nagoya, the growing impact of the Woodward-Hoffmann rules urged me to continue studies on the photochemistry of medium-sized cyclic cross-conjugated dienones connected to the Nazarov cyclization (Fig. 3.3.).

Fig. 3.3. Nazarov electrocyclization.

We observed a remarkable dependency of the reaction course on the ring size and environment. Irradiation of 2,6-cycloheptadienone 1 in cyclohexane forms [2 + 2] cyclodimers (not shown), whereas, in methanol, it gives the simple β-methoxy adduct 2 (and then β,β'-dimethoxy product) as observed with 2-cycloheptenone. In sharp contrast, in less nucleophilic media such as acetic acid or the slightly acidic *tert*-butyl alcohol, this dienone undergoes a Nazarov-type β,β'-bond formation under irradiation to give the bicyclo[3.2.0]heptan-3-one adducts 3 (Fig. 3.4.. Eq. 1) [16]. The eight-membered dienone 4 behaves similarly, leading to 2-substituted bicyclo[3.3.0]octanones 5 even in methanol (Eq. 2). The reaction occurs by way of the strained, ground-state *cis*,*trans* dienones like 6, as firmly proven with the higher analogue 2,8-cyclononadienone [17, 18].

[16] H. Nozaki, M. Kurita, and R. Noyori: Photochemical Reactions of 2,6-Cycloheptadienone. *Tetrahedron Lett.* **1968**, *9*, 3635 – 3638.

The reaction course in strong acids is intriguing. The protonated cross-conjugated dienones are very stable in the dark. However, upon irradiation, they show unique behavior, which again depends on the ring size. In fluorosulfonic acid, protonated 2,6-cycloheptadienone 7

Fig. 3.4. Photochemistry of cross-conjugated cyclic dienones.

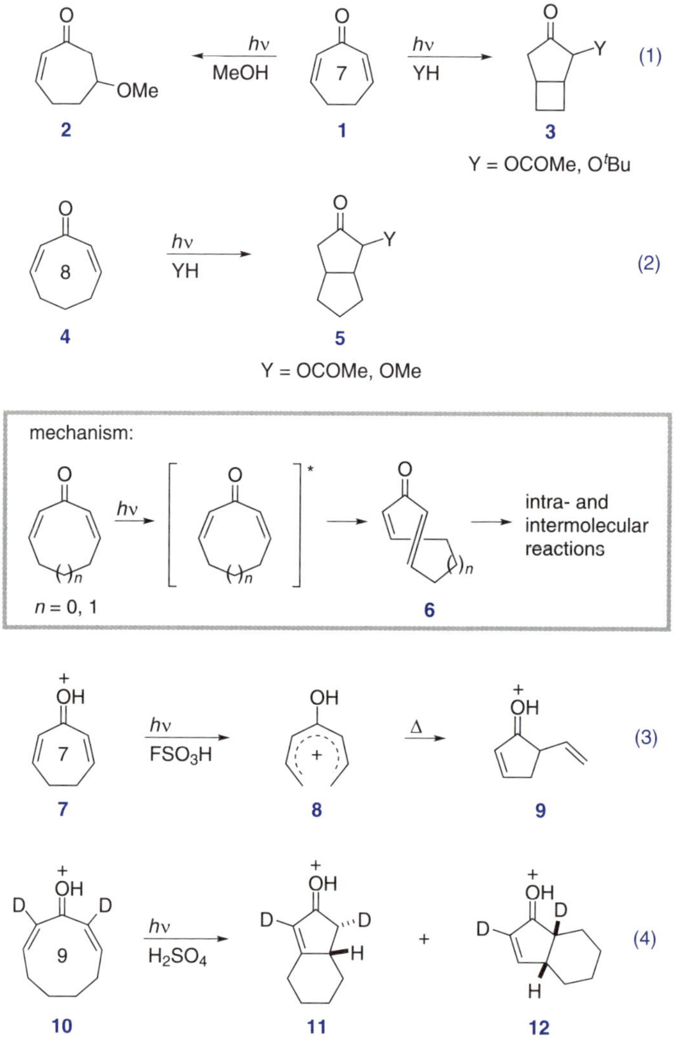

[20] R. Noyori, Y. Ohnishi, and M. Katô: Novel Photorearrangements of Cyclic Cross-Conjugated Dienones in Sulfuric Acid. Transformations Involving a Stereospecific Symmetry-Forbidden Process. *J. Am. Chem. Soc.* **1975**, *97*, 928–929.

does not cyclize, but instead undergoes a ring opening. Its conrotatory C(4)-C(5) single-bond cleavage, forming 4-hydroxyheptatrienyl cation 8, followed by thermal pentadienyl to cyclopentenyl cationic electrocyclization gives protonated 5-vinyl-2-cyclopentenone 9 (**Fig. 3.4.**, Eq. 3) [19]. In contrast, irradiation of *cis,cis*-2,8-cyclononadienone 10 causes disrotatory Nazarov ring closure, and the resulting *cis*-fused 5/6-bicyclic 2-hydroxyallyl cation undergoes the thermal, symmetry-forbidden [1s,4s] sigmatropic hydrogen shift to give 11 (**Fig. 3.4.**, Eq. 4) **[20]**. In addition, 10 experiences photochemical skeletal changes to give the bicyclic ketone 12. *cis,trans*-2,11-Cyclododecadienone undergoes the same skeletal changes with the same stereospecificity.

The Nazarov chemistry, studied together with Masao Katô, taught us many things. Most importantly, as suggested by a giant in molecular orbital theory M. J. S. Dewar, allyl, pentadienyl, and heptatrienyl cations are destabilized by the presence of an oxygen substituent at the *even-numbered* positions, thereby providing various unique chemical consequences. This directive effect becomes conspicuous in our Fe-carbonyl-polybromoketones chemistry.

3.6 Lessons of transition-metal effects on reactive intermediates (1968)

Although the aforementioned photochemistry was quite challenging and in vogue at the time, the major direction of our research group was shifted to OMCOS. My general conception, which later became a sensible approach, was that structure and reactivity of organic molecules or their fragments are strongly and diversely affected through interaction with metallic species, as we experienced in the Cu-catalyzed carbene cyclopropanation. Thus, we intended to explore efficient synthetic methods by full use of the characteristics of transition metals. It was not as easy as we initially thought because of the lack of fundamental knowledge at the time. "What is essential is invisible to the eye" (Antoine de Saint-Exupery, *The Little Prince*). I consider that bold imagination and creativity are the common denominators.

We first realized the wide utility of metal carbonyls. Since the discovery of $Ni(CO)_4$ by Mond in 1890, there had been numerous elaborated metal carbonyls with versatile reactivity applicable as homogeneous catalysts. In addition, metal carbonyls had been used as stoichiometric entities in organic synthesis via activation of unreactive compounds and also trapping of otherwise unstable species. Metal carbonyls can be used as reducing agents, and they have a high nucleophilicity because of the low oxidation state of the central metal atoms. In this regard, since such reductions generally require stoichiometric use of reagents, we considered low-cost Fe carbonyls as the most convenient ones.

Pursuing trimethylenemethane-metal complexes (Unrealized idea leading to a surprising consequence)

In 1968, immediately after arriving at Nagoya, we discovered an unusual chemistry of Fe carbonyls. I must answer the frequent question of why I began this research. In my early days, among the most astonishing chemistry was the discovery of cyclobutadiene-$Fe(CO)_3$ complex 13 in 1965 by Rowland Pettit (1927–1981) of the University of Texas at Austin (Fig. 3.5.) [A]. The $Fe(CO)_3$ complexation converts the otherwise unstable, rectangular, antiaromatic, singlet cyclobutadiene molecule [B] into the stable, aromatic complex with the square organic ligand. Surprisingly, its Ce^{4+} oxidation generates free cyclobutadiene that undergoes various unique reactions as a strained diene, but not as the triplet, anti-aromatic species.

[A] (a) G. F. Emerson, L. Watts, and R. Pettit: Cyclobutadiene- and Benzocyclobutadiene-Iron Tricarbonyl Complexes. *J. Am. Chem. Soc.* **1965**, *87*, 131–133. (b) For more on cyclobutadiene chemistry, see the autobiography of Günther Maier: *Das war's – Erinnerungen eines Doktorvaters*. GNT-Verlag, Berlin, **2021**.

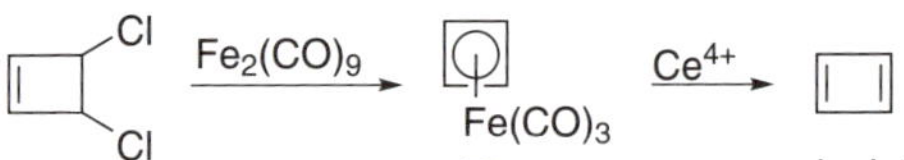

13 singlet

Fig. 3.5.
Pettit cyclobutadiene chemistry.

Inspired by this fascinating chemistry, I became interested in the $Fe(CO)_3$-fixation of the cross-conjugated trimethylenemethane (TMM) molecule 14 which exists in a ground-state triplet with the maximum bond order at the central carbon (**Fig. 3.6.**) [C]. We hoped that oxidation of the complex likewise generates a three-carbon reactive species, whose cycloaddition to alkenes or 1,3-dienes could construct odd-numbered carbocyclic compounds. In fact, G. F. Emerson, a former student of Pettit, had already succeeded in the synthesis of the parent TMM-$Fe(CO)_3$ complex 18 (R = H) from the dichloro precursor 16 and $Fe_2(CO)_9$, and we obtained the aryl-substituted derivatives 18 (R = phenyl) from the corresponding methylenecyclopropanes 17 and $Fe_2(CO)_9$ [21]. However, all actions of oxidants or donor compounds failed to generate the three-carbon diradicals 19 in our hand.

Some years later in 1973, I was invited to the Robert A. Welch Foundation Conference in Houston as a discussion leader for Pettit's lecture on this serendipitous story. I also talked with this colorful hero about our shared opinion about the worthless TMM-$Fe(CO)_3$ complexes. Pettit had a charismatic appearance with big eyes, and he usually made dynamic gestures in his lectures. His chemistry was straightforward and beautiful, and he was an intuitive scientific hero of our young days. This encounter was truly impressive and unforgettable.

Fig. 3.6.
Trimethylenemethane fixation.
R = H, Ph.

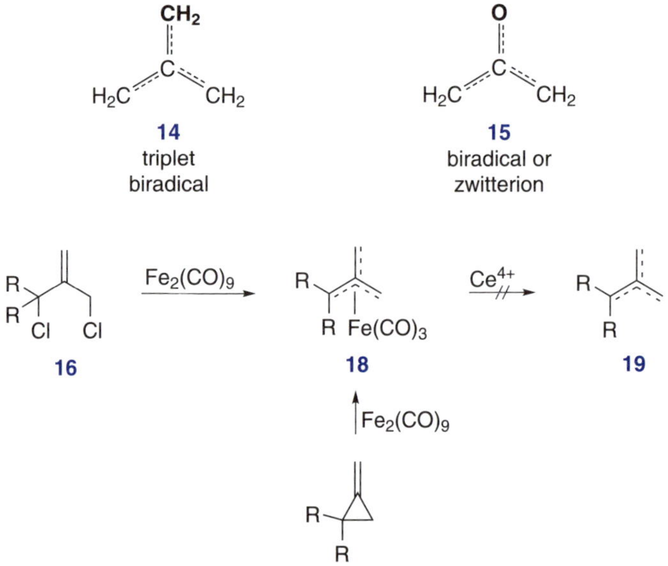

14 triplet biradical

15 biradical or zwitterion

16 18 19

17

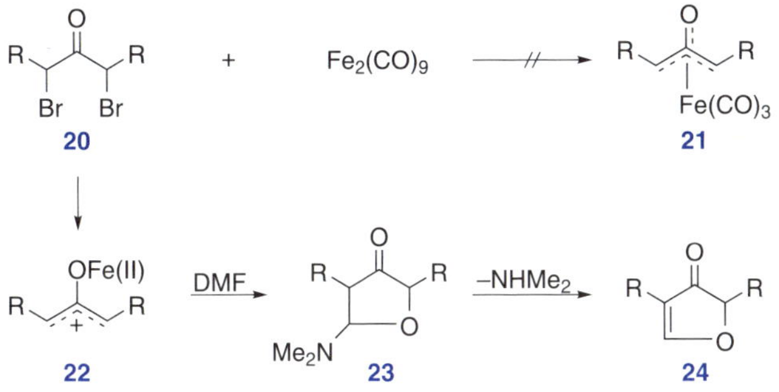

Fig. 3.7. An unusual cyclization reaction promoted by $Fe_2(CO)_9$. R = ^{i}Pr.

We then anticipated that replacement of its carbon elements by an oxygen atom might form the oxa analogue 15 (**Fig. 3.6.**), which would be less stable and thus more reactive. I asked my colleague Hidemasa Takaya to try the reaction of α,α'-dibromo ketone 20 and $Fe_2(CO)_9$ in dimethylformamide (DMF) aiming for the synthesis of 21 (**Fig. 3.7.**). This idea did not work. Instead, it unexpectedly resulted in furanone 24. What surprised us most was that the C–O unit came not from $Fe_2(CO)_9$ but from the DMF solvent! In fact, the initial reductive cycloaddition of dibromo ketone 20 and amido carbonyl formed 23, and subsequent β-elimination of dimethylamine ended up with the generation of 24 **[22]**. This is not considered a failure but is quite the opposite. We quickly understood that the reactive species was the σ-bonded Fe(II)-oxyallyl 22 instead of the expected π-interacted Fe(0)-oxyallyl complex 21. The reactive C_3 intermediate soon turned out to react with various unsaturated compounds. This unexpected fortune occurred immediately after launching my research group at Nagoya.

[22] R. Noyori, Y. Hayakawa, S. Makino, N. Hayakawa, and H. Takaya: Carbon-Carbon Bond Formations Promoted by Transition Metal Carbonyls. VIII. Iron Carbonyl Promoted Reaction of α,α'-Dibromo Ketones and Carboxamides. A Convenient Route to Muscarines *via* Furanones. *J. Am. Chem. Soc.* **1973**, *95*, 4103–4105.

Unrealized idea II

I would like to describe another puzzling story. We wanted to activate methylenecyclopropane 17 with a Ni or Pd catalyst in the hope of generating reactive TMM complexes [23]. In fact, we found that Schrauzer's $Ni(acrylonitrile)_2$ catalyst effected the [3 + 2] cycloaddition across methyl acrylate, giving the substituted methylenecyclopentane product 26. However, the reaction mode and pathways were not straightforward [24, 25]. The adduct 26 is derived mostly from the C(1)-C(2) cleavage of 17 as proven by the use of CHD=CDCOOMe (**Fig. 3.8.**, Eq. 1). We were clueless. Direct C(1)-C(2) cleavage with Ni(0) is unlikely. The formal [σ2 + π2] addition is now considered to occur via an initial metallacyclopentane species formed from the strained alkene 17, acrylate, and Ni(0), which undergoes cyclopropylcarbinyl to allylcarbinyl skeletal change and reductive elimination of Ni(0) [D]. Analogous vinylic bond fission is seen in Ni(0)-catalyzed isomerization of 17 to 1,3-butadiene [26] and reaction with Fe(0) or Pd(II) [21, 23]. The reaction modes are somewhat different with dimethyl fumarate (**Fig. 3.8.**, Eq. 2). The fumarate adducts 27

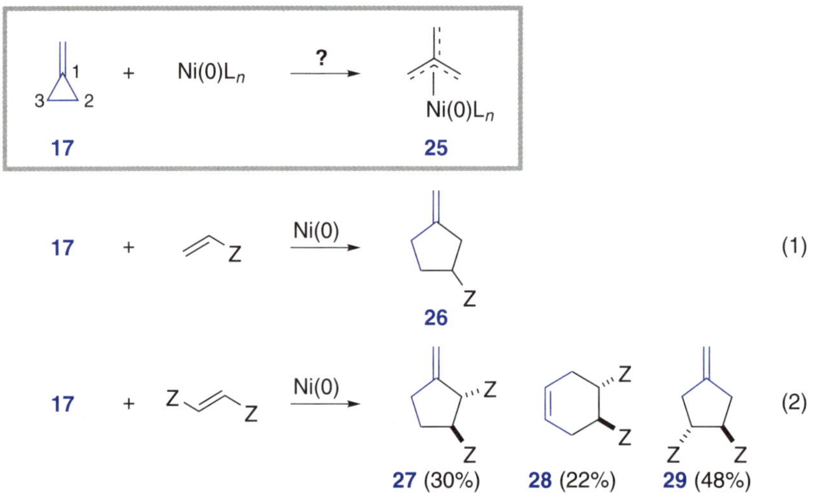

Fig. 3.8. Ni(0)-catalyzed [σ2 + π2] cycloaddition. Z = CO_2Me.

[E] B. M. Trost and D. M. T. Chan: New Conjunctive Reagents. 2-Acetoxy-methyl-3-allyl-trimethylsilane for Methylenecy-clopentane Annulations Catalyzed by Palladium(0). *J. Am. Chem. Soc.* **1979**, *101*, 6429–6432.

and 28 are produced similarly via the C(1)-C(2) cleavage of 17, but the symmetrically disubstituted adduct 29 involves the C(2)-C(3) cleavage as we earlier expected. This was the sole consolation to my young colleagues who engaged in this study. In accordance with a long-sought TMM-Ni(0) intermediate 25 or a rapidly equilibrating methylene-metallacyclobutane isomer, the reaction of deuterium-labeled 17 afforded 29, in which the three methylene units are fully scrambled [27].

My TMM chemistry remained unfinished. It was with great reluctance that I left this intriguing subject behind. Later on, useful Pd-mediated TMM reactions were developed by Barry Trost with the use of different, functionalized starting units **[E]**.

References for chapter 3

[A] (a) G. F. Emerson, L. Watts, and R. Pettit: Cyclobutadiene- and Benzocyclobutadiene-Iron Tricarbonyl Complexes. *J. Am. Chem. Soc.* **1965**, *87*, 131–133. (b) For more on cyclobutadiene chemistry, see the autobiography of Günther Maier: *Das war's*. GNT-Verlag, Berlin, **2021**.

[B] S. Masamune, F. A. Souto-Bachiller, T. Machiguchi, and J. E. Bertie: Cyclobutadiene is not square. *J. Am. Chem. Soc.* **1978**, *100*, 4889–4891.

[C] P. Dowd: Trimethylenemethane. *Acc. Chem. Res.* **1972**, *5*, 242–248.

[D] P. Binger: Cyclodimerisation von Methylencyclopropan an einem Ni(0)-Katalysator. *Angew. Chem.* **1972**, *84*, 352–353.

[E] B. M. Trost and D. M. T. Chan: New Conjunctive Reagents. 2-Acetoxymethyl-3-allyltrimethylsilane for Methylenecyclopentane Annulations Catalyzed by Palladium(0). *J. Am. Chem. Soc.* **1979**, *101*, 6429–6432.

A POST-DOCTORAL FELLOW AT HARVARD UNIVERSITY

4

 DOI.ORG/10.47261/1564-4

4.1 Joining E. J. Corey's group at Harvard (1969–70)

A year after my arrival at Nagoya, I went to Harvard to work with Professor Elias J. Corey (*1928) as planned with the generous permission of the chemistry department.

On Christmas night of 1968, I stepped on American soil for the first time at the age of thirty. It was minus 20 °C when I first arrived in snowy Boston. The cold weather, which I had never experienced before, and the fact that I could not understand a word of English made me foresee the hardships to come. However, with the help of my cousin Toshiko Kumagai and her husband, Shuku, who worked at the Boston branch of Japan Airlines, I quickly acquired a heavy coat and warm shoes, and began my life in the neighboring city of Cambridge. I rented a small room on Oxford Street, a 20-minute walk from the chemistry buildings. Late at night after finishing my experiments, I would carry a large bag full of laundry like Santa Claus to a nearby laundromat, which was very unusual for me, as they did not exist in Japan. People often ask me what was most difficult about living in Cambridge fifty years ago; it was a completely different environment, and everything was new to me. I felt that no matter how hard it was, I had nothing to lose. What I was able to accomplish as a researcher was limited, but every day during my fourteen-month stay I encountered surprises and observed many differences between the two countries.

I was amazed by the enormous difference in the standard of research environments and in living between the US and my home country. At Harvard, postdoctoral fellows were paid some 600 dollars a month, while my monthly salary as an Associate Professor at Nagoya, even when leading an independent research group, was equivalent to a mere 65 dollars (back then 1 USD = 360 JPY). We could purchase almost anything needed for our research, including a variety of reagents, catalysts, and purified solvents as well as high-quality, precoated thin layer chromatography (TLC) plates and many other disposables which were not easily obtainable in Nagoya. Dry ice and liquid nitrogen as coolants were available anytime free-of-charge for our experiments. The departmental library was open all day and night. Everything was completely different to what I was used to in Japan.

E. J. Corey was appointed Full Professor at Harvard at the age of 31 in 1959. He was a promising organic chemist at a young age, and by the time I arrived this 40-year-old professor was already eminent across the world. He was extremely knowledgeable, able to answer any questions immediately, and well-organized in the academic matters related to his research. He was diligent and worked tirelessly. At the same time, he was very devoted to his family being a father of two sons and a daughter. He would walk thirty minutes to work at the laboratory from 9:00 a. m., go back home for lunch, come to the lab again and return home at 5:30 p. m. He wrote papers at home at night and was an efficient and great writer. At that time, Corey had eight laboratories in the Converse and Conant buildings and supervised about twenty people, including more than a dozen elite postdoctoral researchers from around the country and abroad, plus seven or eight graduate students. The post-graduate students wore polo shirts and T-shirts, while the postdoctoral fellows usually wore ties in the labs. When I started as a postdoctoral fellow, Corey was beginning his sabbatical year but had decided to stay in Cambridge. Relieved from his teaching duties, he concentrated even more on his research.

The main research topic in Corey's laboratory was the chemical synthesis of bioactive compounds, such as naturally occurring compounds and artificial drugs with complex structures (**Fig. 4.1.**). Detailed planning was crucial, but the rest is experiment after experiment, and each step was a process of trial and error. Corey's devotion to research was second to none. First, Corey presented each student with an individual research theme and a detailed experimental plan. He then instructed them on the specific methods of the experiment and would go around the lab once or twice every day to ask for results. Although just working hard did not necessarily mean that research would move forward.

Fig. 4.1.
E. J. Corey lecturing on prostaglandin synthesis, Kyoto, 1972.

Sometimes, I felt slightly homesick and there were days when I would look at the planes in the sky leaving from Logan Airport and think that if I could just get on one of them, I could go home to Japan.

During the evenings, enthusiastic young researchers, the logic-loving "debaters," used to gather around the small blackboard in the corridor and begin discussions on research strategies. A certain degree of English skills was required to have debates, but above all, a passion for chemistry and persuasiveness, and I enjoyed the discussions with my younger lab mates.

The American "total synthesis school" of the time, targeting complex terpenes, steroids, alkaloids, and other medicinally important compounds, was led by R. B. Woodward and E. J. Corey, and included Gilbert Stork, Eugene van Tamelen, William Johnson, George Büchi, and many others. Corey, though, was an intelligent all-round chemist. He had a strong belief of the power of organic synthesis but at the same time recognized its inherent logical intricacy. He ambitiously pioneered the computer-assisted design of synthetic routes to complex organic molecules, resulting in the original LHASA (Logic and Heuristics Applied to Synthetic Analysis) program [A].

[A] E. J. Corey and W. T. Wipke: Computer-Assisted Design of Complex Organic Syntheses: Pathways for Molecular Synthesis can be Devised with a Computer and Equipment for Graphical Communication. *Science* **1969**, *166*, 178–192.

4.2 Prostaglandin synthesis and hydrogenation

My bench was in the Converse Basement, room 4 (**Fig. 4.2.**), where I worked with my lab mate, Ned M. Weinshenker from Frederick D. Greene's group at MIT and later Haruhisa Shirahama from Hokkaido University in Japan.

We were engaged in the research of prostaglandins (PGs), a series of polyoxygenated C_{20} fatty acids, discovered by Ulf von Euler (1905–1983) in the 1930s, which control a wide range of physiological responses in the human body. The chemical synthesis was crucial because: (1) the biosynthesis did not meet the increasing demand for the significant, but naturally scarce substances; (2) the structural modification was needed to properly tune the tissue selectivity and metabolic stability for the therapeutic use. In the late 1960s, Corey had pioneered this significant research field. After completing several works including the total synthesis of $PGF_{2\alpha}$ [28, 29], I was asked to selectively hydrogenate a $PGF_{2\alpha}$ derivative that has two C=C bonds to a $PGF_{1\alpha}$ derivative possessing a single C=C bond (**Figs. 4.3.–4.4.**) [30]. This was also successful. Some people tease me and say that I am not a good experimentalist as I tend to think more in my mind, but since I was able to co-author three papers during my 14-month stay, I would like to defend myself a bit and say that I am not so bad at experiments as people think.

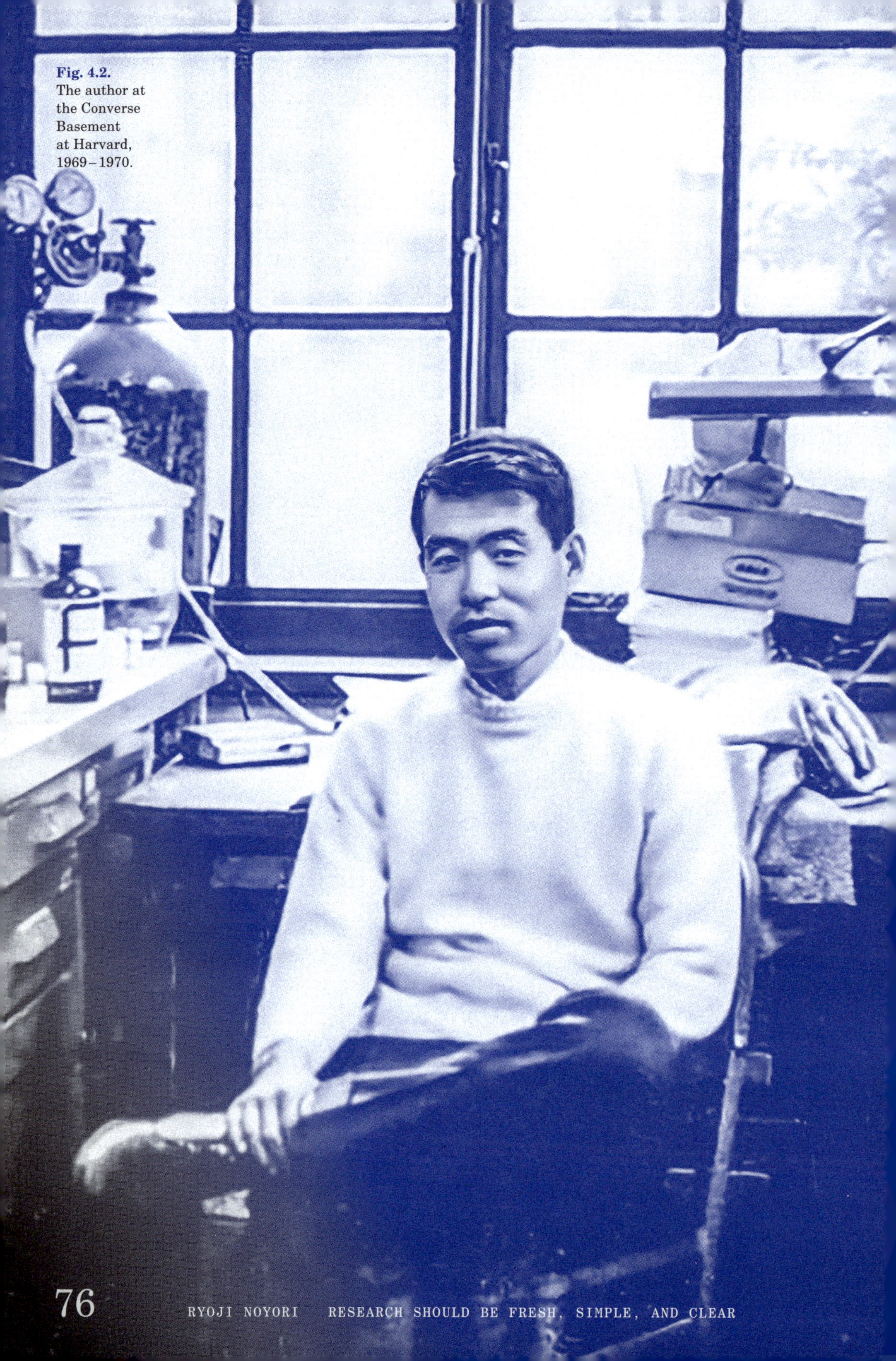

Fig. 4.2.
The author at the Converse Basement at Harvard, 1969–1970.

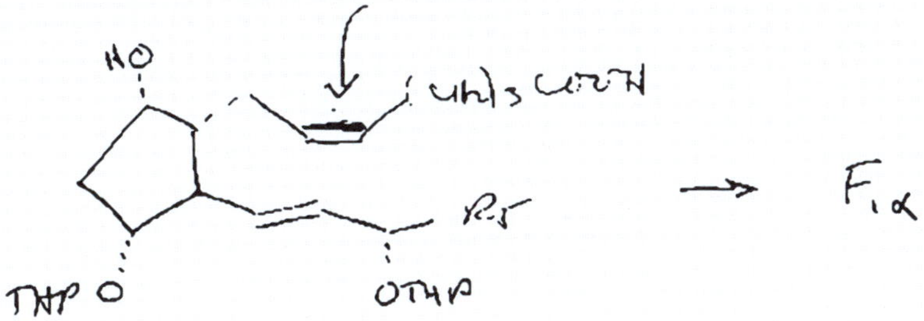

Fig. 4.3. Corey's handwritten suggestions to the author. Corey proposes the selective conversion of $PGF_{2\alpha}$ to $PGF_{1\alpha}$ by homogeneous Rh-catalyzed hydrogenation.

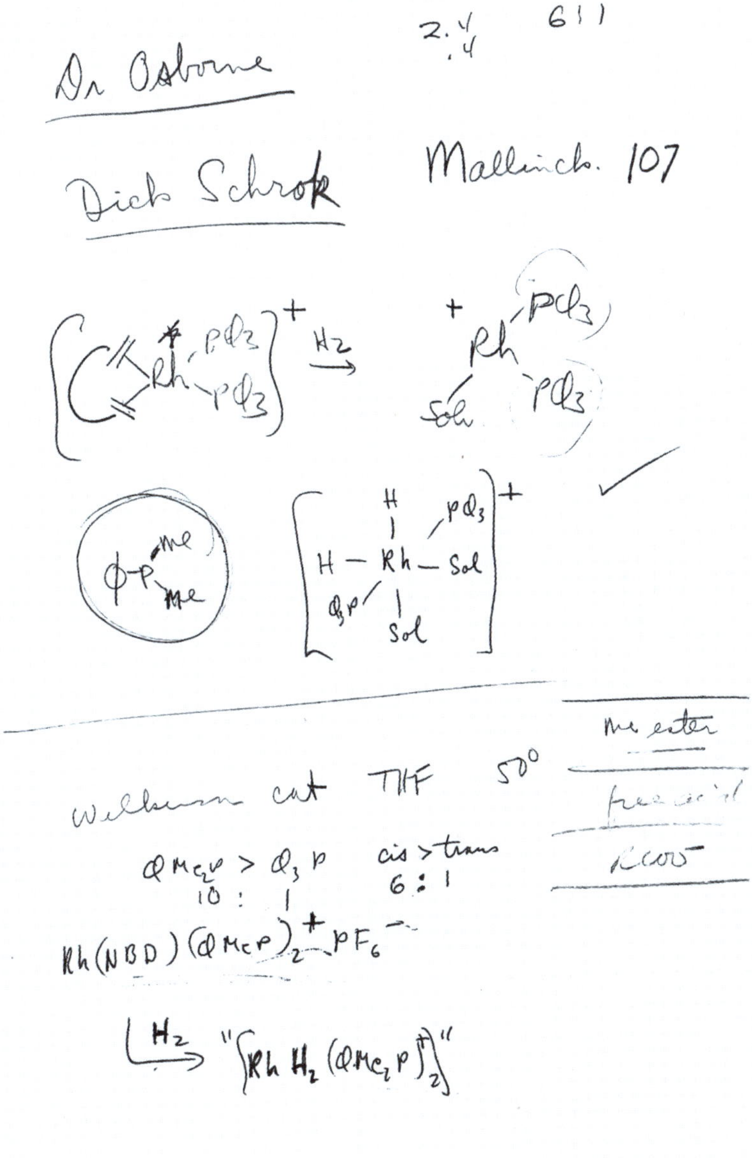

Fig. 4.4. Corey also suggests that the author meets with John Osborn's student Richard Schrock (based at the Mallinckrodt Building, room 107), who kindly provided a new cationic Rh catalyst (Fig. 4.5.).

[B] J. F. Young, J. A. Osborn, F. H. Jardine and G. Wilkinson: Hydride Intermediates in Homogeneous Hydrogenation Reactions of Olefins and Acetylenes using Rhodium Catalysts. *Chem. Commun. 1965*, 131–132.

I recall this research suggestion was the start of my three-decade work on hydrogenation at Nagoya. Reading almost all literature on this very new and challenging topic enhanced my interest in homogeneous hydrogenation. Harvard was also a place to meet with young researchers. I was primarily involved in the PG chemical synthesis under Corey, while studying organometallic chemistry in parallel, and this was essential for my future plans for conducting research at Nagoya University. I often discussed with John Osborn (1939–2000), who discovered the phosphine-Rh complex catalyzed hydrogenation in Geoffrey Wilkinson's laboratory at Imperial College London **[B]**. Moreover, I was personally invited by this British inorganic chemist to attend his regular lectures on fundamental organometallic chemistry. Later, he often told his friends, "Noyori is my student in organometallic chemistry." Osborn's student Richard R. Schrock (*1945, 2005 Nobel Laureate in Chemistry) kindly provided me with a precious sample of a newly created cationic Rh catalyst for our hydrogenation research. Corey seemed to have contacted Osborn in advance to ask for the catalyst and wrote me a note to go and see Schrock, who was then working in the inorganic chemistry building at the Harvard campus. I remember that I was extremely motivated by this note, which led to my journey in asymmetric Rh chemistry.

Although the final solution to saturate the *cis* double bond was obtained by using a simple heterogeneous Pd/C catalyst (**Fig. 4.6.**) [30], I tested many kinds of homogeneous systems as well, perhaps more than hundred times using 1 mg each of the precious PG sample. During this task on selective C=C bond hydrogenation, I was helped by K. Barry Sharpless (*1941), a fellow 2001 Nobel Laureate in Chemistry, who was then a postdoctoral fellow in the biochemistry laboratory of Konrad Bloch (1964 Nobel Laureate in Physiology or Medicine). Since high performance liquid chromatography (HPLC) was yet undeveloped, his

Fig. 4.5. In 2011, R. Schrock recalls his old interaction with the author. Interview with R. Eisenberg. Voices of Inorganic Chemists. Celebration of the 50 Year of Inorganic Chemistry, the American Chemical Society, 2011 (›Links [8].).

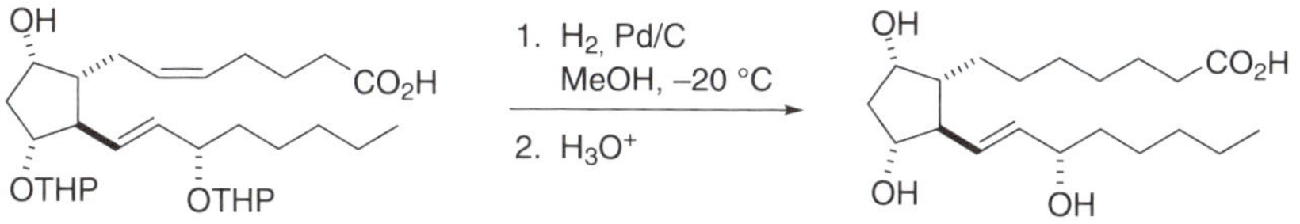

Fig. 4.6. Selective hydrogenation of a $PGF_{2\alpha}$ derivative. THP = 2-tetrahydropyranyl.

special TLC technique using an $AgNO_3$ / acetonitrile-impregnated silica-gel plate was extremely effective for analyzing the structurally very similar olefinic compounds.

Fig. 4.7. Author with John Osborn at the 10th International Symposium on Homogeneous Catalysis at Princeton in 1996.

The most valuable gift I received from my stay at Harvard was the warm friendship of many promising students and postdoctoral fellows who supported and guided me both at the time and throughout my entire academic career. The members of the years 1969 – 70 in the Corey group who later played significant academic roles include Kazuo Achiwa, David E. Cane, Ken Harding, Louis S. Hegedus, John A. Katzenellenbogen, Isao Kuwajima, and Haruhisa Shirahama. Dieter Seebach, Martin F. Semmelhack, and Gary H. Posner were a little ahead of me; Alain Krief, Koichi Narasaka, K. C. Nicolaou, and Masakatsu Shibasaki joined the group a little later. I share a special bond with Hisashi Yamamoto, a student of Nozaki and then Corey, who also pursued OMCOS at Nagoya (CHAPTER 9). Nearly seventy Japanese chemists worked under Corey's supervision. They and other the alumni of the Corey School spread across the world, in both academia and industry, and have made tremendous contributions to the establishment of modern organic synthesis.

Fig. 4.8.
The author (left) and Hisashi Yamamoto, a graduate student, at Harvard Yard, 1969.

4.3 My astonishment in encountering world-leading chemists

The Harvard chemistry department was a colorful collection of 20th century heroes of organic chemistry, including Robert B. Woodward, Louis F. Fieser (emeritus), Paul D. Bartlett, Frank H. Westheimer, and William von E. Doering, in addition to the rising star E. J. Corey. Here, I felt that structural / mechanistic organic chemistry and synthetic chemistry, and perhaps other sub-disciplines, were not separable. The regular early-evening department seminars were highlighted by the active participation of these academic giants. In addition, I attended the famous Woodward seminar, which lasted until midnight several times.

There are many anecdotes about Woodward, the charismatic chemist who was described as the greatest organic chemist of the 20th century. His perfectionism may have been the result of his European view of academia, which respects dignity and classical style. He was a well-dressed gentleman who liked to exaggerate and be dramatic. His name was written in gold letters on the beautiful wooden door of his specially appointed professor's office, which also had its own washroom and napping room. Everything was blue, and the laboratory was equally neat and tidy. His private parking space was also painted sky blue, a prank by a student.

This special professor did not give regular lectures and would hold the Woodward seminars on an irregular basis. Around 7 p.m., one or two graduate students would present their research results, and the audience would freely express their opinions. However, this was just the beginning. Eventually, around 8 p.m., the blackboard would be wiped clean with a wet cloth. Then, the main actor would appear slowly, wearing a dark blue suit and the standard sky blue tie, and take a seat in the center of the front row. The classroom would be quiet for a moment, as he picked up his favorite cigarette, a Benson & Hedges. After a while, he would write a quiz-type question in colored chalk on the blackboard. He had about five colors of chalk, which he carefully wrapped in a cloth and brought with him. This custom did not change when he came to give a lecture at Nagoya in later years. Graduate students and post-doctoral researchers who wanted to show their presence and ability in front of the professor were the first to come forward to the blackboard to answer the questions. The professor would look on, chatting with those around him, and when the answer was correct, he would smile and put a check to the answer with his usual red chalk. This was when the seminar really got started. The explanation of the true meaning of the question would begin. The differences between the common explanation and their own ideas were shown. Of course, we were free to debate, but there was little chance of winning. Since the seminar was informal and open to the public, even those of us outside his research

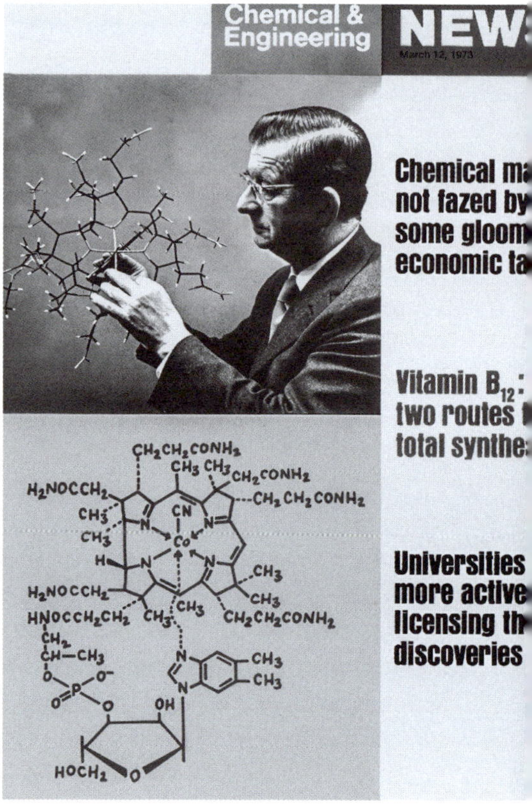

Figs. 4.9.–4.10. R. B. Woodward (and R. Hoffmann) on the covers of *Chemical & Engineering News* in 1972 and 1973.

group attended, and there were also participants from MIT and other universities in the Boston area who wanted to enjoy the lectures and challenge the great professor.

Inspired by the Woodward seminar, the Hirata group at Nagoya University also introduced evening seminars in the mid-1960s, with guests from Meijo University and Nagoya City University as well. I joined the Nagoya seminar as soon as I arrived there, but I cannot compare the two as the characters of the participants were so different.

Woodward's seminar was not conducted so frequently when I was at Harvard. He had already received his long-awaited Nobel Prize in Chemistry for "The Art of Chemical Synthesis" in 1965, but his highly anticipated project, "Chemical Synthesis of Vitamin B_{12}," which started in 1961 and then proceeded jointly with Albert Eschenmoser of the Swiss Federal Institute of Technology beginning in 1965, had approached the most critical stage (**Figs. 4.9.–4.10.**) **[338]**. Moreover, he was busy compiling a grand theory of conservation of molecular orbital symmetry to be published in *Angewandte Chemie*. This theoretical work, first

[338] R. Noyori: Fakten sind der Feind der Wahrheit – Betrachtungen über Zufallsentdeckungen und ungeahnte Entwicklungen der asymmetrischen Katalyse. *Angew. Chem.* **2013**, *125*, 83–98; R. Noyori: Facts are the Enemy of Truth—Reflections on Serendipitous Discovery and Unforeseen Developments in Asymmetric Catalysis. *Angew. Chem. Int. Ed.* **2013**, *52*, 79–92.

published in 1965, was expected to grant him his second Nobel Prize, but he died suddenly in 1979 at the age of 62. Two years after his death, his young collaborator, Roald Hoffmann, and Kenichi Fukui were jointly awarded the Nobel Prize in Chemistry. Corey has long believed that their theory originated from a conversation with Woodward in May 1964, as described in his 2004 Priestley Medal address **[C]**.

[C] (a) E. J. Corey: Impossible Dreams, *J. Org, Chem.* **2004**, *69*, 2917–2919; (b) R. Hoffmann: A Claim on the Development of the Frontier Orbital Explanation of Electrocyclic Reactions. *Angew. Chem., Int. Ed.* **2004**, *43*, 6586–6590.

I learned much by watching the behavior of Corey and Woodward. I deeply respected them, but I felt that they could not be my role models, as I personally felt that my capability remained far below theirs and my mentality was totally different from the ones of these chemistry giants. The US and Japan are also very different in terms of the society; the US is diverse, competitive, and values individuals, whereas Japan is more homogeneous, collaborative, and values groups over individuals, factors that strongly affect the academic culture of both countries. The economical gap between the countries was enormous, though the gap slowly closed later on. Furthermore, the Japanese university faculty system and funding system were so different to those in the US. Japan had no postdoctoral system until around the 1990s, and scholarships and fellowships provided to students were and even still are insufficient. We had to find our own ways to cleverly survive in such difficult circumstances, which is probably not easy for many to imagine. My lab experience at Harvard with Corey and learning the Woodward style has had a great influence on my research and on leading my own group.

Although Corey and Woodward are both Americans, I felt that their personalities and behaviors were quite different. In contrast to Woodward, Corey never drank or smoked. He was very serious in every aspect. Their ways of laboratory management were also very different. Corey is a well-prepared, efficient scientist with remarkably organized knowledge; he firmly remembers every major chemistry event that occurred since 1950. I heard this directly from him. He saw each lab member at least once a day, asking "how is your experiment going?" and gave many suggestions accordingly.

Woodward's lab, which focused on the synthesis of vitamin B_{12}, was different to Corey in many ways. The students and postdoctoral researchers hardly found time to discuss with Woodward (only several times a year), who spent much of his time in Switzerland and also writing a review for *Angewandte Chemie* with R. Hoffmann on molecular orbital (MO) theory. Woodward usually talked with a limited number of members who were engaged in imminent problems to be solved. This was his style, which I heard from a Japanese friend of mine who worked in Woodward's group. I learned that Yoshito Kishi became highly recognized in this way when he was a postdoctoral researcher in Woodward's group in the late 1960s.

Overall, both Corey and Woodward were an extreme, and none can be so ambitious. I have tried to create unique science, but never liked to compete with others. Being a (hopefully) reasonable and modest Asian scientist, I stood on the opposite side of these charismatic American figures.

4.4 Experiencing the stimulating American society

I was astonished by the enormity of the US and all the ways to get around it, like a long-distance bus service called the Greyhound and a bus-like airplane called an air shuttle, which frequently flew to other regional cities. I had never even taken a domestic flight within Japan before, so the US seemed like a huge place to me. My first stay in the US would be meaningless if it was just for research. I felt that it was also important to have fun and explore the society.

In the late 1960s, while the men were working, housewives in Japan went to the market every day to buy food for the day. In the US, a country with advanced automotive technology, couples drove to the market on Fridays after work to buy everything for the whole week. It was a testament to the difference between the social advancement of women and the ubiquity of large refrigerators. The sheer size of milk and juice cartons, wide variety of fruits, and inexpensive alcohol stood out to me. In Japan, milk and fruit juices were usually sold in small bottles (180 ml), but in the US they were sold in large paper cartons, which made me realize the wealth of the country. Oranges and grapefruits from California and Florida were also a symbol of American food culture. In Japan, alcoholic beverages were clearly separated from other beverages, and I used to buy a large bottle of beer from the liquor store. In the US, however, people were buying a whole case of 24 small bottles together with other foods at the supermarket. Johnny Walker Black Label was such a luxury in Japan that I even kept a gifted bottle on my altar. But in the US, I delighted in the fact that I could drink it casually.

For lunch, I often went with Hisashi Yamamoto (**Fig. 4.8.**) to the Faculty Club on campus where Harvard professors gathered with friends. I was told that the club served horse steak, a traditional Boston dish, so I tried it, but it was very tough and the taste was rather disappointing. I went through most of the restaurants in the international neighborhood of Harvard Square. I wonder if the German-style pub-restaurant, Wursthaus (Sausage House), where I had my first meal in Cambridge, still exists.

Being a bachelor at that time, I had plenty of time to spare, and enjoyed a good relationship with my new friends. For example, on the evening of July 20, 1969, I was invited to my lab mate's house, to watch the live TV broadcast of the moon landing of Apollo 11 together with his family. At that time, bright American students did not know much about the East, so I tried to teach them things about Japan, but they often did not understand and seemed to look upon our customs with wonder.

On weekends, even though I was exhausted, I refreshed my mind and went out with my cousin's family and friends to play golf, drive to Cape Cod and around New Hampshire, and participate in car rallies. The natural scenery in New England is vast and beautiful. From time to time, I also attended meetings of the Japanese Association. Even with the many opportunities the US had to offer, many Japanese people were forced to return due to sensitive nerves. It was not an easy life for me either, but it seems that I was in a very fortunate situation compared to them. The American society was full of dynamism, and it was an intensive experience. There was no way we could compete with the US, but I was determined to use this experience as a springboard for the development of chemistry in Japan.

4.5 A memorable one-month vacation

After my memorable fourteen-month stay at Harvard, I wanted to learn more about the US by visiting some of my acquaintances. At the University of Maryland, Ned Weinshenker from Corey's lab had been appointed as Assistant Professor, and he invited me to present my research that I was conducting in Nagoya at his university. It was my first time to give a lecture overseas, and I was in a cold sweat as I had no idea how to communicate with the faculty members who gathered around me. I also visited Japanese researchers at their laboratories at Stanford University and the University of California, Los Angeles. It was also at this time that I visited George A. Olah (1927 – 2017, Nobel Laureate in Chemistry in 1994) and his wife, Judith, at Case Western University, where I was amazed by the most advanced NMR facility at that time to study stable carbocation structures. After visiting the home of Koji Nakanishi in New York, who had just moved to Columbia from Tohoku University, I departed from J. F. K. Airport for a European tour.

In the UK, I was welcomed by my former lab mate John Hutton, who had been involved in prostaglandin synthesis at Corey's lab, at his pleasant home and garden in Oxford. When he met me at the train station, he told me off for travelling in a second-class train carriage. The pride of a British scientist is indeed impressive, and the Japanese chemist, used to poverty, learned one more thing. In Switzerland, I had a chance to visit the Department of Chemistry at ETH Zurich, a place where young people around the world aspire to study, and where my friend from Japan was working in Eschenmoser's lab at that time. I was deeply impressed by the historical elegance of this school. I travelled to some other European cities for sightseeing before returning to Japan. I spent all my savings on this trip. I thought it would be my only chance to travel abroad, but this assumption turned out to be totally incorrect.

4.6 My lab in Japan

Thus, my first international adventure fruitfully ended thanks to the thoughtful consideration and kindness of many people. I recall that this was a difficult time for both American and Japanese universities for different reasons. The US got bogged deeper in mud of the Vietnam War, which profoundly affected the morale of students and the academic activities on the campus. At the same time, Japan's universities were in the midst of a school dispute with the All-Communist Party among other sectors, including the occupation of the Yasuda Auditorium of the University of Tokyo. I was worried about the state of the laboratories at Nagoya University. Every time I met a visitor from Japan in Cambridge, they would warn me that my laboratory might be shut down. During this time, I was directing the new laboratory while communicating in writing with Takaya and Katô in my group and other faculty members in the department. Even if I wrote back immediately, it would take two weeks for the air mail to go back and forth. International telephone calls were too expensive for us. I deeply acknowledge my generous colleagues at Nagoya who assumed my managerial responsibility during my leave and also apologize for any inconvenience caused by my absence.

References for chapter 4

[A] E. J. Corey and W. T. Wipke: Computer-Assisted Design of Complex Organic Syntheses: Pathways for Molecular Synthesis can be Devised with a Computer and Equipment for Graphical Communication. *Science* **1969**, *166*, 178 – 192.

[B] J. F. Young, J. A. Osborn, F. H. Jardine and G. Wilkinson: Hydride Intermediates in Homogeneous Hydrogenation Reactions of Olefins and Acetylenes using Rhodium Catalysts. *Chem. Commun.* **1965**, 131 – 132.

[C] (a) E. J. Corey: Impossible Dreams, *J. Org, Chem.* **2004**, *69*, 2917 – 2919;
(b) R. Hoffmann: A Claim on the Development of the Frontier Orbital Explanation of Electrocyclic Reactions. *Angew. Chem., Int. Ed.* **2004**, *43*, 6586 – 6590.

THREE DECADES OF OMCOS RESEARCH AT NAGOYA

5

 DOI.ORG/10.47261/1564-5

In April 1970, I returned to my own laboratory at Nagoya. Having learned many things in the US, I was full of fresh ideas. In particular, Rh-catalyzed asymmetric hydrogenation (AH) was a challenge that needed to be tackled immediately. But I already had a range of exciting research subjects lined up in my group, preventing us from changing our research direction. Each day, more and more things came up that needed to be done. We were so busy maintaining an active front that we could not devote our scarce human resources to new projects.

5.1 Fe carbonyl-promoted [3 + 4] and [3 + 2] cyclo-coupling reactions (1968 – 1984)

While at Harvard, I had been supervising my laboratory at Nagoya through frequent air mail correspondence. Following the discovery of the unexpected reaction shown in Fig. 3.7. (CHAPTER 3), we immediately developed an Fe carbonyl-aided [3 + 4] cyclo-coupling of α,α'-dibromo ketones and 1,3-dienes in hydrocarbon solvents forming 4-cycloheptenones (Fig. 5.1.) [31]. As in the Diels-Alder reaction, dienes with a high equilibrium concentration of the *s*-cis conformer serve as efficient receptors of the bifunctional C_3 units. The use of diene-$Fe(CO)_3$ in place of free dienes resulted in a significant increase in yield. As expected, the reaction of cyclic dienes, furans, and some pyrrole derivatives afforded the bicyclic ketones in high yields [32, 33].

[31] R. Noyori, S. Makino, and H. Takaya: Carbon-Carbon Bond Formations Promoted by Transition Metal Carbonyls. I. Reaction of α,α'-Dibromo Ketones with Iron Carbonyls in the Presence of 1,3-Dienes. A New Route to Troponoid Compounds. *J. Am. Chem. Soc.* **1971**, *93*, 1272 – 1273.

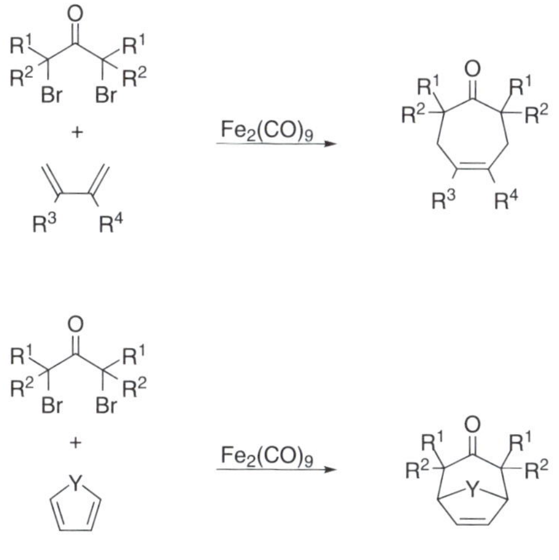

Fig. 5.1. Fe carbonyl-promoted [3 + 4] cyclo-coupling. Y = CH_2, O, NCOR, NCO_2R.

This discovery marked a milestone in my early career, since only very few examples were known of the carbocycle formation using a C_3 carbon unit [A]. The novelty of this method, which allowed a direct troponoid synthesis [31, 33 – 36], was recognized by Professor Tetsuo Nozoe at Tohoku University, a father of non-benzenoid aromatic

Fig. 5.2. Tetsuo Nozoe (1902 – 1996).

[B] J. I. Seeman: The Nozoe Autograph Books: Stories behind the Stories. *Chem. Rec.* **2013**, *13*, 483–514.

chemistry who had discovered troponoid compounds. From then on then, this creative and influential scientist kindly guided me for many decades **[B]**. This finding made me, then a totally unknown young Japanese scientist, visible in the global chemistry community. Following my publication in 1971, Hans M. R. Hoffmann of Imperial College London reported a similar Zn/Cu-promoted process [C]. NaI may be used in certain cases, but I insisted in using $Fe_2(CO)_9$ for this reaction which originated from my rather adventurous idea (**CHAPTER 3**).

We were poised to explore the potential of this exciting chemistry to the greatest extent. In 1971, as Yoshihiro Hayakawa joined our group from the Hirata group, the synthesis of carbocyclic compounds using Fe carbonyls made great progress. Mechanistically, from the beginning, I was convinced that the reactive intermediate is the Fe(II)-oxyallyl species 1 (**Fig. 5.4.**) and viewed the seven-membered ring formation as a concerted [2π + 4π] cycloaddition. The reaction starts with a two-electron reduction of the bromo ketone forming an enolate, as proven by the reaction of *endo*-α-bromocamphor in DMF-D_2O which gives *exo*-α-deuteriocamphor. With a dibromo ketone, the initially formed enolate eliminates the bromide ion from the allylic position.

Fig. 5.3. Reaction of *endo*-α-bromocamphor in DMF-D_2O.

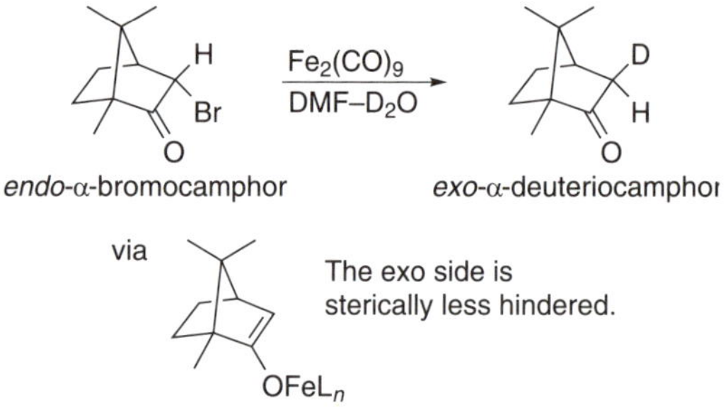

The oxyallyl intermediates 1 with a configuration of either W-, sickle-, or U-form depending on the substituents, undergo various intramolecular skeletal changes [37–39]. Some unique transformations include conrotatory electrocyclization of a five-carbon 4π-electron system involving the aromatic ring (**Fig. 5.4.**, Eq. 1), cationic [1a, 4s] sigmatropic rearrangement (Eq. 2), the neopentyl cation rearrangement via 1,2-methyl shift (Eq. 3), *tert*-butylallyl to cyclopropylcarbinyl cation rearrangement (Eq. 4), and the cationic [3, 4] sigmatropic rearrangement (Eq. 5).

The Fe(II)-oxyallyl species 1 acts as both a uni- and bifunctional C_3 electrophile. The allylic cation 1 is susceptible to nucleophilic trapping with methanol or sodium acetate and sometimes deprotonation in basic media such as THF or DMF. Particularly noteworthy is the endowment of an *electrophilic* character to a carbon α to the original keto function, which is otherwise *nucleophilic*, as is seen in enolate chemistry. The allylic

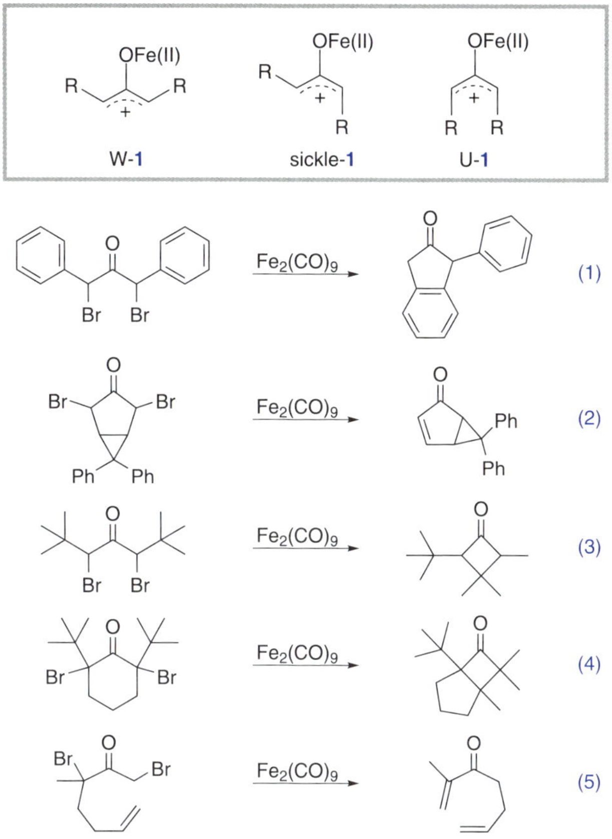

Fig. 5.4. Skeletal changes via Fe(II)-oxyallyl intermediates. R≠H.

cation is stabilized by electron-releasing substituents at the C(1) and C(3) positions and, in turn, is destabilized or uniquely activated by the presence of the central oxygen atom masked with Fe(II). This phenomenon was already observed in our study on photochemistry (**CHAPTER 3**). Due to this mechanistic reason, the reaction does not allow the use of α,α'-dibromoacetone, the simplest dibromo ketone, as a C_3 unit.

The oxyallyl species adds across not only 1,3-dienes but also to certain suitably substituted olefins (**Fig. 5.5.**). For example, the [3 + 2] reaction with aryl olefins directly provided the 3-arylated cyclopentanones [40, 41]. The cyclo-coupling with morpholino enamines, followed by β-elimination of morpholine, provides a single-pot procedure for cyclopentenone synthesis [42 – 44]. Likewise, as seen in the reaction with DMF (**Fig. 3.7.**), 3(2*H*)-furanones are produced via the hetero [3 + 2] annulation involving the C=O linkage of carboxamides, followed by elimination of dialkylamines [22, 45].

A Complete Neglect of Differential Overlap (CNDO) calculation using model reactions suggests that the [3 + 4] reaction occurs by a symmetry-allowed [2π + 4π] cycloaddition mechanism, in which the regioselectivity is controlled by the frontier molecular orbitals of

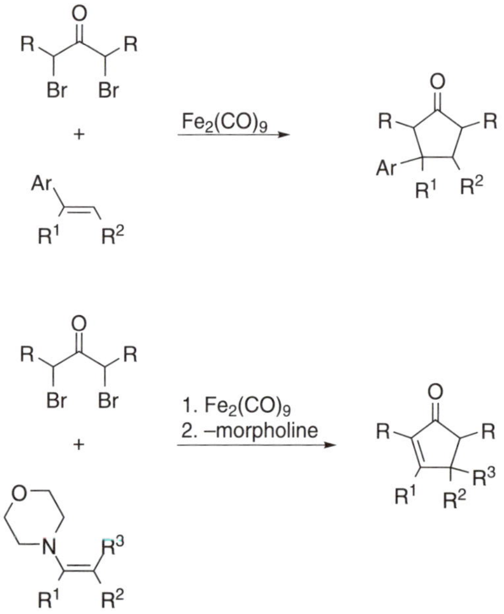

the two components (**Fig. 5.6.**) [46]. In contrast, the symmetry-forbidden [3 + 2] reaction is best interpreted by assuming a stepwise but stereospecific mechanism involving a zwitterion, stabilized by the substituents of the acceptor and donor components. Thus, a strong carbocation-stabilizing group is needed in the olefinic substrates [47, 48]. The hetero [3 + 2] reaction is understood in a similar manner. Overall, the reactivity of the C_3 oxyallyl intermediates is reminiscent of that of tetracyanoethylene undergoing both [2 + 4] and [2 + 2] cycloadditions.

The chemistry is highly general. As illustrated in **Fig. 5.7.**, these new reactions allowed the flexible construction of five- and seven-membered carbocycles and others, depending on the structures of substrates. This method can be extended to certain intramolecular skeletal transformations as well.

Fig. 5.5. Fe carbonyl-promoted [3 + 2] cyclo-couplings.

The Fe carbonyl-aided cyclo-coupling reactions enabled us to construct a wide variety of organic frameworks, which are of theoretical and practical interest. The conditions are mild, and a range of functional groups, including carbonyl, ester, amide, cyano, ether linkage, and others are tolerated. The usefulness was demonstrated by straightforward syntheses of numerous naturally occurring bioactive compounds after further mechanism-based elaborations [49, 50].

In addition to the natural / unnatural troponoids and other terpenoids (**Fig. 5.9.** a) [51 – 54], we accomplished the general synthesis of tropane alkaloids via the reductive [3 + 4] cyclo-coupling (**Fig. 5.9.** b). In order to achieve this goal, the following three problems had to be solved: (1) incapability of dibromoacetone to generate 2-oxyallyl species; (2) *N*-methylpyrrole's predominant aromatic, instead of 1,3-diene, character undergoing electrophilic substitution instead of the desired cycloaddition; and (3) instability of the azacyclic structure of the adduct causing a retro-Mannich reaction under the reaction conditions. These problems could be overcome by replacing dibromoacetone with *sym*-tetrabromoacetone as a C_3 unit and *N*-methylpyrrole by *N*-methoxycarbonylpyrrole as a substrate [55 – 57]. Tetrabromoacetone serves as an equivalent of unsubstituted 2-oxyallyl, coupled with debromination from the

Fig. 5.6. Mechanisms of the [3 + 4] and [3 + 2] cycloadditions. EDG = electron-donating group.

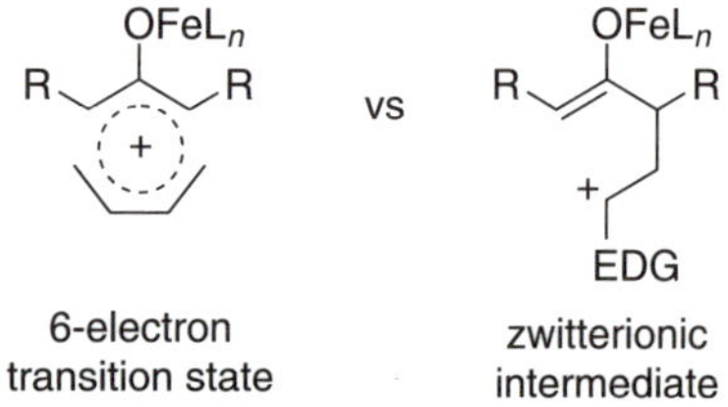

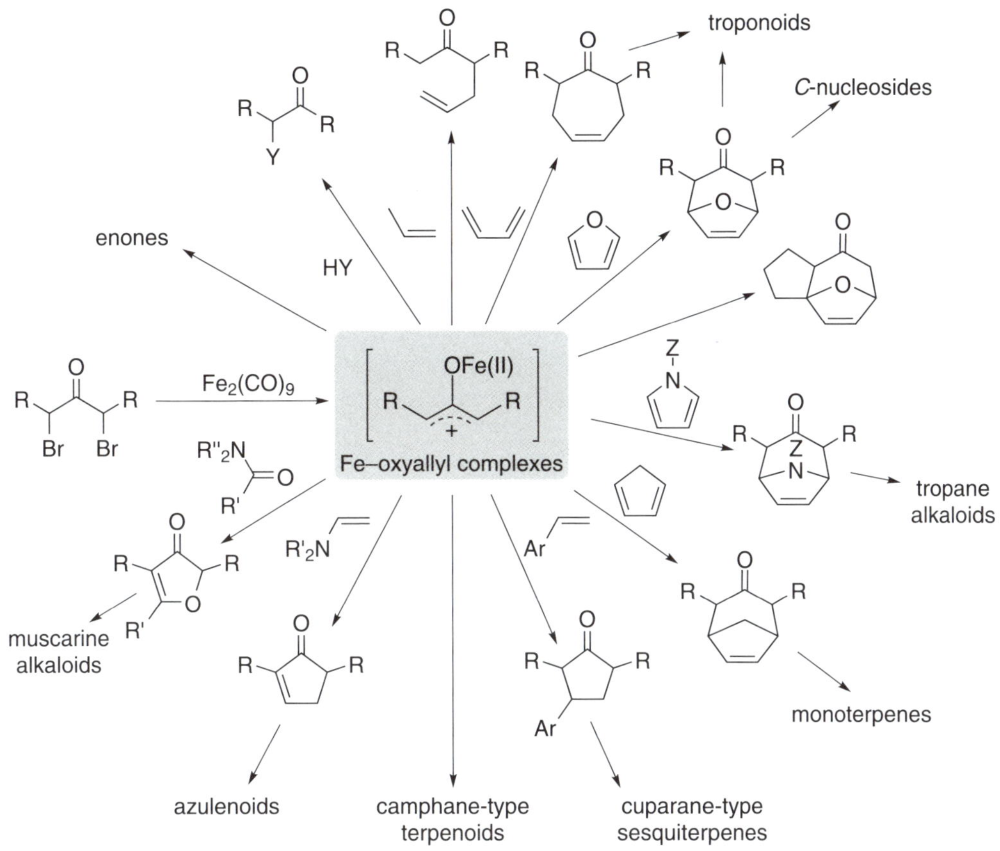

Fig. 5.7. Diverse pathways of the Fe carbonyl-promoted reactions.

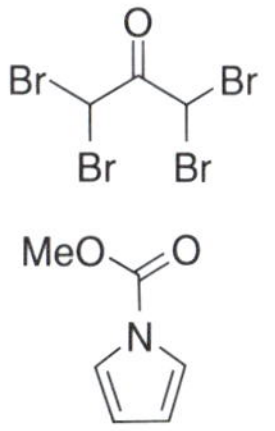

Fig. 5.8. Top: *sym*-tetrabromoacetone; bottom: *N*-methoxycarbonylpyrrole.

products, thereby expanding the reaction scope significantly [58]. This method is applicable to α,α,α'-tribromides derived from other methyl ketones. This finding in 1973 inspired Professor Eiji Ochiai (1898 – 1974). I still well remember that this magnate in Japan's pharmaceutical science and expert of alkaloid chemistry took the time to participate shortly in a natural products conference in Tokyo to encourage me and listen to our presentation.

In a similar manner, we were able to open doors to the general pathway to synthesize *C*-nucleosides starting from simple non-carbohydrate precursors, acetone and furan (**Fig. 5.9.** c) [59 – 61]. This class of compounds possess a multitude of biological properties such as antibiotic as well as anticancer and antiviral activities. Various artificial analogues became available as well, although we were unable to discover useful pharmaceuticals [62 – 73]. Notably, pseudouridine is the most abundant RNA modifier in cellular RNA, critically contributing to the discovery of mRNA vaccines by Katalin Karikó and Drew Weissman in 2020 [D].

[D] E. Dolgin: The Tangled History of mRNA Vaccines. *Nature* **2021**, *597*, 318 – 324.

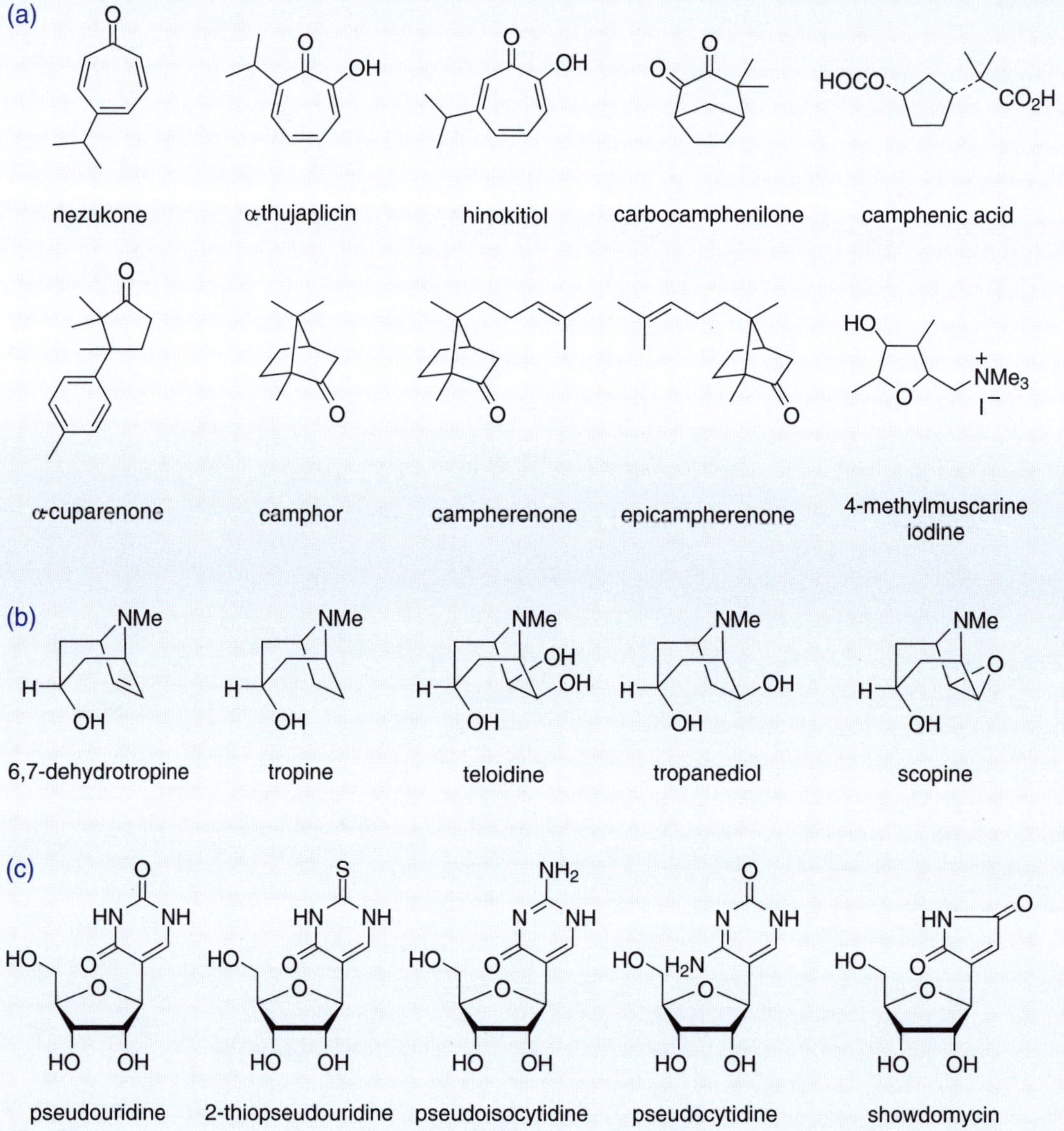

Fig. 5.9. Synthetic applications of the Fe carbonyl-polybromo ketone reactions. (a) Terpenoid synthesis. (b) General tropane alkaloid synthesis. (c) General *C*-nucleoside synthesis.

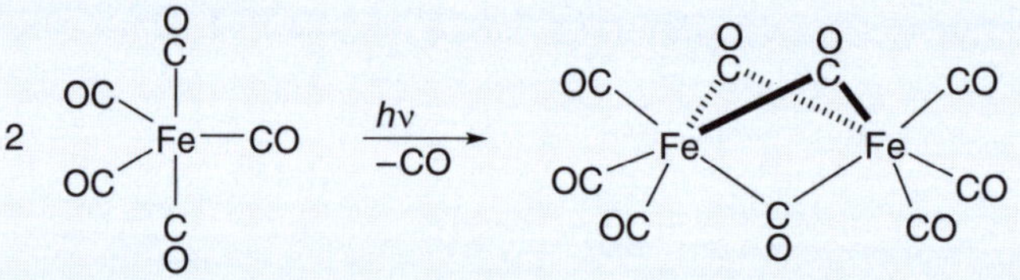

Fig. 5.10. Left: $Fe(CO)_5$ straw yellow to orange liquid; right: $Fe_2(CO)_9$ orange crystals.

I had long been held captive by this intriguing chemistry [38]. In 1978, we summarized our eight-year endeavor by publishing six contiguous full papers in the *Journal of the American Chemical Society* [35, 36, 41, 43, 57, 74] Furthermore, in 1983, this work was compiled as a chapter in *Organic Reactions* **[75]**. As a reducing agent for polybromo ketones, both $Fe(CO)_5$ and $Fe_2(CO)_9$ are applicable, but the non-volatile crystalline $Fe_2(CO)_9$, m.p. 100 °C, is better due to the higher reducing ability and stability as well as its lower toxicity. Zn/Ag couple, [76] Zn/Cu couple, and NaI [B] could be used for this purpose, but the reaction pathway is profoundly affected by the properties of the reducing agents and reaction media. We used Fe carbonyls in aromatic hydrocarbons as a standard reducing agent because of their high reliability and predictability.

[75] R. Noyori and Y. Hayakawa: *Reductive Dehalogenation of Polyhalo Ketones with Low-Valent Metals and Related Reducing Agents in Organic Reactions* (ed: W. G. Dauben), Vol. 29, Chapter 2. John Wiley & Sons, New York, **1983**.

5.2 Transition metal catalysis of strained hydrocarbons: implications in olefin metathesis (1968 – 1982)

The Woodward-Hoffmann rules were the first general concept to predict the pathway of organic reactions including stereochemistry. The presupposition is based on sound molecular orbital logics and is highly reliable. As for the violations of the rule, "*There are none!* Nor can violations be expected of so fundamental a principle of maximum bonding," the inventors convincingly concluded in their full article, published in *Angewandte Chemie* in 1969 **[E]**. However, for transition metal chemistry, this dogma does not apply. The power of transition metal catalysts presented a range of "formal violations" to this fundamental rule, expanding the realm of molecular transformations. Subsequently, the origin of the unique metal effects was argued in diverse ways such as the d orbital participation to ordinary σ and π MOs of organic reactants or, more likely we thought, involvement of short-lived organometallic intermediates.

[E] R. B. Woodward and R. Hoffmann: The Conservation of Orbital Symmetry. *Angew. Chem., Int. Ed.* **1969**, *8*, 781–853.

In this regard, from the launch of my group at Nagoya, I intended to challenge this fascinating subject, in addition to trimethylenemethane chemistry (**CHAPTER 3**). As a simple-minded chemist, I viewed ethylene and methylene (carbene) formally as "cycloethane" and "cyclomethane," respectively. In a similar manner, I believed there was a common chemical feature in canonical metallacylobutane, metallacyclopopane, and metallaethene structures and also across metal-cyclopropane, -alkene, and -carbene complexes. Thus, I was poised to unravel, with Hidemasa Takaya, fundamental chemical functions by examining the behavior of various σ-framed strained hydrocarbons in the presence of transition metal complexes. As a consequence, we were able to discover various intriguing chemistries, which were unrecognizable from classical approaches.

Our attention was directed to the stereoselectivity, regioselectivity, and periselectivity of the reactions of highly strained polycyclic

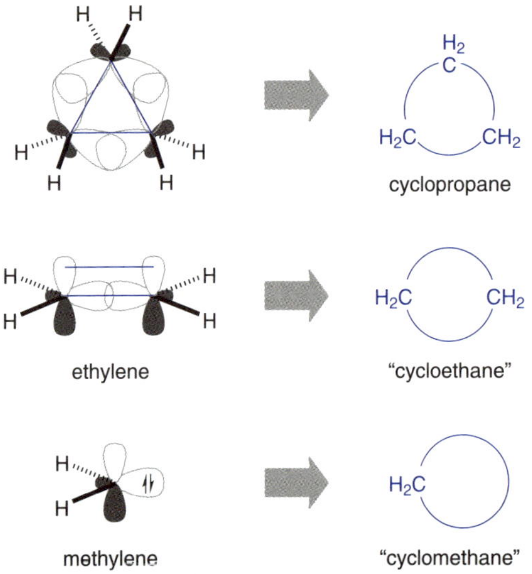

hydrocarbons in the presence of Ni(0) complexes. This choice was correct for our purpose. When I returned from the US to Japan in 1970, our research on the Ni(0)-catalyzed reaction of highly strained hydrocarbons had already begun to make a significant impact. Unlike most other transition metal catalysts that cause mere intramolecular skeletal changes, depending on the reaction system, Ni(0)(acrylonitrile)$_2$ [= Ni(an)$_2$] (red crystals with a polymetallic structure) and Ni(0)(1,5-cyclooctadiene)$_2$ (monomeric and soluble crystals) catalysts have an effect on both intra- and intermolecular reactions, thereby providing more substantial chemical changes [77 – 91]. As illustrated in **Fig. 5.12.**, many strained hydrocarbons undergo unique chemical transformations in the presence of Ni(0) catalysts, strongly supporting the existence of σ-bonded organometallic intermediates.

Fig. 5.11.
I viewed ethylene and methylene (carbene) formally as "cycloethane" and "cyclomethane".

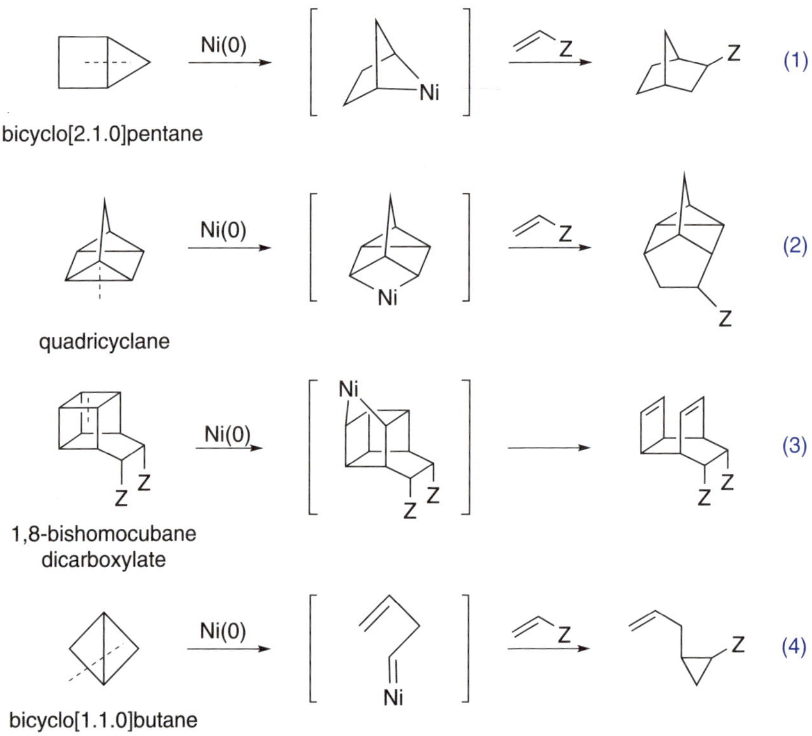

Fig. 5.12.
Ni(0)-catalyzed reactions of strained polycyclic hydrocarbons. Z = CN, CO_2R

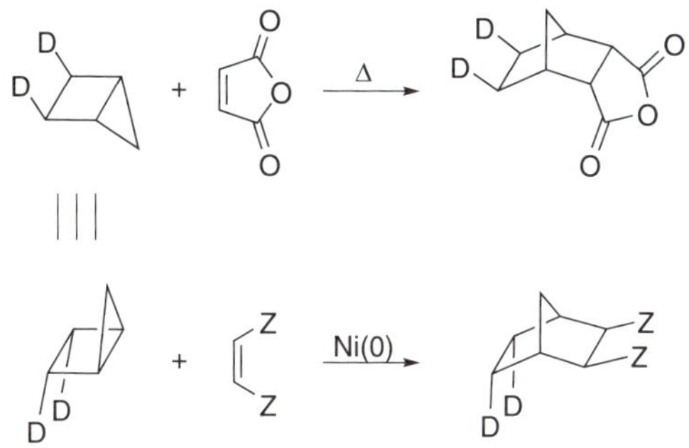

Bicyclo[2.1.0]pentane, despite having a high strain energy of 57 kcal/mol, is thermally stable and decomposes only at temperatures over 300 °C. However, $Ni(an)_2$ catalyzes a formal [2σ + 2π] cycloaddition with acrylonitrile, methyl acrylate, maleate, and fumarate already below 70 °C to give substituted norbornane products (**Fig. 5.12.**, Eq. 1) [79, 84, 88]. Deuterium labeling experiments verified that the olefin approaches the bicyclic envelope from the exo side, which is in contrast to Paul Gassman's purely thermal reaction [F], occurring from the endo side via diradical or zwitterionic intermediates. This difference is not only for steric but also for molecular orbital (MO) reasons. Likewise, quadricyclane, which is viewed as a cyclopropane / cyclobutane / cyclopentane-fused hydrocarbon, behaves as a bicyclo[2.1.0]pentane that forms the [2σ + 2π] adducts (Eq. 2) [86]. The reaction course is again totally different from the purely thermal [2σ + 2σ + 2π] reaction. We postulated the ring expansion mechanism, generalized in **Fig. 5.14.**, which involves oxidative addition of the strained C–C bond to Ni(0) forming a metallacyclobutane, followed by alkene insertion giving a metallacyclohexane and then Ni(0) extrusion, ending up with the cyclopentane products.

However, in the absence of alkenes, quadricyclane behaves as a strained cyclobutane to form norbornadiene. The [2σ + 2σ] valence isomerization also occurs with 1,8-bishomocubanes (**Fig. 5.12.**, Eq. 3) [83], as was already seen with Rh(I) complexes [G]. We interpreted this cyclobutane-diolefin isomerization as a straightforward mechanism via metallacyclopentane intermediates formed via initial edge-on interaction (**Fig. 5.14.**, Eq. 2). Our experiments and MO calculations [87] clearly eliminated the possibility of a concerted mechanism through a face-on, cyclobutene-transition metal complex long claimed by Frank D. Mango (Shell) [H]. Notably, this reaction mode is in sharp contrast to the Ag^+ ion or Pd(II) catalysis causing a bicyclo[2.2.0]hexane to dicyclopropyl conversion categorized as a formal [$2\sigma_a + 2\sigma_a$] process [I]. These strongly σ-accepting but weakly π-donating metals tend to cause the skeletal rearrangements via back-side, electrophilic corner-on interaction. Thus, the reaction course of these strained systems appears to remarkably depend on the properties of transition metals.

Fig. 5.13. Comparison of thermal Ni(0)-catalyzed cycloadditions of deuterated bicyclo[2.1.0]pentane with electron-deficient olefins.

[87] R. Noyori, M. Yamakawa, and H. Takaya: Nickel Catalyzed Reactions Involving Strained Bonds. XIII. Peri- and Regioselectivities of the Nickel(0)-Catalyzed Valence Isomerization of the 1,8-Bishomocubane System. A Molecular Orbital Consideration. *J. Am. Chem. Soc.* **1976**, *98*, 1471–1478.

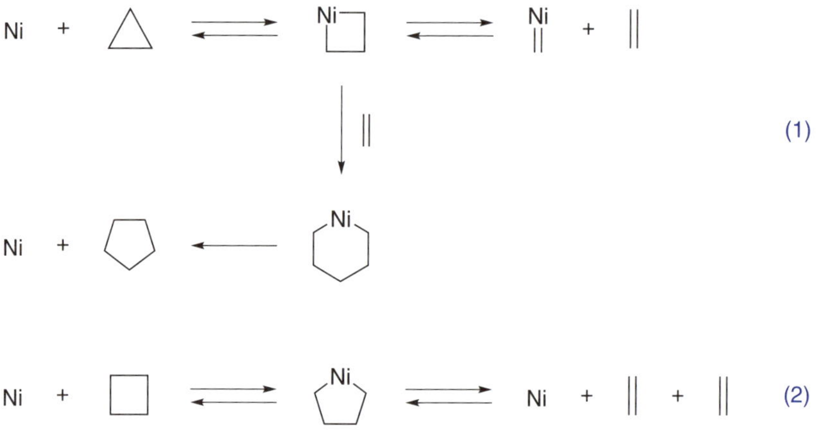

Fig. 5.14. Ni(0)-catalyzed reactions of strained hydrocarbons.

Fig. 5.15. Ni(0)-catalyzed cycloaddition of norbornadiene with methylenecyclopropane (top) and of benzonorbornadiene with electron-deficient olefine.

In principle, the conversion between a cyclobutane and diolefin is bi-directional. We found that Ni(0) complexes induced a formal [2π + 2π] cross-addition of certain strained olefins. Norbornadiene undergoes a cycloaddition to methylenecyclopropane to produce the *endo* [2 + 2] adduct [81], whereas cyclopropanated or benzo-fused norbornadiene and electron-deficient olefins generate the *exo* [2 + 2] adducts [90]. These otherwise thermally forbidden reactions proceed commonly via a five-membered metallacyclic intermediate (**Fig. 5.14.**, Eq. 2).

Most intriguing was the behavior of bicyclo[1.1.0]butanes (**Fig. 5.12.**, Eq. 4) [78, 82, 85, 89, 92]. The parent hydrocarbon, first reported by Kenneth B. Wiberg in 1959, is the smallest bicyclic hydrocarbon. The compound is stable despite its high strain energy of 66 kcal/mol. Thermolysis of the compound at 200 °C breaks two peripheral σ bonds to give butadiene, leaving the central bond intact. In contrast, the Ni(0) catalyst causes simultaneous geminal two-bond cleavage, and the resulting short-lived Ni(0)-complexed allylcarbenes can be trapped by coexistent acrylonitrile, acrylate esters, maleic or fumaric esters to give the cyclopropane products with retention of the configuration.

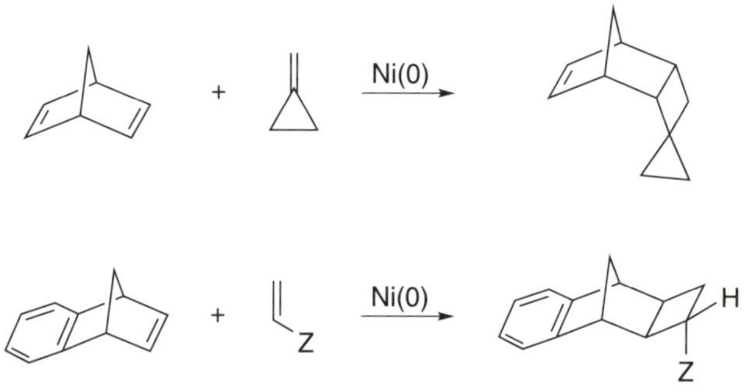

Z = CN, CO_2Me

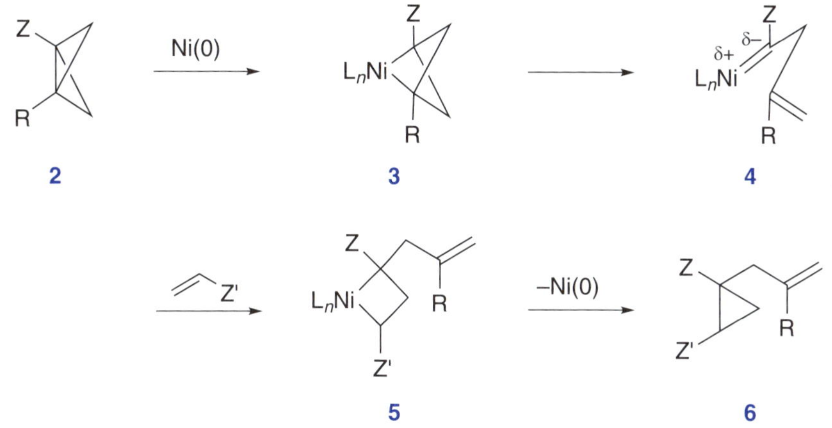

Fig. 5.16. Ni(0)-catalyzed reactions of bicyclo[1.1.0]butanes and electron-deficient alkenes. R = Me, Ph, or H. Z, Z' = CO_2Me, CN, H. L = CH_2=CHZ'.

As illustrated by the top scheme in **Fig. 5.14.**, Eq. 1, the metal promoted reaction of cyclopropane to carbene is reversible. In bicyclobutane chemistry, the highly strained cyclopropane is converted to the less strained cyclopropanes via intramolecular retro-carbene addition followed by intermolecular carbene re-addition. The mechanism of the two-fold carbene reaction is not trivial. As shown in **Fig. 5.16.**, the reaction starts with the oxidative addition of the central strained C–C bond of 2 on Ni(0) via an edge-on interaction. The resulting bicyclic four-membered metallacycle 3 causes cleavage of one of the four peripheral C–C bonds via intramolecular bond-switching, giving the Ni(0)-allylcarbene complex 4. Since the carbene center has eminent nucleophilicity due to the back-donation from Ni(0), the site of bond cleavage is decisively directed by electronic properties of the substituents, i. e. electron-donating or -withdrawing, for example, Me *vs* CN or CO_2Me [92]. The Ni(0)-bound carbene or Ni(II) metallaethene 4 in turn reacts with a Ni-coordinated electron-deficient alkene and the resulting metallacyclobutane intermediate 5 gives rise to the final cyclopropane product 6. Our general conclusion is that, depending on the property of the nature of metal and substituents, metallacyclobutanes can undergo both fragmentation forming an olefin and metal-carbene species and reductive elimination giving a cyclopropane and a metallic complex (**Fig. 5.14.**).

These experimental and theoretical approaches may no longer be appreciated nowadays, but they guided later advancement of organometallic catalysis, particularly olefin metathesis mechanism. I still remember an amazing short *Chemical & Engineering News* article in 1967 describing a W-catalyzed metathesis developed by the Goodyear Tire and Rubber Company **[J]**. In May 1974, we first presented our preliminary findings at the Japan-US Seminar on Prospects in Organotransition-Metal Chemistry in Honolulu (organized by Minoru Tsutsui, James P. Collman, Jack Halpern, Yoshio Ishii, and Sei Otsuka). Among the topics that were

[J] *Chem. Eng. News* **1967**, *45*, 51.

discussed at this seminar were the chemistry of strained hydrocarbons perturbed with transition metals, and I enjoyed the rigorous discussion between Halpern, Mango and Rowland Pettit. Robert H. Grubbs, at that time a 32-year-old assistant professor at Michigan State University, proposed the mechanism of the emerging olefin metathesis reaction. We listened to the talks of many senior academics from both countries, but I was also deeply impressed by highly motivated young Americans including George M. Whitesides (MIT), John A. Osborn (Harvard), John Faller (Yale), Tobin J. Marks (Northwestern), and Robert Bergman (Caltech), who all eventually became world-class leaders in the field.

Later in 1978, with the aid of MO studies, I provided an overview of this organometallic concept in Athens, Georgia at the American Chemical Society's inorganic chemistry symposium organized by R. Bruce King [93]. Again, I encountered J. Halpern and R. Pettit at this symposium. During those old days, this intriguing subject attracted numerous US fellow chemists including Philip Eaton, William G. Dauben, Paul G. Gassman, Leo Paquette, Charles P. Casey, Thomas Katz, and others in addition to the colleagues mentioned above.

5.3 Starting with binaphthol chemistry: BINAL-H asymmetric reductions (1976–1980)

We began to focus on asymmetric synthesis in the mid-70s, when practical methods remained very rare. Since the central metal in organometallic compounds is intrinsically achiral, suitable molecular design of chiral ligands is essential to achieve effective asymmetric transformations. At the time, natural amino acids monopolized the chiral chemistry at large, while carbohydrates and C_2 chiral tartaric acid played some roles as well. While others were using readily available auxiliaries, I made a bold decision to explore the general use of an atropisomeric 1,1'-binaphthyl scaffold for this purpose. To me, this artificial molecular architecture is simple and distinctly beautiful [94, 95].

Any molecular design must be rational, thus satisfying to the scientific mind. Unlike the commonly utilized natural product-derived auxiliaries or ligands possessing non-symmetrical, central C_1 chirality, the C_2 symmetrical 2,2'-functionalized 1,1'-binaphthyls simplify the chemistry by halving the number of possible diastereomeric species present during the reaction. Superiority of the C_2 structures had already been demonstrated by others, for example, tartaric acid-derived auxiliaries and transition metal ligands such as DIOP (Kagan), DIPAMP (Knowles), or CHIRAPHOS (Bosnich and Fryzuk). The binaphthyls are conformationally more flexible and can accommodate a range of metallic elements by rotating about the $sp^2C(1)$-$sp^2C(1')$ pivot without seriously increasing

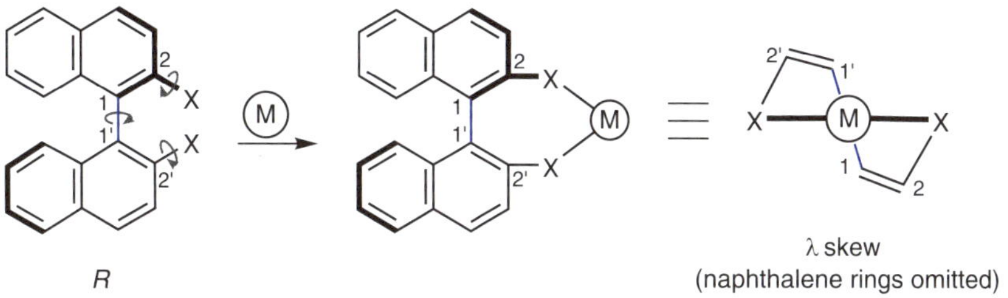

Fig. 5.17. Formation of chiral seven-membered metal chelates from functional binaphthyl compounds. X = heteroatom or group. M = metal.

torsional strain, and, in turn, the resulting seven-membered chelate rings containing only sp^2 carbon atoms are skeletally unambiguous (**Fig. 5.17.**). The highly skewed δ or λ geometry is expected to generate high chirality-discriminating ability in the reaction transition state. It was my belief that the combination of selected metals and such binaphthyl derivatives as chiral elements display a high degree of enantio-selection in various stoichiometric and catalytic transformations. Since organic synthesis is inherently artificial in nature, I intended to logically control the molecular chirality by the use of purely unnatural compounds. From a practical point of view, this may not have been so sensible. Nevertheless, to a certain extent, our approach was distinctive.

Our first trial was simple: We used 2,2'-dihydroxy-1,1'-binaphthyl (2,2'-binaphthol) for the chiral modification of lithium aluminum hydride (LAH). Others had also dealt with the subject, including Stephen R. Landor, Harry S. Mosher, J. P. Vigneron, Kenji Koga, Albert I. Meyers, Teruaki Mukaiyama, Gabriel Saucy, and Dieter Seebach, mostly with the aid of naturally occurring substances or their derivatives. But truly practical procedures were still lacking.

The usefulness of our method remained uncertain. Since pure binaphthol was commercially unavailable, we had to resolve the racemate via its phosphate. In 1979, our hard work paid off [96 – 103]. The BINaphthol-modified ALuminum Hydride reducing agent, named BINAL-H, exhibited exceptionally high enantio-selection in the reduction of a wide range of prochiral carbonyl compounds. For example, acetophenone and butyrophenone provided the chiral alcohols with 95 % and near 100 % *ee*, respectively. **Fig. 5.19.** (a, b) illustrates the empirical formula of BINAL-H 7 and the general sense of asymmetric induction. Notably, the presence of unsaturated groups such as aryl, alkenyl, and alkynyl is crucial for the stereo-selection; in other words, the two substituents flanking the carbonyl groups are distinguished by the difference in their electronic properties. Our transition state model shown in **Fig. 5.19. c** explains the general auxiliary / product configurational relationship, *S*/*S* or *R*/*R* [103]. Thus, with (*S*)-BINAL-H, the *S*-generating transition structure 8 is favored over the diastereomeric *R*-generating structures 9 that is destabilized by the *n*/*π* type electronic repulsion between a binaphthol

Fig. 5.18. Book cover of Fieser and Fieser's *Reagents for Organic Synthesis, vol. 9*, Wiley, **1981**.

(a)

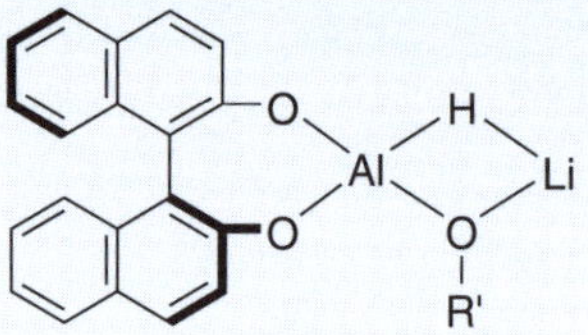

(*S*)-**7**, [(*S*)-BINAL–H]

(b)

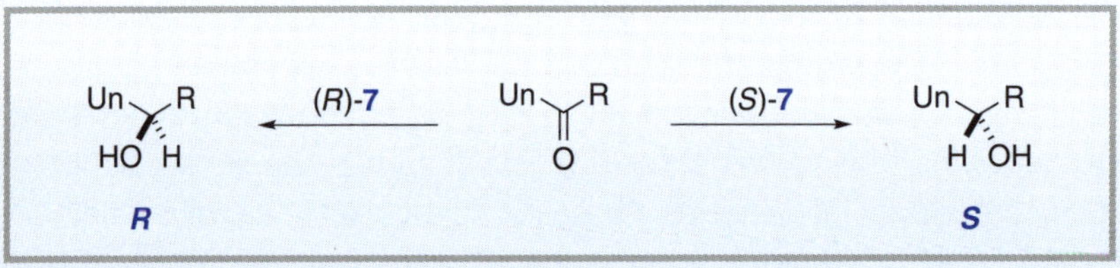

(c)

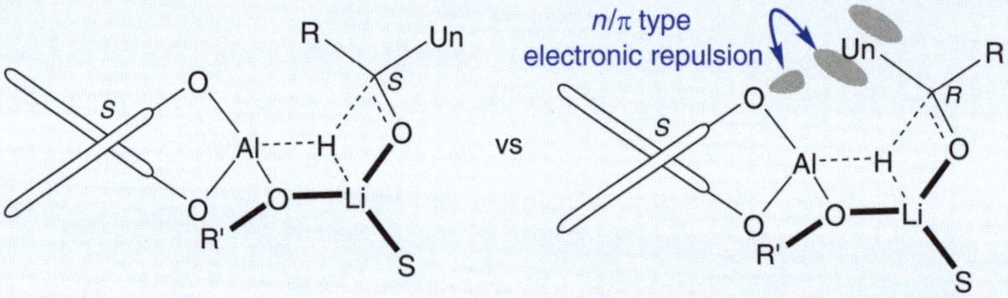

8, favored *S*-generating transition structure

9, unfavored *R*-generating transition structure

(d) Application in the Corey prostaglandin synthesis:

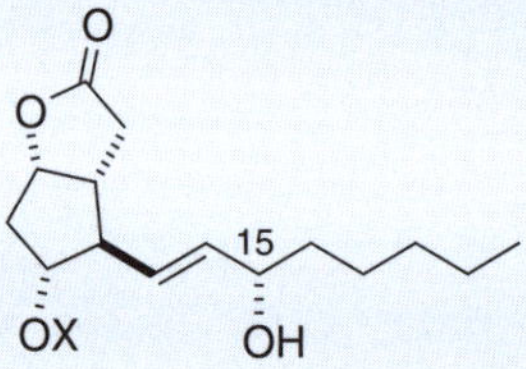

10

X	15*S*/15*R*
COMe	99.4:0.6
THP	99.5:0.5
H	100:0

Fig. 5.19. The BINAL–H asymmetric reduction. (a) Empirical structure. (b) Generalized scheme. (c) Transition state model. (d) Use in prostaglandin (PG) synthesis. R = alkyl, H. R' = Et. Un = aryl, alkenyl, alkynyl, and other unsaturated groups. THP = 2-tetrahydropyranyl.

oxygen and the unsaturated moiety. The oxygen / R non-bonded repulsion in the former becomes significant by increasing the bulkiness but cannot overbalance the electronic influence.

Originally, we intended to utilize this method for the synthesis of some insect pheromones due to the interest of our collaborator Mugio Nishizawa (1945 – 2010) who had a natural products chemistry background. In fact, the asymmetric reduction of ynones effectively produces chiral propargylic alcohols which act as synthetic intermediates of the pheromones of various beetles. Chiral allylic alcohols obtained from conjugated enones are versatile intermediates for the construction of chiral organic frameworks through, for instance, the Claisen rearrangement, the Pd(II)-assisted sigmatropic reaction, or intramolecular S_N2' displacement of the derivatives. Obviously, certain chiral allylic alcohols are important in their own right.

Fortunately, our BINAL-H reagent was proven remarkably effective in generating the 15*S* configuration of the prostaglandin series (PGs), greatly facilitating Corey's commercial synthesis (**Fig. 5.19. d**) [97, 102]. The desired 15*S* creation in 10 is a result of double asymmetric induction and is significantly assisted by the chiral framework of the precursory lactonic enones. For example, the 9-*O*-THP substrate gives the desired 15*S* alcohol with a near perfect, 99.5 : 0.5 selectivity. (*S*)-BINAL-H well differentiates between the 15-keto faces (reagent control (*S* : *R*) = 20.6 : 1), while the preexisting chiral bicyclic units further provides a favorable bias producing the α alcohol (substrate control (α : β) = 9.7 : 1).

Thus, we were able to eventually overcome the difficulty in the availability of chiral binaphthol. This endeavor marked the first example of its practical synthetic use, although Donald Cram's chiral molecular recognition (1987 Nobel Prize, with Jean-Marie Lehn and Charles J. Pedersen) had already utilized binaphthol-based templates. Now this chiral motif has become a first choice for various asymmetric syntheses.

5.4 Determination of the enantiomeric ratio in the 1970s

Measurement is crucial in addition to the rational design of enantioselective reactions. Research on asymmetric synthesis depends heavily on methods for determining the optical purity or enantiomeric excess (*ee*) of chiral compounds. As we struggled earlier in asymmetric carbene chemistry (CHAPTER 2), the progress in this field was very slow until the end of the 1970s because analysis of enantiomers still relied on the use of a polarimeter. In our BINAL-H investigation, direct comparison, when necessary, was done carefully with the synthetic and authentic resolved materials which had been purified by tedious preparative gas chromatography (GC); both samples were subjected alternately to the rotation

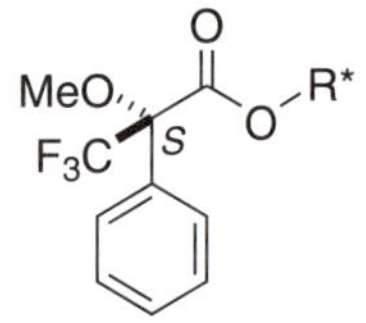

MTPA ester

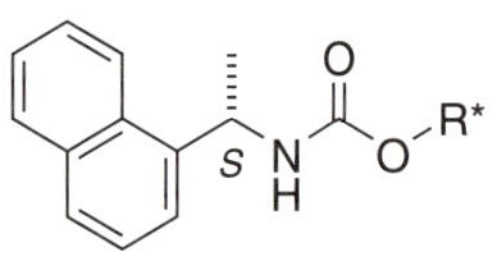

1-(1-naphthyl)ethylcarbamate

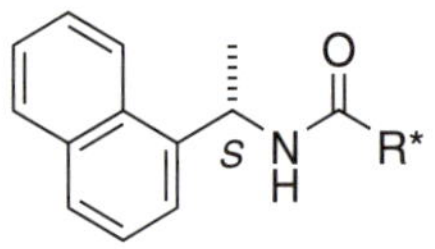

1-(1-naphthyl)ethyl amide

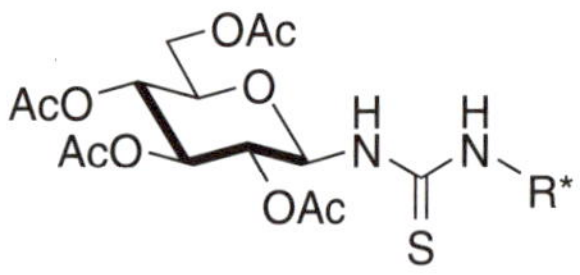

GITC-derived thiourea

R* = chiral, *R* or *S*

measurement under the entirely same conditions (solvent source, concentration, temperature, cell, etc.) with the GC-purified synthetic and authentic resolved materials. The pure reference samples, actually, were not easily available, and the comparison of the observed rotations with the literature values was not highly reliable. As a consequence, the lack of generally useful analytical methods had long hampered the advancement of this significant field.

Subsequently, synthetic chemists worldwide gradually began to determine the optical purity by using ^{1}H-NMR, 60 to 100 MHz at the time (400 or 600 MHz are used nowadays), with chiral lanthanide shift reagents, or more conveniently, by producing a diastereomeric mixture after derivatization with a chiral auxiliary. Some examples are seen in **Fig. 5.20.** For example, chiral alcohols were converted to the diastereomeric Mosher α-methoxy-α-trifluoromethyl-α-phenylacetic (MTPA) esters or 1-(1-naphthyl)ethylcarbamates; carboxylic acids were transformed to 1-(1-naphthyl)ethyl amides; amines were turned by condensation with 2,3,4,6-tetra-*O*-acetyl-D-glucopyranosyl isothiocyanate (GITC) into the corresponding thiourea. Chiral gas liquid chromatography (GLC) could also be used for certain volatile compounds. Later, development of ^{19}F- and high-field ^{1}H-NMR significantly accelerated the analysis of diastereomeric compounds.

Fig. 5.20. Diastereo-derivatization of chiral alcohols, carboxylic acids, and amines.

Obviously, direct HPLC analysis, if available, is much more convenient. In this regard, I was very pleased to know that our primitive *n*-butyllithium / (−)-sparteine chemistry (**CHAPTER 2**) [6] inspired Yoshio Okamoto at Nagoya to pioneer the helicity-specific polymerization of trityl methacrylate in 1979 [K]. This unique helical polymer allowed him to prepare a chiral high performance liquid chromatography column that could directly separate various enantiomeric substances including non-polar hydrocarbons [104]. Then, in the late 1980s, he further explored an HPLC technology utilizing a polysaccharide-based chiral stationary phase, allowing for the general, direct analysis of UV-active chiral compounds without derivatization. This accomplishment set a new standard for synthetic chemistry and was eventually recognized by the prestigious Japan Prize in 2019.

5.5 Prostaglandin synthesis (1976 – 1989)

Prostaglandins (PGs) exhibit diverse activities in body tissues and cells and are recognized as local hormones for maintaining homeostasis of circulatory, respiratory, and digestive organs, as well as other biological activities. In nature, PGs are produced from C_{20} fatty acids, as outlined

Fig. 5.21. Structures of prostaglandins.

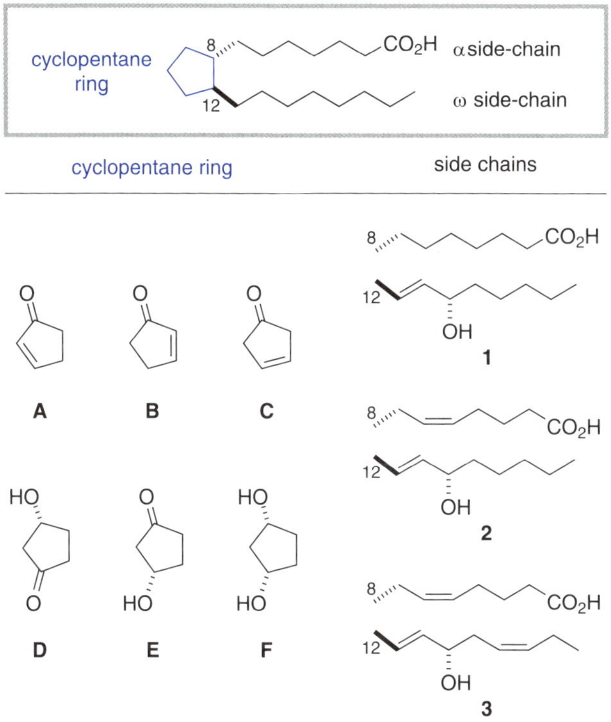

in **Fig. 5.22.**, by the action of cyclooxygenase yielding the common endo-peroxide intermediates, PGG and PGH, which undergo isomerization and reduction by various enzymes in microsomes. Major difficulties in their therapeutic applications include scarce natural occurrence and nonselective widespread action on most tissues and cells. In the late 1960s to 1970s, academia and industry exerted immense efforts to advance chemical synthesis of PGs. E. J. Corey solved these problems in his achievement of stereoselective, stepwise synthesis. I was very pleased to able to contribute to his commercial production of PGs at Ono Pharmaceutical Co. Ltd. by inventing the asymmetric BINAL-H

Fig. 5.22. Biosynthesis of prostaglandins from C_{20} fatty acids.

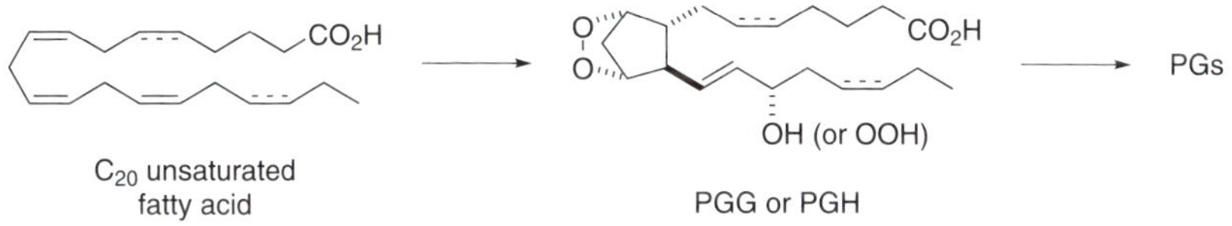

[105] R. Noyori und M. Suzuki: Prostaglandin Synthesis. 9. Prostaglandin-Synthesen durch Dreikomponenten-Kupplung. *Angew. Chem.* **1984**, *96*, 854–882; R. Noyori and M. Suzuki: Prostaglandin Synthesis. 9. Prostaglandin Syntheses by Three-Component Coupling. *Angew. Chem. Int. Ed.* **1984**, *23*, 847–876.

reduction (**Fig. 5.19. d**). As such, several natural PGs and synthetic analogues are now used as drugs for various purposes. Later in 1982, the Nobel Prize in Physiology or Medicine was jointly awarded to Sune K. Bergström, Bengt I. Samuelsson, and John R. Vane for their discoveries concerning PGs and related biologically active substances. Fortunately, Corey received the Nobel Prize in Chemistry in 1990 for his development of the theory and methodology of organic synthesis.

Biology and chemistry of natural and unnatural PGs remain important. My experience at Harvard with Corey in 1969–1970 (**CHAPTER 4**) naturally spurred me to develop an efficient PG synthesis in a different way. This family consists of an oxygenated cyclopentane ring and C_7 and C_8 side chains, which are named α and ω, respectively, with *trans* configuration. Therefore, we simply aimed to realize the compact three-component PG synthesis by combining suitably functionalized cyclic and side-chain units **[105]**. This straightforward, flexible route allows the synthesis of various artificial PG analogues possessing the desired tissue selectivity and higher metabolic stability. In relation to this, we explored a series of selective metal-mediated synthetic methods.

Long sought, general three-component coupling synthesis

This project began as a collaboration with Masaaki Suzuki, who was educated by Yoshimasa Hirata at Nagoya and Woodward at Harvard and had experience with natural products synthesis. Among numerous possibilities, the three-component coupling synthesis that we sought consists of conjugate addition of a nucleophilic ω side-chain unit to an O-protected (*R*)-4-hydroxy-2-cyclopentenone followed by electrophilic trapping of the enolate intermediate by an α side-chain organic halide (**Fig. 5.23.**).

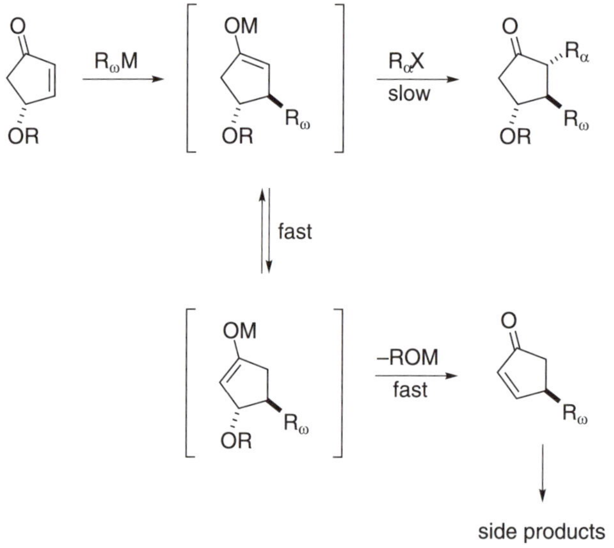

Fig. 5.23. Problem of the three-component synthesis of prostaglandins. M = metal. X = halogen. R_α = C_7 side chain. R_ω = C_8 side chain.

However, we were too optimistic. The route was not as simple as it seemed. Standard Gilman's lithium diorganocuprates or related mixed cuprates undergo conjugate addition to the enones, but the resulting enolates cannot be alkylated with α side-chain halides. The failure is probably due to the facile double-bond migration in the enolate causing β-elimination to form cyclopentenone products, hampering the desired vicinal carba-condensation.

We then concentrated on the generation of an enolate species, which is susceptible to direct alkylative trapping by use of stoichiometric amounts (not excess) of an ω side-chain unit and (*R*)-4-*tert*-butyldimethylsiloxy-2-cyclopentenone 11. Eventually in 1985, organocopper chemistry combined with a special alkylation procedure furnished the first solution to the long-standing synthetic problem (**Fig. 5.25. a**) [106 – 109]. We found that an organocopper reagent formed *in situ* from equimolar amounts of CuI and the ω side-chain alkenyllithium, and 2 – 3 equivalents of tributylphosphine, undergoes smooth conjugate addition to the chiral siloxy enone 11 using a 1 : 1 reagent / substrate molar ratio [110, 111]. The resulting enolate intermediate can be trapped by an alkyl iodide with the aid of organotin chemistry. Thus, a one-pot, sequential treatment of the organocopper reagent with an enone, HMPA, triphenyltin chloride, and a side-chain *Z*-allylic iodide leads to the desired three-component linking product 12 in ca. 80 % yield. However, the workup is not simple, and purification is required to obtain the final product as the reaction mixture contains various inorganic and organic side products that need to be removed.

$Li[CuR_2]$

Empirical formula of Gilman's lithium diorganocuprate

R = alkyl, alkenyl, aryl.

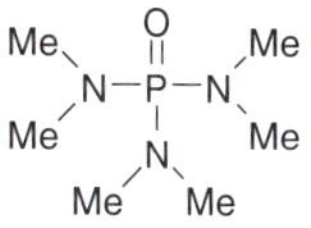

Fig. 5.24. Hexamethylphosphoric triamide (HMPA).

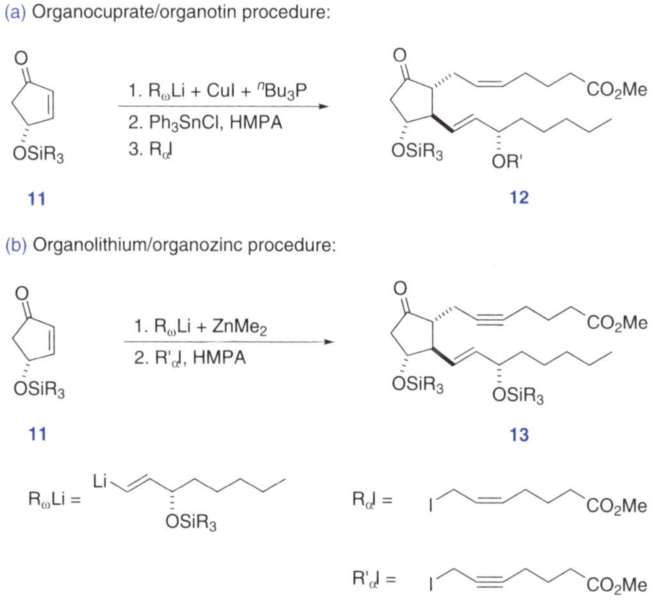

Fig. 5.25. The three-component PG synthesis. (a) Organocuprate / organotin procedure. (b) Organolithium / organozinc procedure. SiR_3 = $SiMe_2{}^tBu$. R' = SiR_3, or THP.

"Green chemistry" has become a key issue in chemical processes, which will be discussed further in **CHAPTER 7**. In earlier years, the synthesis of 1 kg of pharmaceutical drugs with a complex structure was estimated to produce 25 – 100 kg unwanted waste. Now, efficient organic syntheses must be achieved in short steps without forming unnecessary or toxic byproducts. Precious catalysts, if necessary, are desired to be recovered and recycled.

Fig. 5.26.
Author giving a lecture on prostaglandin synthesis at Nagoya. December 1980.

Eventually, in 1989, we reached the near-ideal Cu-free solution for this reaction **(Fig. 5.25. b)**. The clean procedure utilizes organozinc chemistry, in which neither the phosphine-complexed Cu reagent nor triphenyltin chloride is needed [103, 112, 113]. Thus, when an equimolar mixture of dimethylzinc and the ω side-chain alkenyllithium in THF was treated sequentially with 4-siloxy 2-cyclopentenone 11, propargylic iodide and HMPA, the desired condensation product 13 was formed in good yield. Under these reaction conditions, all the inorganic residues are removable by aqueous workup, making the operation simple.

As summarized in **Fig. 5.28.**, the invention of the tandem conjugate addition / alkylation sequence accomplished a very short synthetic route to PGs. The absolute configuration of the C(11) and C(15) positions is secured at the stage of the starting building blocks. (*R*)-4-Hydroxy-2-cyclopentanone is obtainable by its very practical optical resolution by condensation with a hemiacylal derived from (1*R*,3*R*)-*trans*-chrysanthemic acid [114] or via kinetic resolution by the (*R*)-BINAP-Rh catalyzed 1,3-hydrogen shift [115] or (*S*)-BINAP-Ru catalyzed hydrogenation [116]. The C(15) configuration in the ω side chains is determined by the (*S*)-BINAL-H reduction of the unsaturated ketones [102]. The desired *trans*, *trans* relative configuration in the functionalized cyclopentanone is generated by asymmetric induction in the three-component linking

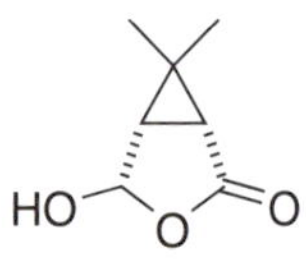

Fig. 5.27.
A hemiacylal derived from (1*R*,3*R*)-*trans*-chrysanthemic acid.

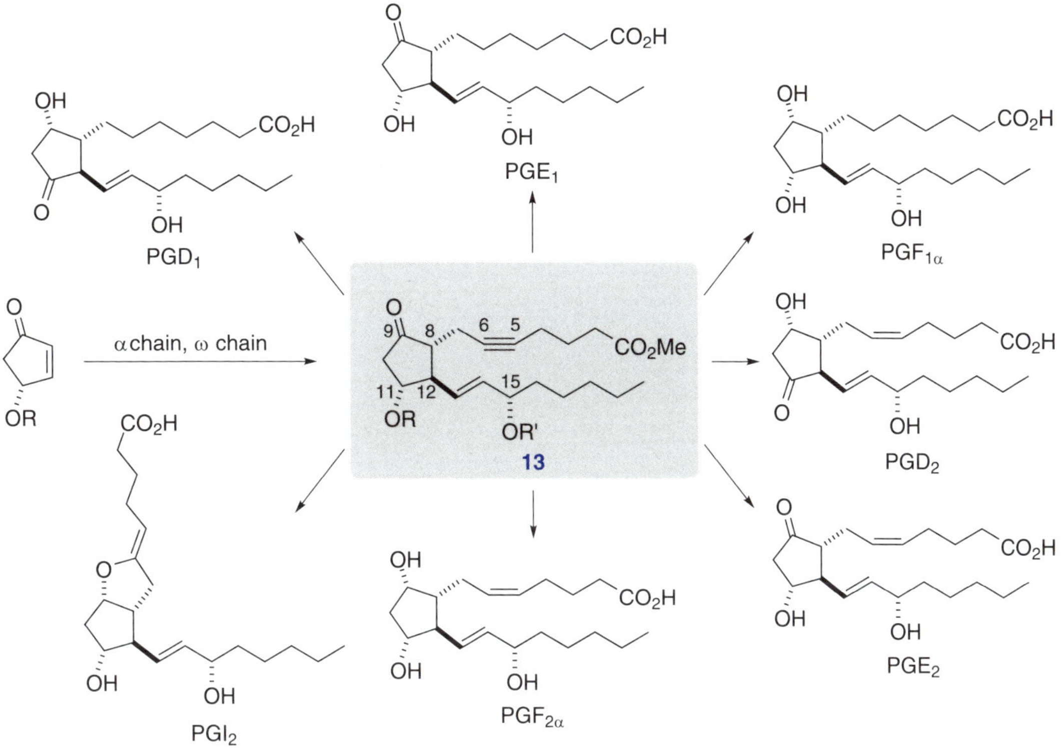

Fig. 5.28. General PG synthesis.

process. The 11,12-*trans* relationship is induced simply by a nonbonded interaction between the 11-siloxy group and the organometallic reagent that is introduced, and similarly, alkylation of the enolate generates the 12,8-*trans* geometry based on the newly created C(12) stereogenic center.

The central acetylenic intermediate 13 serves as a common intermediate for the general synthesis of naturally occurring PGs of series 1 and 2 [117]. The selective half-hydrogenation of the 5,6-triple bond of 13 to the *Z* double bond is best achieved over a 5% Pd / $BaSO_4$ catalyst, whereas the controlled hydrogenation over a Pd/C catalyst gives a PGE_1 derivative. The conversion of the E-type intermediate to prostaglandin Ds (PGDs) requires the reversal of the oxidation state at the C(9) and C(11) positions, which is achieved by using different protecting groups in the starting alcohol blocks [118]. L-Selectride has proven to be the best reagent for stereoselective conversion of the 9-keto compound to the 9α-alcohol. In this regard, an unprecedented kinetic discrimination is seen in the reduction of the chiral cyclopentanone by the BINAL-H reagent 7. Although (*S*) BINAL-H [(*S*)-7, R' = Et] is almost inert to this cyclopentanone, the reduction with the enantiomeric (*R*)-7 reagent proceeds smoothly ($k_R/k_S \sim 130$) to form the 9α-alcohol with sufficient diastereoselectivity (99 : 1), which is attributed to the nonplanar, chirally skewed conformation of the cyclopentanone. Furthermore, physiologically significant

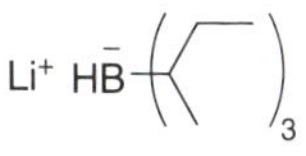

Fig. 5.29. L-Selectride.

but very labile PGI_2 (half-life 3 min under physiological conditions) with a unique 2-alkylidenetetrahydrofuran structure is also obtainable from the 5,6-didehydro-$PGF_{2\alpha}$ derivative [119] by its *anti*-5-*exo*-*dig* cyclization aided by $PdCl_2(PhCN)_2$ followed by stereo-retentive, reductive depalladation with ammonium formate, and deprotection [107].

Versatile vicinal carba-condensation of α,β-unsaturated ketones

The PG synthesis described above capitalizes on organometallic chemistry. The invention of unique reagents and catalysts solved a wide range of synthetic problems regarding efficiency and selectivities in skeletal construction and functional group transformations.

One of the key operations is the conjugate transfer of sp^2-hybridized carbon to enones, which requires the use of a 1 : 1 enone / side-chain stoichiometry. This is best attained by using 1 equivalent of the reagent, formed from the *E*-alkenyllithium and dimethylzinc in a 1 : 1 molar ratio [112, 113]. The reaction would proceed via lithium zincate, but without methyl transfer. Notably, dimethylzinc effectively suppresses a potential proton abstraction of the enolate intermediate from coexisting ketones. This characteristic allows efficient trapping with alkyl iodides in the presence of HMPA [112]. Another useful reagent effecting the "stoichiometric" conjugate addition to enones is an organometallic formed *in situ* from sp^2- or sp^3-hybridized organolithium, CuI, and tributylphosphine in a 1 : 1 : 2 – 3 molar ratio. In this conjugate addition, the enolate becomes a sole strong nucleophile present in the reaction system and thus reacts cleanly with electrophiles in high yield. In this regard, this method is superior to the classical organocuprate procedure. The reaction of the enolate with alkyl halides, however, needs triphenyltin chloride and HMPA in order to suppress the basicity, but retain their nucleophilicity [106]. In order to understand the stannyl chloride effect, we studied the factors controlling the stability of the pentacoordinate stannate complex [120].

The utility of these regio-controlled three-component reactions is very wide. As illustrated in **Fig. 5.31.**, enolate 14 without coexisting nucleophiles can be trapped with various electrophiles, providing a useful one-pot, two-step method for vicinal carba-condensation with α,β-unsaturated ketones [105, 121, 122] Much later, in the mid-1990s, we discovered a rapid, productive catalyst system promoting conjugate addition of dialkyl- or diarylzinc compounds to α,β-unsaturated ketones; a mixture of CuCN and protic *N*-benzylbenzenesulfonamide catalyzes the reaction with a Cu : Zn ratio of up to 1 : 10,000 to give the β-substituted ketones in nearly quantitative yields [123]. A range of enones with *s-cis*, *s-trans*, and flexible geometry are usable. The dimer of the *in situ*-formed enolate 14 (M = ZnR), prior to aqueous workup, further undergoes an aldol reaction or Pd(0)-assisted coupling with allyl acetate, resulting in the regio-defined vicinal carba-condensation products.

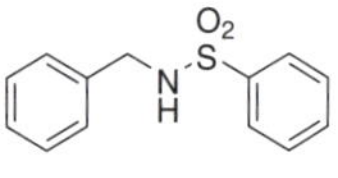

Fig. 5.30. *N*-benzylbenzene-sulfonamide.

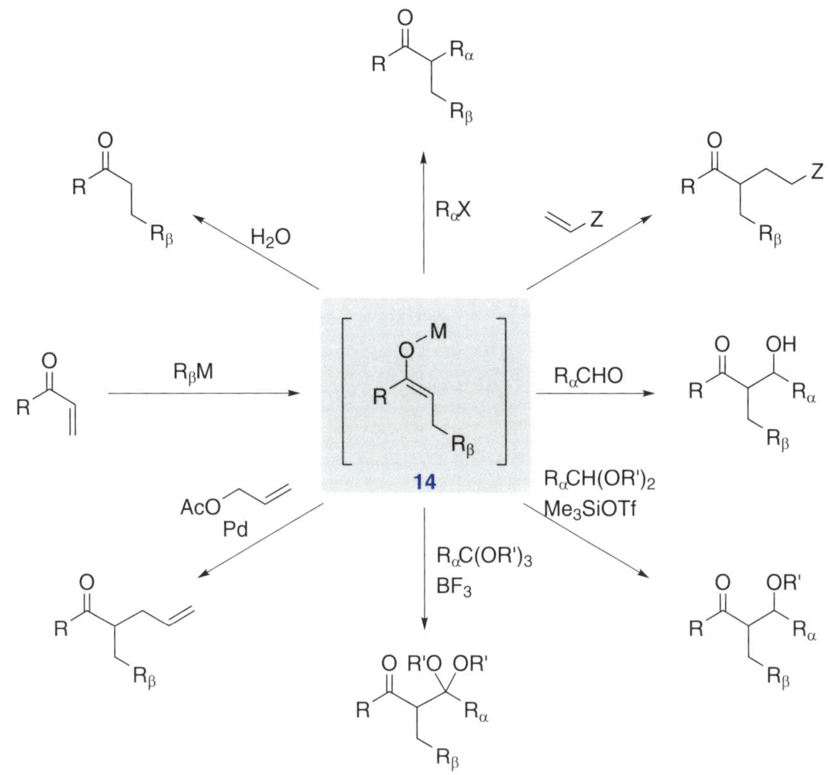

Fig. 5.31. Vicinal carbacondensation of α,β-unsaturated ketones. M = metal. X = halogen. Z = electron-withdrawing group.

Prostaglandin analogues

We also opened doors to a three-component aldol route to PGs. As shown in **Fig. 5.32.**, a one-pot, high-yield construction of the PG skeleton was accomplished by our Cu-mediated conjugate addition of an ω side-chain unit followed by trapping of the *in situ* generated enolate with an α side-chain aldehyde (for trapping with formaldehyde, see **[L]**). Subsequent removal of the 7-hydroxyl group from the adducts and the protective groups provide natural PGs of the E series [114, 124, 125].

[L] G. Stork and M. Isobe: Simple Total Synthesis of Prostaglandins from 4-Cumyloxy-2-cyclopentenone. *J. Am. Chem. Soc.* **1975**, *97*, 6260–6261.

Unnatural PGs are important as well. In the 1980s, PG research entered a new stage with the addition, to the widespread local hormone activities, of new actions relating to vital defense processes such as inflammation, tissue repair, immune response, the regulation of cell proliferation and differentiation, and a sleep-inducing effect. Our general method

Fig. 5.32. Aldol route to prostaglandins. $SiR_3 = SiMe_2{}^tBu$. $R' = SiMe_2{}^tBu$ or THP.

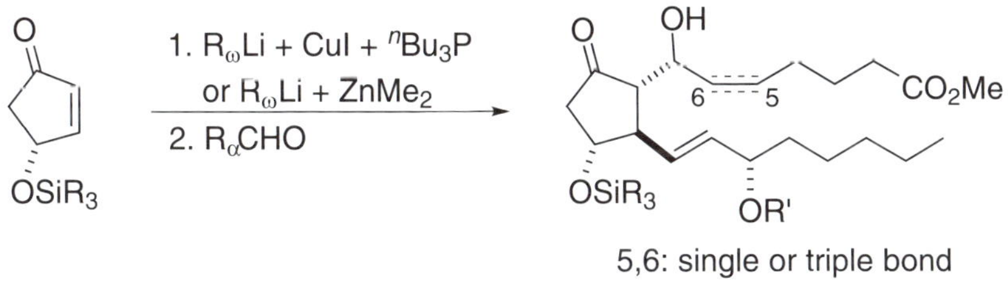

allowed the expeditious synthesis of all natural and various artificial PGs for these studies [126]. Arising from this research, we had a collaboration with a laboratory of the Pharmaceutical Division of Teijin Limited. This group was led by Seizi Kurozumi, a friend of mine from Kyoto, who had trained with Barry Trost at the University of Wisconsin. At that time, academia-industry collaborations were fruitful in the open, basic research phase. I felt that exchanging ideas between researchers from different backgrounds, such as chemical versus pharmaceutical, stimulated not only the progress in research but also the growth of the younger generation.

With our procedure, we could synthesize isocarbacyclin, a stable carbocyclic analogue of physiologically significant prostacyclin (PGI_2) (**Fig. 5.28.** and **Fig. 5.33. a**) [127, 128]. [^{3}H]APNIC (19-(3-azidophenyl)-20-noriscarbacyclin), an azido-functionalized PGI_2 analogue, acted as an efficient photoaffinity probe for a PGI_2 receptor [126, 129, 130]. Moreover, we were surprised to find that (15*R*)-TIC ((15*R*)-16-*m*-tolyl-17,18,19,20-tetranorisocarbacylin), an artificial analogue with an unnatural 15*R* configuration, and 15-deoxy-TIC have a very high binding affinity and selectivity for the PGI_2 receptor in the central nervous system [131 – 133]. In a fruitful chemical-medical collaborative research project with Bengt Långström at Uppsala, Yasuyoshi Watanabe at Osaka City University, and Masaaki Suzuki at Gifu University that implemented positron emission tomography (PET) using ^{11}C-incorporated tracers (**Fig. 5.34.**), the invention of these specific molecular probes led to the discovery of a novel PGI_2 receptor in the central nervous system of the monkey as well as the human brain (**Fig. 5.35.**) [134]. Here, a powerful catalytic tool was needed, because the $^{11}CH_3$ group must be incorporated in the final step of the synthesis of the (15*R*)-TIC methyl ester, and the total time for synthesis, workup, purification, and sterilization should be less than 40 min because of the short half-life time of ^{11}C (20 min). Notably, our Pd-mediated rapid coupling of [^{11}C]methyl iodide with aryltributylstannane played a pivotal role in this biomedical study [135 – 137]. Thus, the designed TICs acted as a new type of potential chemotherapeutic agents for applications in neurodegeneration.

It was Professor Osamu Hayaishi (1920 – 2015), an influential biochemist and discoverer of oxygenase enzymes, who intensively studied the bioinformation of PGs at the time. During his work with Masanori Fukushima (Aichi Cancer Center) on the sleep-wakefulness regulation by PGD_2, he claimed in a newspaper article that his team had discovered its antitumor effect. As an organic chemist, I immediately thought the real active compound is its dehydration product, PGJ_2, which is susceptible to thiol nucleophiles. Its structural analogues PGAs are more readily accessible. In fact, the three-component aldol method provides a simple access to antitumor Δ^7-prostaglandin A_1 (Δ^7-PGA_1) methyl ester (**Fig. 5.33. b**) [126, 138]. We presented the chemical implications of the antitumor and antiviral activities of Δ^7-PGA_1 and PGA_1 methyl ester by

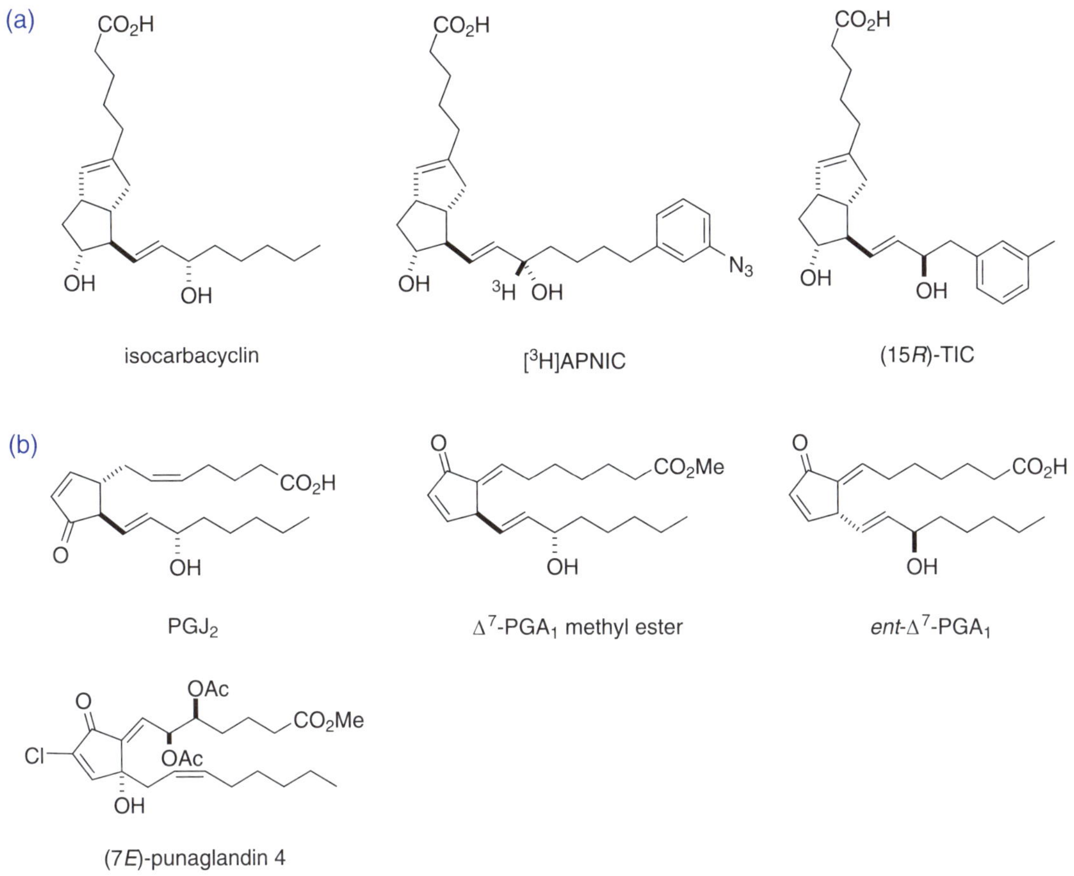

Fig. 5.33. Synthesis of physiologically important PG analogues. (a) Isocarbacyclin and its analogues. (b) Antitumor PGs.

examining the reactions with thiols [139 – 141]. Molecular design can overcome the metabolic instability of Δ^7-PGA_1 [142], while maintaining its antitumor potency. Thus, the 12*R* derivatives markedly increase the induction of p21, a Cdk inhibitor, leading to sharp cell cycle arrest at the G1 phase at a very low dose level [143].

In relation to the anti-neoplastic activity of PGs, we exploited a convergent synthesis of punaglandins (PUGs), marine eicosanoids

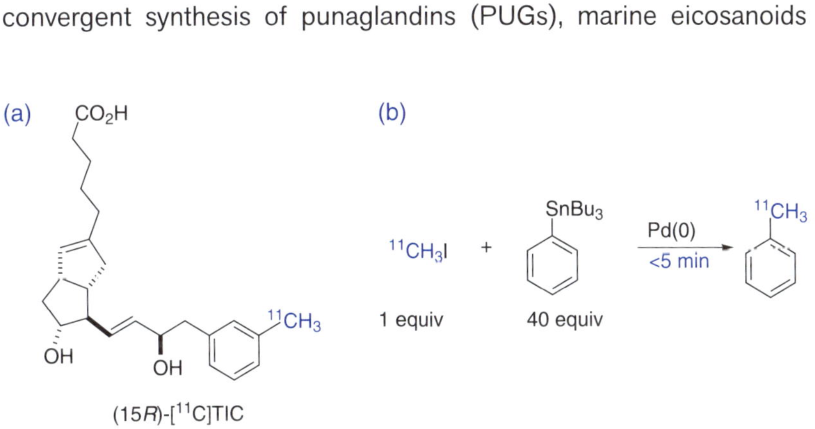

Fig. 5.34. (a) Structure of (15*R*)-[^{11}C]-TIC. (b) Rapid coupling protocol for PET study using $^{11}CH_3$-labelled tracers.

Fig. 5.35.
The uptake of $(15R)$-[^{11}C]-TIC (Fig. 5.34.a) in human brain. The PET images of 18 horizontal slices from the lower to the upper portions of the brain (volunteer: M. Suzuki: June 13, 2000, PET Center of Uppsala University) and the interdisciplinary collaborative team. From the left, M. Suzuki, R. Noyori, Y. Watanabe, and B. Långström.

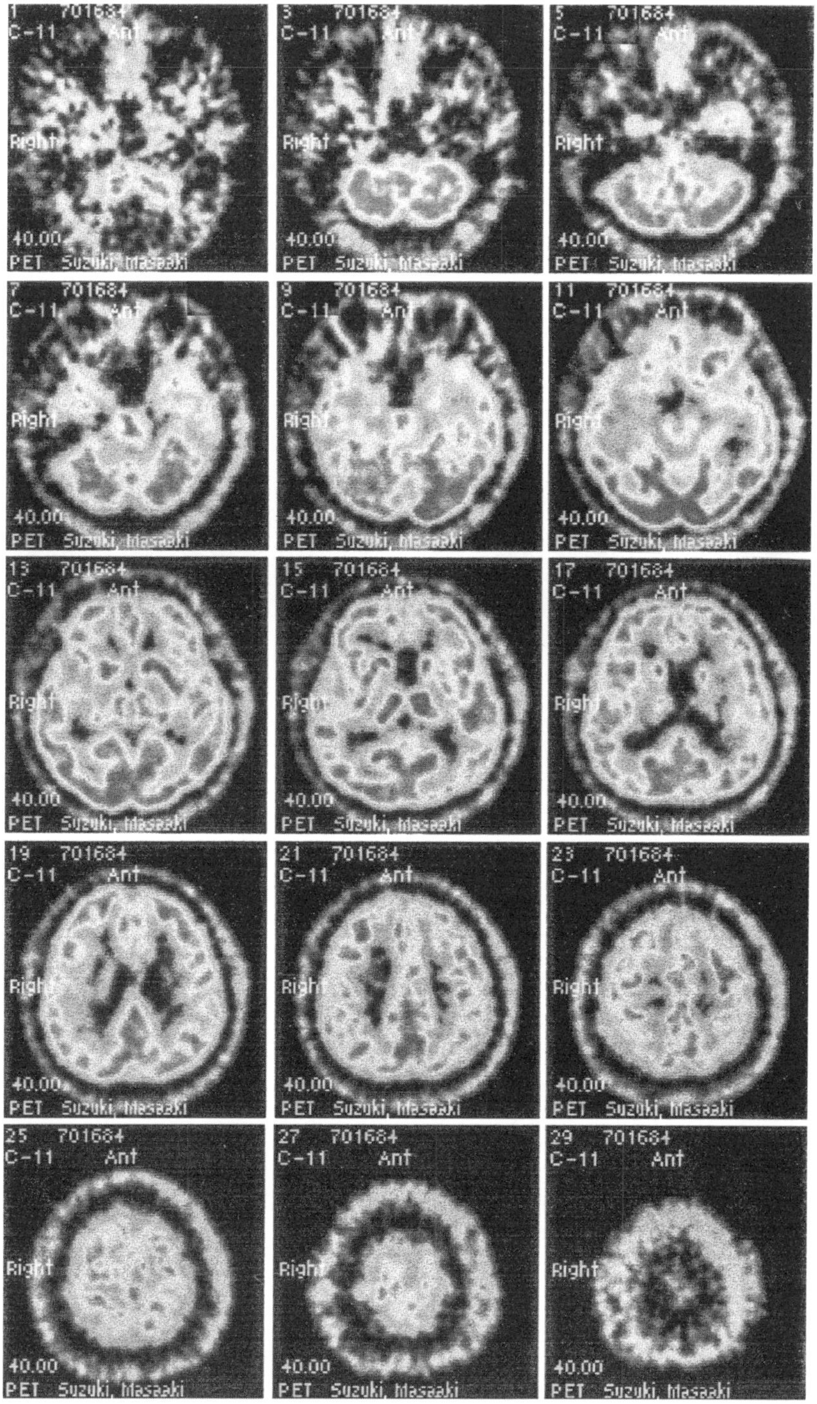

possessing potent inhibitory effects [144, 145]. Careful comparison of our stereo-authentic synthetic products and naturally occurring materials, collected by Paul Scheuer's group at the University of Hawaii, dictated revision of the originally postulated structures.

5.6 Transition metal catalyzed transformations of endoperoxides and epoxides (1979 – 1990)

Involvement of the enzymatic transformation of PG endoperoxides, PGG and PGH, in the PG biosynthesis (**Fig. 5.22.**) prompted us to investigate the behavior of 1,4-epiperoxides in the presence of man-made, well-defined transition metal catalysts having a one- and two-electron exchange ability. We also hoped that these catalytic systems could provide synthetic opportunities for position-specific oxygenation of 1,3-dienes.

Consequently, as shown in **Fig. 5.36.**, $RuCl_2(PPh_3)_3$ appeared to induce the selective, high-yield isomerization of bicyclic 2,3-didehydro

Fig. 5.36. Stereochemical relationship in the diene to diepoxide conversion.

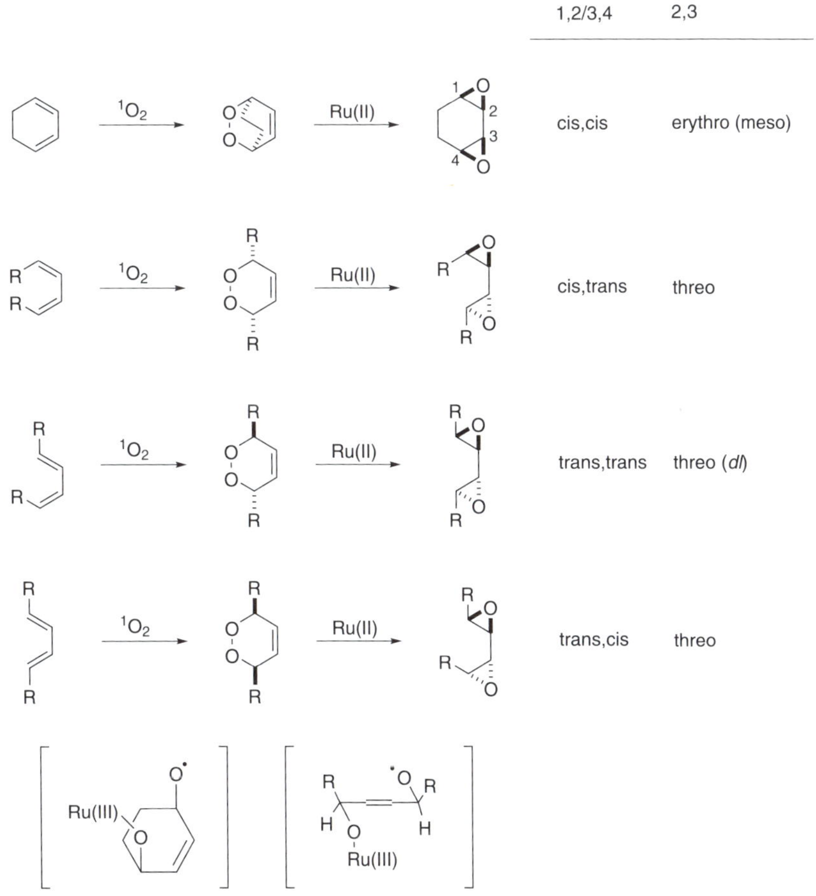

15 **16**

Fig. 5.37. Transition-metal catalysis of a PG endoperoxide: (a) Ru(II)-catalyzed reaction; (b) Pd(0)-catalyzed reaction.

1,4-epiperoxide to the *syn*-1,2 : 3,4-diepoxides [146, 147]. The stereochemical outcome reflects the unique difference between the ground-state geometry of the epiperoxides, C_s (bicyclic) *vs* C_2 (monocyclic). The reaction is interpreted in terms of the intermediacy of inner-sphere radicals 15 or 16 formed by atom transfer of Ru(II) to peroxy substrates.

PGH$_2$ methyl ester (17), a saturated epiperoxide, undergoes clean and stereospecific fragmentation to afford malonaldehyde and

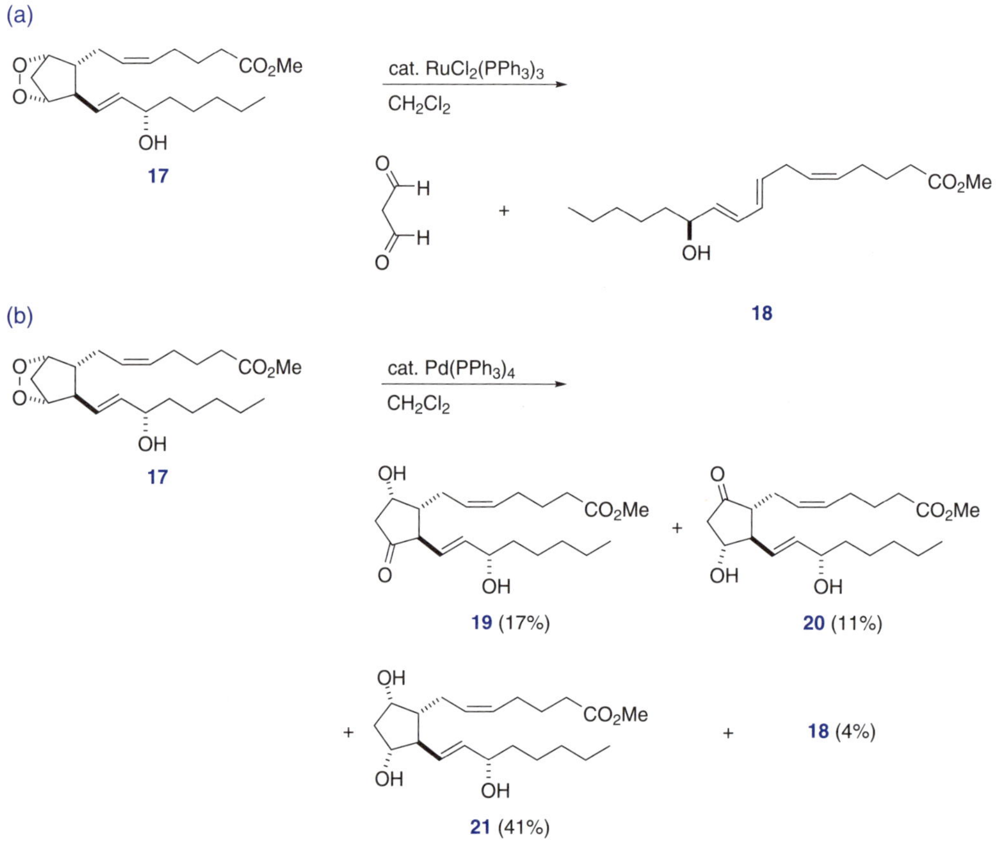

(5Z,8*E*,10*E*,12*S*)-12-hydroxy-5,8,10-heptadecatrienoate (18) if catalyzed by RuCl$_2$ (PPh$_3$)$_3$ (**Fig. 5.37.**) [147]. In contrast, Pd(0) promotes decomposition of the epiperoxide 17, leading mainly to primary PGs including PGD, PGE$_2$, and PGF$_{2\alpha}$ methyl esters (19 – 21), providing a chemical model of the biosynthesis [148, 149].

Ru(II) and Pd(0) catalysts apparently behave differently [148]. Since Pd(0) complexes have a strong nucleophilicity, we hoped to use this property for the selective conversion of activated epoxides including alkenyl epoxides. As illustrated in **Fig. 5.38.**, we found a convenient way to oxygenate 1,3-dienes, where the reaction course is highly dependent on the structures and substitution pattern of the substrates [150]. We consider that the reaction proceeds via either zwitterionic or chargeneutralized Pd alkoxides, 22 or 23, which undergo intramolecular hydrogen transfer or react with carbon or heteroatom nucleophiles. The isomerization of cyclic monoepoxides leads selectively to 3-cycloalkenones, providing a simple way to obtain synthetically useful 3-cyclopentenone from cyclopentadiene. In a similar manner, epoxy ketones are convertible to 1,3-diketones [151, 152].

Fig. 5.38. Site-specific oxygenation of 1,3-dienes.

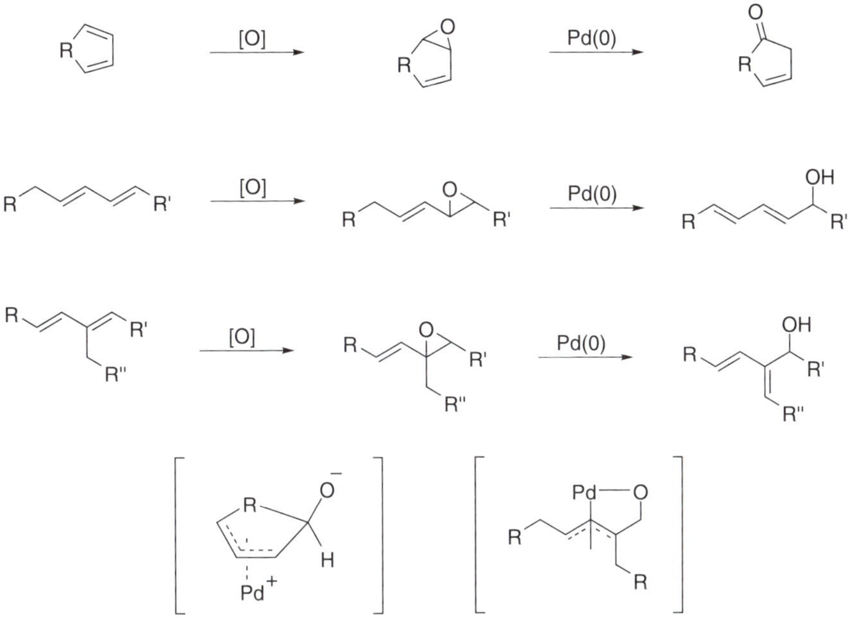

22

23

Launching the IUPAC OMCOS symposium series

Promotion of research and education is not a mission exclusive to senior academics. I have long been engaged in the promotion of organic chemistry nationally and internationally. The symposium series on Organometallic Chemistry directed toward Organic Synthesis (OMCOS) is among the most successful IUPAC ventures, and already four decades have passed since its first meeting was held in Fort Collins, Colorado, in 1981. So how was this brought about? I was heavily involved in it by planting a seed in its prehistory.

In the 1970s, IUPAC already had several relatively large-scale symposia series, for example on natural products chemistry and organic synthesis, held in even-numbered years. It was in the mid-1970s, when Shô Itô (Professor at Tohoku University, a member of the Nozoe school), as the Chair of IUPAC Organic Chemistry Division, was looking for some emerging and promising topics to be promoted by holding a symposium in odd-numbered years. This event was to be a fixed period trial and may or may not continue based on assessing its performance. In response to Itô's inquiry, I suggested a possibility of organizing an OMCOS symposium after some discussion with Gary H. Posner (Corey student) of Johns Hopkins University. Subsequently, Itô immediately agreed with my suggestion, proposed a reasonable plan to IUPAC, and the idea was accepted without any issues. I was pleased to know of this approval. However, at the time in the late 1970s, I was still a young professor who was inclined to fully focus on carrying out research, and I also felt that my administrative experience was not sufficient to organize an international conference in Japan. Therefore, in 1977, at an organometallic conference held at the lake Biwa in Japan, we asked John K. Stille to hold the inaugural meeting at Fort Collins, together with Louis S. Hegedus (another Corey student), and they happily accepted our proposal. Thus, the OMCOS conference series had a good start from both the scientific and the social perspective.

As an "unsung" OMCOS meeting initiator, I am very pleased to see the unexpectedly long and glorious existence of the conference series. Moreover, I would like to express my appreciation of the many senior Japanese colleagues who led this unique, promising research field in the yet difficult environment during the 1960s. Nothing comes from nothing.

Fig. 5.39. From the left: Robert H. Grubbs, Hiroshi Yamazaki, and the author at the Lake Biwa Hotel in 1977.

Fig. 5.40. The inaugural OMCOS meeting at Fort Collins in 1981. From the left: the author, Jiro Tsuji, John K. Stille, Sumire Negishi, Albert I. Meyers, Eiichi Negishi, and Shunichi Murahashi.

Fig. 5.42. Japan's organometallic leaders gathered in 2000. Front row, from the left: Akio Yamamoto, Sei Otsuka, Nobue Hagihara, Hideki Sakurai, and author. Back row, from the left: Shigetoshi Takahashi, Shunichi Murahashi, Mitsuo Ishikawa, Akira Nakamura, Hiroshi Yamazaki, Kiitiro Utimoto, and Hideo Kurosawa.

Fig. 5.41. At the first OMCOS meeting at Fort Collins in 1981. From the left: Gary H. Posner, the author, and Howard Alper at Stille's residence.

5.7 Chemistry of enolates: Stereoselectivity in aldol reactions (the mid-1970s – the late 1990s)

During Woodward's career, the logic of stereo-selection in a relative (not absolute yet) sense relied largely on the distinction of convex- and concave faces of a cyclic structure. This concept remained valid in enolate chemistry, as we observed the key role it played in our PG synthesis requiring the *trans* configuration of the α and ω side chains (**Fig. 5.25.**).

An intriguing problem in the mid-1970s was that of "acyclic stereoselection", i. e., control of relative or absolute stereochemistry in acyclic systems. The major challengers to this newly emerged concept were as follows: Paul A. Bartlett, Clayton H. Heathcock (both at the University of California, Berkeley), David A. Evans (Caltech and then Harvard), Satoru Masamune (MIT), and Yoshito Kishi (Harvard). Our colleagues focused on enolate chemistry among others as it plays an important role in organic synthesis. Their utility is attributed to the capability of undergoing a variety of nucleophilic reactions with desired selectivity. Traditionally, enolates were generated by the exposure of ketones to basic reagents. Unfortunately, the resulting enolates undergo notorious α to α' isomerization in the presence of coexisting proton donors such as solvents or substrates, hampering the selective transformations, as shown in **Fig. 5.23.** Then, in the 1970s, a number of more elaborate methods appeared to generate positionally and geometrically well-defined metal enolates. Teruaki Mukaiyama (University of Tokyo) played an important role in this regard by inventing his original aldol reaction using alkenyl silyl ethers promoted by $TiCl_4$. Under such circumstances, the "directed aldol reaction" was a major issue those days.

Aldol is the generic name for β-hydroxy carbonyl compounds normally formed by the addition of enolates to aldehydes or ketones. This grouping is widely found in organic compounds that show a range of biological, chemical, and physical functions. *Erythro* and *threo* (or *syn* and *anti*) diastereomers are possible when the compounds possess two consecutive stereogenic carbons at the α and β positions. The major issue was whether or not the *threo* / *erythro* or *syn* / *anti* selection in the aldol reaction reflects the enolate geometry, either *E* or *Z*. In fact, the kinetic stereo-regulation in a Lewis acid-mediated reaction is clearly defined by the enolate configuration; the *Z* enolates give the *erythro* aldols, whereas the *E* enolates afford the *threo* aldols predominantly (**Fig. 5.43.**). The stereoselectivity is considered to be governed by the diastereomeric, metal-linked six-membered structure 24 called the Zimmerman-Traxler type transition state. The metallic species M assembles both active enolate and aldehyde substrate, facilitating the reaction.

Meanwhile, we went an opposite direction, studying the behavior of new, ion-paired, naked enolates in an aprotic solvent. The true anion

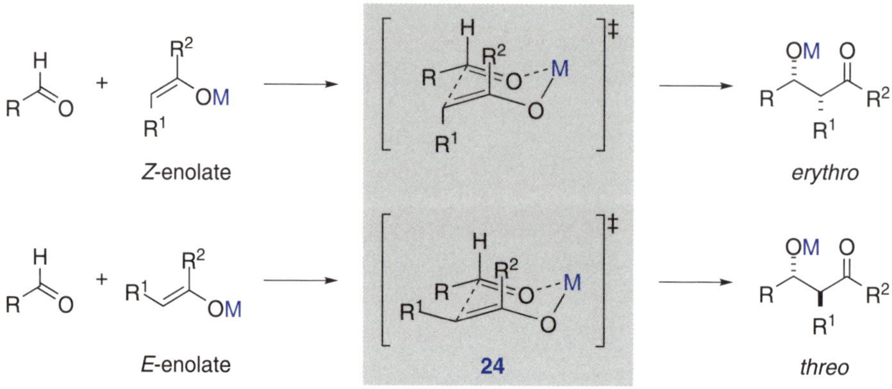

Fig. 5.43. Lewis-acid aided stereoselective aldol reaction via Zimmerman-Traxler type transition states.

must behave differently from the normal Lewis-acid bound species. In view of the great strength of Si–F bonds (159 kcal/mol in Me_3SiF) [M], the reaction of alkenyl trimethylsilyl ethers and the fluoride ion is anticipated to produce the unprecedented, metal-free enolate species and Me_3SiF. We shared the same idea with Isao Kuwajima and his student Eiichi Nakamura at the Tokyo Institute of Technology, and in 1977 we jointly published an article on the aldol reaction catalyzed by tetraalkylammonium fluoride, instead of Lewis acids **[153]**, [154].

[153] R. Noyori, K. Yokoyama, J. Sakata, I. Kuwajima, E. Nakamura, and M. Shimizu: Fluoride Ion Catalyzed Aldol Reaction between Enol Silyl Ethers and Carbonyl Compounds. *J. Am. Chem. Soc.* **1977**, *99*, 1265–1267.

In principle, the reactivity and selectivity of ambient enolates should reflect electronic characteristics; charge density is greater at the oxygen and the HOMO coefficient is larger at the carbon atom. Thus, it is generally considered that hard nucleophiles react preferentially at oxygen under charge control and soft nucleophiles tend to react at carbon under MO control. However, the actual behavior is highly influenced by the nature of the structures and reaction conditions. Enolates formed from ketones and a base in protic media are hydrogen-bonded, whereas preformed, metal-coordinated enolates in aprotic media exist as aggregates and the nature of metals and aggregation states play a significant role in controlling their reaction. Thus, we were intrigued by the properties of truly ion-paired tris(dialkylamino)sulfonium (TAS) enolates under aprotic, rigorously anhydrous conditions. The TAS cation is sterically demanding and is stabilized greatly by electron delocalization. When TAS difluoro(trimethyl)siliconate 25 (1976 US patent from the DuPont company), which is in equilibrium with TAS fluoride and fluorotrimethylsilane, reacts with the alkenyl silyl ether of benzyl methyl ketone 26 in THF, after removing the volatiles under vacuum, a yellowish air-sensitive, crystalline TAS enolate 27 is produced (**Fig. 5.44.a**) [155–158]. The electro-conductivity measurement and the ^1H and ^{13}C NMR analysis substantiated its ionic nature. In going from the starting silyl ether 26 to TAS enolate 27, we observed a 20.0 ppm up-field shift for the vinylic ^{13}C signal which is greater than any of the reported systems including the crown ether-complexed Na enolate, consistent with MO investigations. The isolated TAS

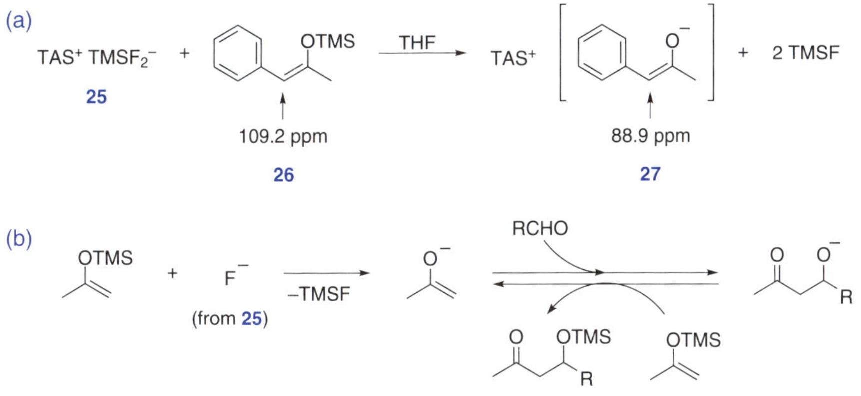

Fig. 5.44. Chemistry of TAS enolates. (a) Synthesis of a TAS enolate. Chemical shifts of ^{13}C NMR are noted. (b) Fluoride ion catalyzed aldol reaction. $TAS^+ = (Et_2N)_3S^+$, $TMS = SiMe_3$.

enolate was found to undergo regiospecific C-alkylation with alkyl iodides including cyclopropyl and 5-hexenyl iodide, without skeletal change, and O-acetylation with acetic anhydride. TAS phenoxide shows similar NMR characteristics and undergoes O-alkylation with organic halides [159].

Aldol addition is reversible in nature. The reaction of isolated TAS enolate 27 and benzaldehyde fails to form an adduct for a thermodynamic reason. The reactive β-oxido ketone must be trapped efficiently by an electrophile. Thus, the fluoride source 25 acts as an initiator of the autocatalytic reaction of an alkenyl silyl ether and aldehyde giving the β-siloxy ketone (**Fig. 5.44. b**). The initial addition of the ion-paired enolates to an aldehyde is reversible but not always in full equilibrium, the extent being dependent on aldehyde structures, aromatic or aliphatic, for example. The steric course is characterized by a unique *erythro* (or *syn*) diastereo-selection which is independent of enolate geometry (*Z* or *E*) (**Fig. 5.45.**, Eq. 1) [160]. This stereo-selection was ascribed to the acyclic,

Fig. 5.45. Aldol reactions via extended transition states.

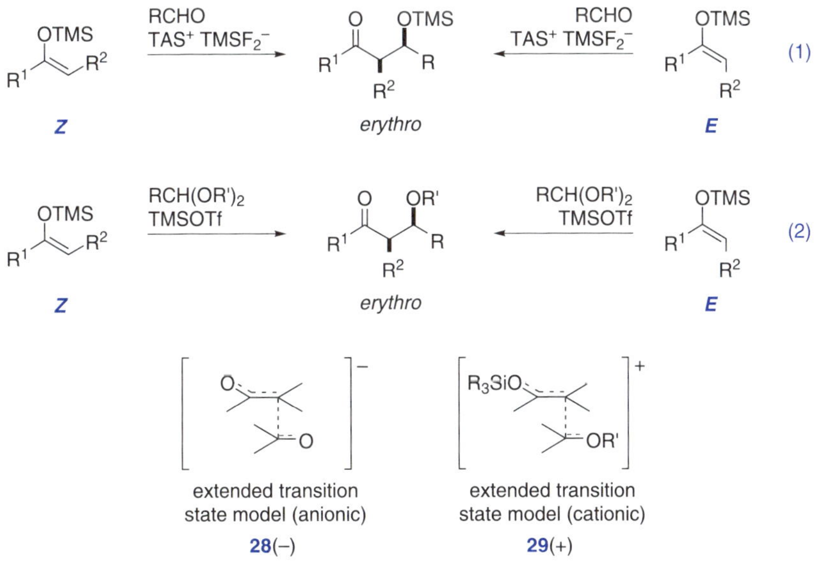

open transition states 28(–) [N]. This notion is in contrast to the conventional pericyclic transition structure 24 often postulated for the *Z/E* dependent chemistry of metal enolates. The same transition state model is applicable to the trimethylsilyl triflate-promoted aldol reaction of alkenyl silyl ethers and acetals (Eq. 2) which exhibits similar *Z/E*-irrespective *erythro* selectivity via 29(+) [161, 162].

We were involved in aldol chemistry over two decades. Our detailed structural study on cycloalkanone-based aldols reached a conclusion that single crystal X-ray diffraction analysis is the only reliable method to determine the relative *threo* and *erythro* configuration [163]. NMR analysis by the Stiles-House empirical method can be used in only very limited cases. In hydrocarbon solvents, many aldols exist as monomers with an intramolecular OH•••O=C hydrogen bond, but some compounds are in an equilibrium with higher aggregates, where the stabilities are controlled by many steric and electronic effects. Furthermore, the preferred structure in solution often differs from the crystalline-state geometry, showing that great care must be taken in dealing with this matter.

5.8 Catalysis with organic and inorganic silicon compounds (1979 – 1988)

Since 1968, when I launched my laboratory at Nagoya, I pursued research topics in OMCOS in the broadest sense. The achievements include among others: Ni-catalyzed reactions of strained hydrocarbons, Fe carbonyl-aided carbocyclic synthesis, and concise prostaglandin synthesis.

Our curiosity was not limited to transition metals. Silicon was another unexplored, but attractive element for catalysis in our days. Organic and inorganic silicon compounds display a multitude of functions in organic reactions, depending on the relative bond strength, relative electronegativity, and involvement of the valence 3p or vacant 3d atomic orbitals, participation of high-lying σ or low-lying σ^* molecular orbitals, etc. Of particular interest are the great bond dissociation energies for Si–O (122 kcal/mol in Me_3SiOMe) and Si–F bonds (159 kcal/mol in Me_3SiF) [M]. The structures and properties of Si compounds had been thoroughly investigated. Thus, organosilicon compounds are essential ingredients in synthetic chemistry but remained to serve as stoichiometric components until the late 1970s.

As an all-purpose catalyst did not exist, we became interested in the unique catalytic roles of Si compounds in synthesis. Unlike conventional Lewis acid catalysts, which are often aggregated and are capable of assembling more than two functional groups onto the same metal center, organosilicon compounds are monomeric and the electron-deficient Si center, like protons but in aprotic media, accepts only one heteroatom functionality. Furthermore, organosilicon catalysts can be structurally and functionally tunable and are usually less harmful than other Lewis acids.

Trimethylsilyl triflate (TMSOTf) was the simplest compound of our choice. It is a readily accessible derivative of trifluoromethanesulfonic (triflic) acid, a superacid. It forms a colorless, fluid liquid, which boils at 45 – 47 °C at 17 mmHg and dissolves in aliphatic or aromatic hydrocarbons and haloalkanes without causing any reactions. Although the Si–O bond is covalent, rather than ionic, in nature, the Si atom is endowed with eminent electron deficiency by the presence of the electron-withdrawing OTf group, as proved by the ^{29}Si NMR signal. As a consequence, the Si atom can interact strongly with various heteroatoms, particularly oxygen, in organic compounds, to generate the corresponding oxonium ions. The situation is much simpler than ordinary Lewis acid chemistry. In the 1970s, however, it remained to act as an excellent, but stoichiometric silylating agent. At Nagoya, we started to develop new possibilities in collaboration with Masaaki Suzuki and Shizuaki Murata (who later became a Professor at Nagoya University). It appeared that Helmut Vorbrüggen of Schering AG had the same idea utilizing TMSOTf in the area of nucleoside synthesis [0].

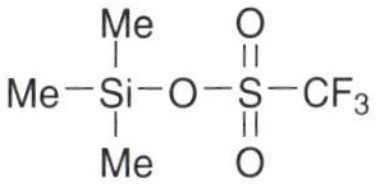

Fig. 5.46. Trimethylsilyl triflate (TMSOTf).

We discovered that TMSOTf with or without hindered tertiary amines such as 2,6-di-*tert*-butylpyridine or dicyclohexylmethylamine serves as a versatile catalyst in numerous organic reactions, including aldol-type reactions [161, 162, 164, 165], glycosylation, hydride reduction [166], allylation [167], Baeyer-Villiger type oxidation [168] as well as other functional group transformations (review: [169]). Acetals, orthoformates, or 2-acetoxytetrahydrofuran or -pyrans readily undergo the reaction with enol silyl ethers via a carboxonium triflate ion-pair intermediate giving the aldol-type products (**Fig. 5.45.**), while parent ketones and aldehydes are inactive. Labile ketones such as 3-cyclopentenone or -hexenone can be readily acetalized using alkyl trimethysilyl ethers without protic compounds [170]. This is a simple and useful procedure. Since our demonstration of this catalytic activity, silyl triflates have gained wide popularity in synthetic organic chemistry. The catalyst can be immobilized by perfluorinated resinsulfonic acid [171]. As a result, a large body of useful nucleophilic reactions have evolved, as generalized in **Fig. 5.47.**

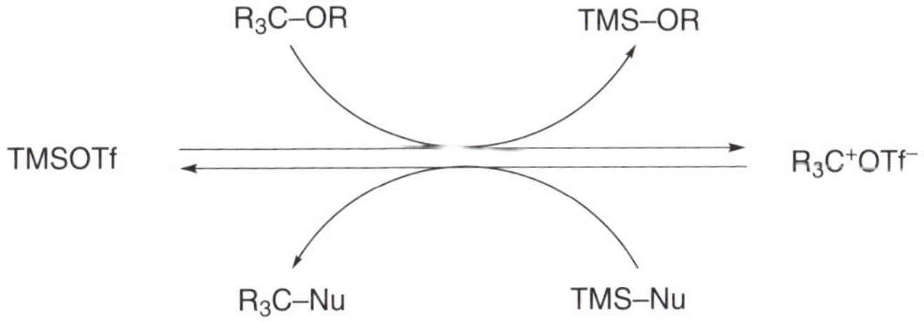

Nu = H, $CH_2CH=CH_2$, CN, OR, OCR=CR'$_2$, SR, etc.

Fig. 5.47. General scheme of trimethylsilyl triflate-catalyzed organic reactions. R_3C^+ = trityl or a heteroatom-stabilized carbocation.

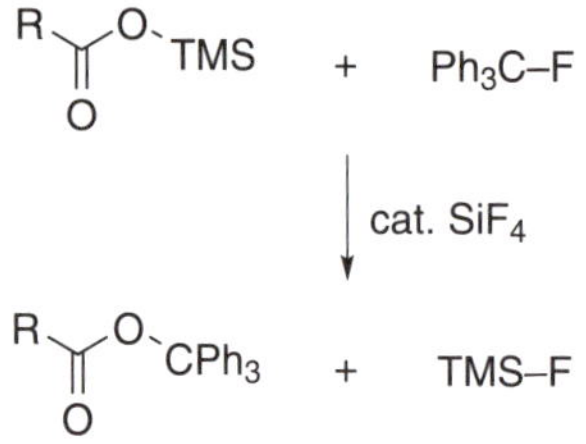

Fig. 5.48. SiF_4-catalyzed reaction of TMS carboxylates with trityl fluoride to give trityl carboxylates and TMS fluoride.

Many of them proceed under very mild conditions in aprotic media without harmful proton exchange, displaying desirable selectivities. TMSOTf has now become an indispensable tool in organic synthesis. Furthermore, a silyl triflate / tertiary amine combined agent was found useful for selective epoxide-allyl alcohol conversion [172–175].

The utility of the eminent oxophilicity of the silicon element was thus amply demonstrated. Then, we noted the affinity between Si and F, which is even stronger than that of Si to the O atom. Thus, certain electron-accepting Si species are anticipated to interact with organofluorine compounds, thereby providing a new means of cationic activation of organic molecules [176]. As expected, SiF_4 catalyzes the reaction of trityl fluoride with TMS carboxylates in acetonitrile giving trityl carboxylates through elimination of TMS fluoride [177]. Phenol, primary alcohols or their TMS ethers are equally tritylated [178]. Secondary alcohols are oxidized to ketones by hydride abstraction with the trityl cation intermediate.

During this study with Shunichi Hashimoto (who later became a Professor at Hokkaido University) and Masahiko Hayashi (now a Professor at Kobe University), we were able to develop a convenient procedure to prepare glycosyl fluorides by treatment of 1-unprotected or 1-*O*-acetylated sugars with an HF-pyridine mixture (pyridium poly(hydrogen fluoride) [179]. We found that SiF_4 or TMSOTf effectively catalyzes glycosylation using appropriately protected glycosyl fluorides and TMS ethers (1:1 molar ratio) (**Fig. 5.49. a**) [180]. Various fluorosugars, but not chlorosugars, and TMS ethers can be used. Tetraalkoxysilanes or even unprotected alcohols can be used as nucleophiles. Various functional groups are tolerated under the reaction conditions using Si-based promoters. We argued that the α/β stereoselectivity significantly depends on solvents, acetonitrile versus ether. This procedure is operationally simple and suitable for large-scale reactions, and is advantageous over the classical Königs-Knorr glycosylation that uses 1-halosugars and the precious

Fig. 5.49. Utility of 1-fluorosugars. (a) Clean glycosylation. Bn = benzyl. (b) Nucleoside synthesis. Bz = benzoyl.

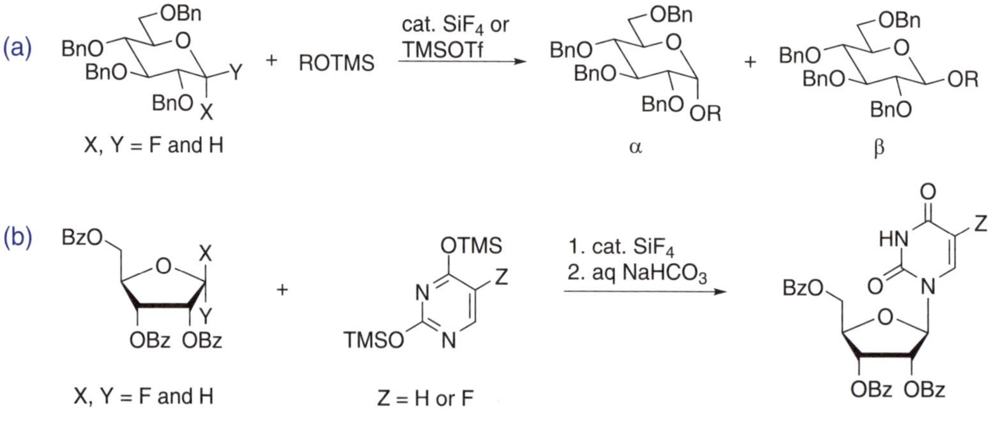

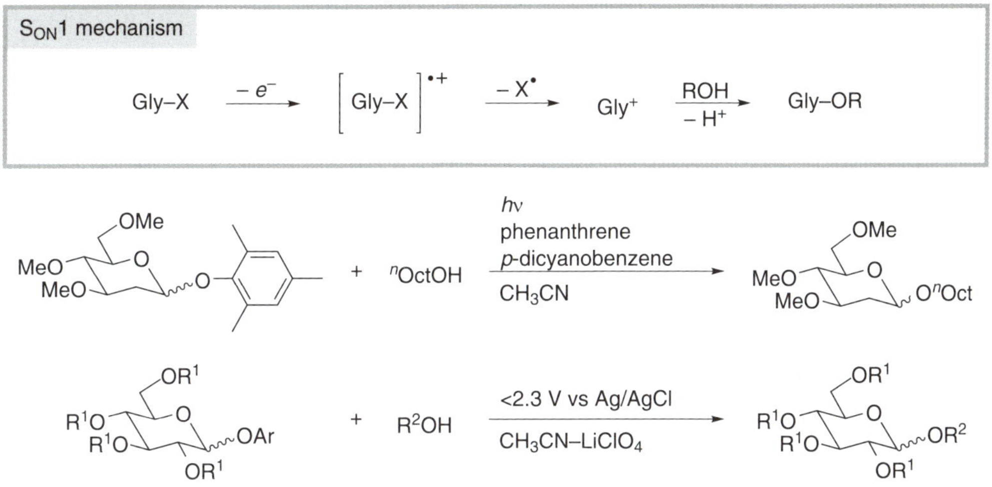

Fig. 5.50. Photochemical and electrochemical glycosylation via an $S_{ON}1$ mechanism. Gly = 1-glycosyl. X = substituents with a high oxidation potential. R^1 = Ac, Bn, Me, H. R^2 = Me, Et, ^{i}Pr, Cy, ^{t}Bu, etc. Ar = Ph, 2,4,6-$Me_3C_6H_2$.

and explosive Ag salts or toxic Hg compounds. In a similar manner, we explored a simple SiF_4-catalyzed method to condense 1-fluorofuranoses and silylated nucleobases, well compared with the silyl-Hilbert-Johnson procedure using a protected 1-*O*-acetylribose with TMSOTf or other Lewis acids (**Fig. 5.49. b**) [181] [O].

I am afraid that my attention easily shifts from one thing to another. During the glycosylation study, we hoped to generate a cationic species without the use of Lewis acids. Learning that in solution, radical cations can dissociate into the solvent-separated cations and radicals, we achieved a photo-sensitized transacetalization between protected 2-deoxyglycosyl *aryl* ethers and alcohols. Combination of an aromatic hydrocarbon with a strong reducing ability and a cyano aromatic with a strong oxidizing ability under light irradiation afforded satisfactory results. This finding marked the first $S_{ON}1$-type nucleophilic reaction proceeding via radical cation intermediates [182]. (**Fig. 5.50.**) Moreover, hydroxyl-protected and -unprotected glycosyl aryloxides were found to react with alcohols under mild oxidative electrolytic conditions leading to the corresponding glycosides [183, 184]. We enjoyed such mechanistic dawdling on the roadside as well.

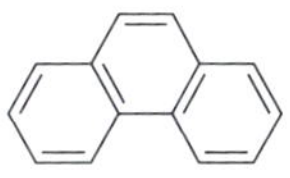

$E_{ox}^{p/2}$ = 1.64 V vs Ag/AgCl

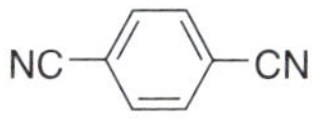

$-E_{red}^{p/2}$ = 1.62 V vs Ag/AgCl

Fig. 5.51. Top: Phenanthrene: bottom: *p*-dicyanobenzene.

Revisiting the strong affinity between silicon and the fluorine atom, we applied the unique property in a reverse manner to activate nucleophiles. As described already, we were able to develop the chemistry of truly ion-paired enolates and phenoxide using alkenyl and phenyl trimethylsilyl ethers with the aid of tris(diethylamino)sulfonium (TAS) difluorotrimethylsiliconate as a fluoride ion source (**Fig. 5.44. – 5.45.**).

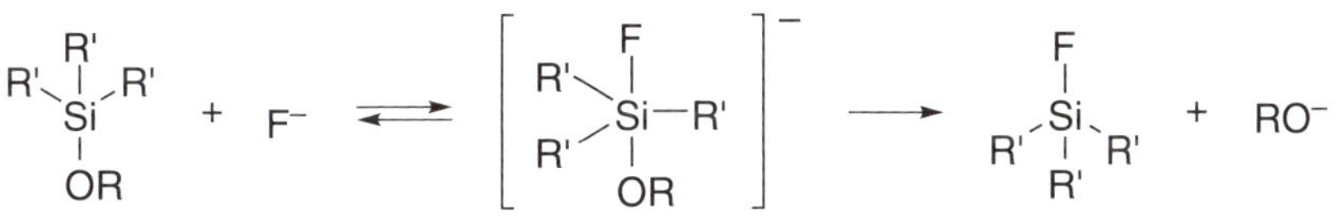

Fig. 5.52. Substitution reactions via penta-coordinate siliconate intermediate.

5.9 Nucleic acid synthesis employing organometallic chemistry (1983–1992)

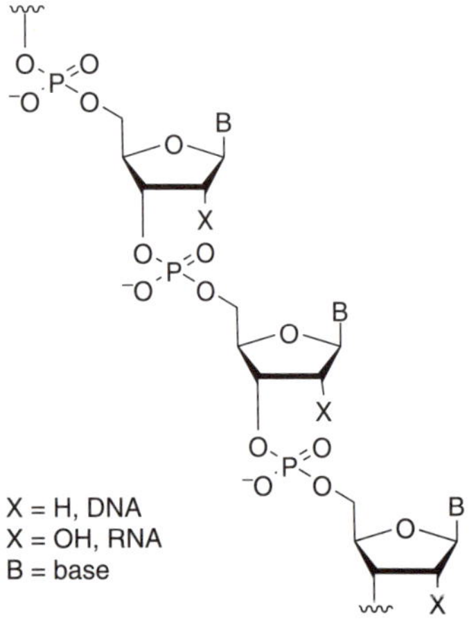

Fig. 5.53.
Nucleic acids.

Nucleic acids, either natural or unnatural, are among the most important class of compounds together with peptides and carbohydrates in the biological area. However, the chemical synthesis posed an array of problems in the transformation and isolation of products, as well as achieving high overall efficiency. I had often complained that their synthetic methods remained rather classical. As a result, my colleagues often criticized, "Noyori does not understand biochemistry." I must admit that this is true, but I believed that my expertise could help biochemists. Organometallic chemistry had been utilized largely for the transformation of non- and semi-polar organic compounds such as saturated or unsaturated hydrocarbons, organic halides, and carbonyl compounds, but long conceived not compatible with highly polar, functionalized substances.

In the mid-1980s, driven to challenge this trend by devising useful OMCOS tools for nucleic acid synthesis, I began a collaboration with Yoshihiro Hayakawa who played a major role in Fe carbonyl chemistry. Fortunately, we achieved efficient solution-phase synthesis of some short-chain oligomers that have unusual structures and a solid-phase preparation of relatively long-sequence DNAs.

We accomplished a series of convenient synthetic transformations utilizing both main-group elements and transition metals. First, O-selective phosphorylation of nucleosides without NH_2 blocking was achieved. This was considered impossible because the NH_2 function is more nucleophilic than the OH group. We solved this problem by reversing the O/N reactivity simply by treatment of the substrates with an equimolar amount of a metallo-organic base such as an alkyllithium, KO^tBu, ^tBuMgCl, $Al(NMe_2)_3$, and $^iBu_2AlH$. Resulting alkoxides are condensed with phosphorochloridate or *p*-nitrophenyl phosphate readily to afford the desired O-phosphorylation products (nucleotide) in high yield [185, 186]. This method provides a simple way to form internucleotide linkages [187].

We explored the utility of a $CH_2{=}CHCH_2$ (allyl) protector for the phosphate and oxygens in guanine and thymidine, which can be deblocked under mild conditions by a Pd(0)-catalyzed reaction (**Fig. 5.54.**) [188, 189] [P]. Moreover, $CH_2{=}CHCH_2OCO$ (allyloxycarbonyl, AOC) serves as an excellent protector for sugar hydroxyls as well as amino and imide moieties of nucleoside bases, which can be deblocked again by combining a Pd(0) catalyst and various nucleophiles [190]. Notably, the three allyl and two AOC protectors in the dinucleotide are

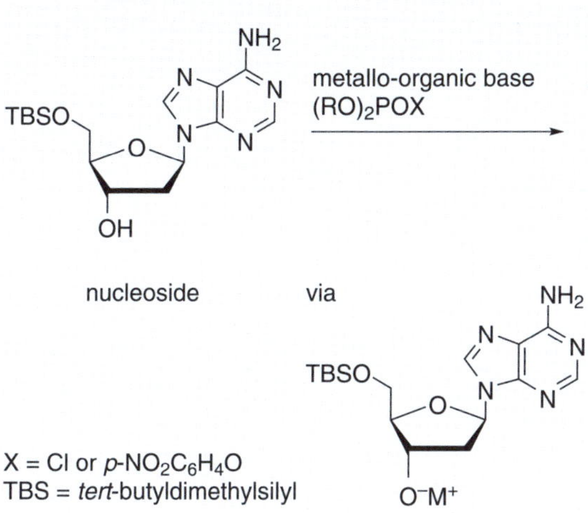

Fig. 5.54. O-selective phosphorylation of nucleosides without NH_2 blocking.

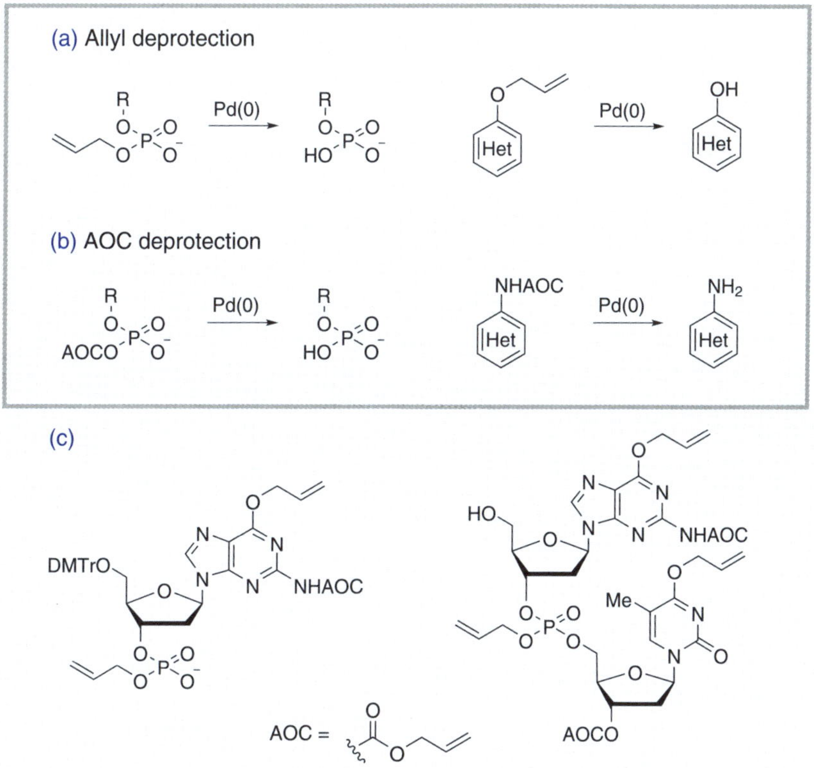

Fig. 5.55. Pd(0)-sensitive (a) allyl and (b) allyloxycarbonyl (AOC) protectors in nucleotide synthesis. (c) Examples of allyl- and AOC-protected nucleotides. Het = aromatic heterocycle.

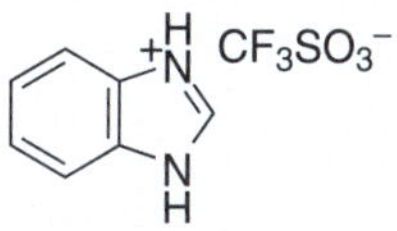

Fig. 5.56. Benzimidazolium triflate.

deblocked all at once by Pd(0) with the aid of diethylammonium hydrogencarbonate, giving the unprotected dGpT (2'-deoxyguanylylthymidine dinucleotide) in 90 % isolated yield. These allylic protectors are stable under the conditions for removal of the conventional *O*-(dimethoxytrityl) (DMTr) or *O*-(*tert*-butyldimethylsilyl) (TBS) groups at the C(3') and C(5') positions [191]. The AOC group thus acts as both specific and general protectors, providing a powerful tool in nucleotide synthesis. In addition, my earlier interest in organosilicon catalysis led to the invention of non-aqueous oxidation of nucleoside phosphites to the phosphates by using bis(trimethylsilyl)peroxide and TMSOTf or TMS-Nafion as a catalyst [192]. Condensation of a nucleoside phosphoramidite and a nucleoside is an essential step in oligonucleotide synthesis, which was greatly improved by using benzimidazolium triflate as a promoter [193].

By combining these methods, we could efficiently access biologically significant 2',5'-linked oligoadenylates (2–5 A's) [194–196], branched-type oligoribonucleotides [197, 198], and other artificial analogues (**Fig. 5.57.**) [199]. Such unique 2–5 A's and related compounds were widely distributed for biochemical studies at the time.

Fig. 5.57. Synthesis of 2–5 A's and branched-type oligonucleotides.

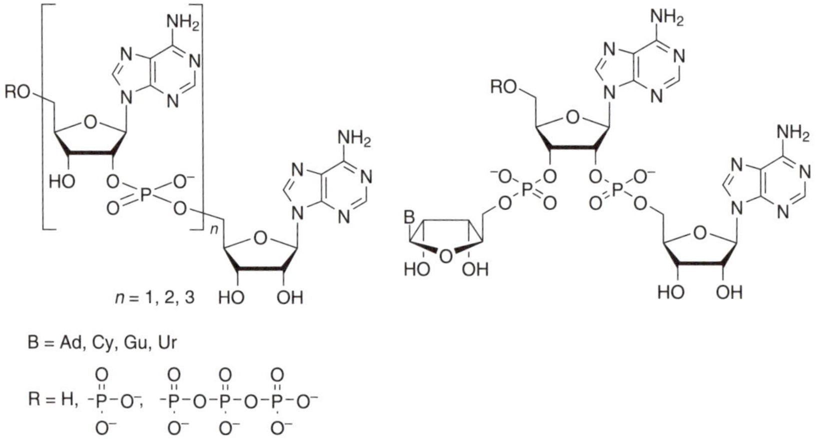

The solid-phase synthesis of oligo-deoxyribonucleotides is an indispensable operation in this field [200], where the phosphoramidite method, first developed by Marvin H. Caruthers, had served as the standard procedure. However, the purity of the DNAs thus synthesized was often variable depending on the sequence and length and often remained unsatisfactory. In 1988–1990, we disclosed a significantly improved automated solid-phase synthesis of deoxyribonucleotide (DNA) oligomers [201–204]. The requisite phosphoramidite monomer units are obtainable by condensation of the corresponding 3'-*O*-free nucleosides and allyloxy(diisopropylamino)phosphine assisted by 1*H*-tetrazol and diisopropylamine (**Fig. 5.58. a**). **Fig. 5.58. b** shows our building blocks: all NH_2 moieties of 2'-deoxyadenosine (dA), 2'-deoxycytidine (dC), and 2'-deoxyguanosine (dG) units are blocked by single AOC groups, while the internucleotide bonds were protected by allyl groups.

To demonstrate the validity of this procedure, we selected as a target 60mer known as a part of a DNA sequence of yolk sac tumor proteoglycan cDNA pPG1. The chain elongation was achieved on an Applied Biosystems Model 381 DNA synthesizer starting with a thymidine which was covalently attached at the 3'-hydroxyl to controlled pore glass (CPG) supports (500 Å pore size) via long-chain alkylamine spacer arm. The synthetic cycle required 10.8 min, in which the phosphite intermediates were oxidized by *tert*-butyl hydroperoxide [202] instead of the conventional aqueous iodine-pyridine mixture. The average coupling yield in this method was 99.3 %, resulting in an overall yield of 66 %, compared well with the standard procedure. Eventually, the application of Pd(0) (2.5 equivalents / allyl), PPh_3, and a formic acid / butylamine mixture allowed a single-step removal of all the allylic protective groups from the CPG-anchored product. In this 60mer synthesis, 104 allylic

Fig. 5.58. Building blocks for oligonucleotide synthesis.

(a)

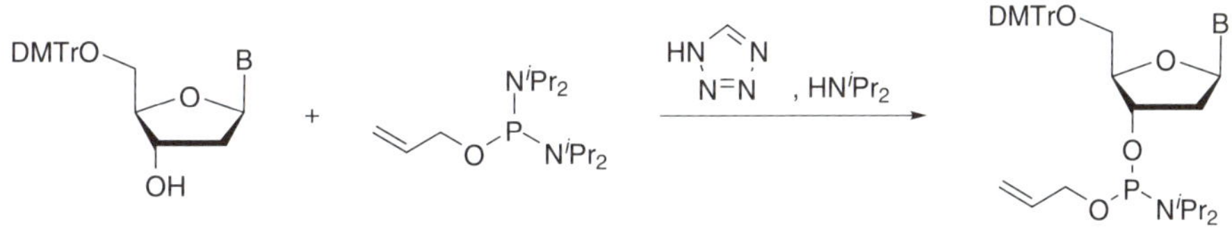

(b)

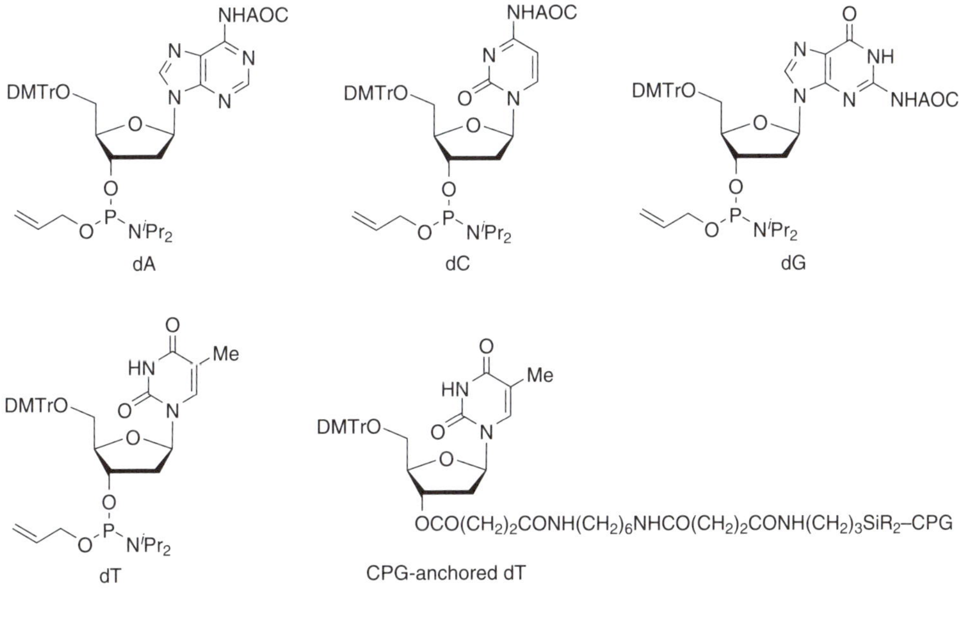

protectors were removed in a single action in almost 100 % yield. The product was easily purified on a solid support by simple washing. The target DNA was detached from CPG by exposure to concentrated ammonia, whereby full removal of the protective groups was achieved as confirmed after digestion by phosphodiesterase and phosphatase. After detritylation, decyanoethylation and support-detachment were conducted under standard ammonium hydroxide treatment. Finally, heating the product with concentrated ammonia accomplished the synthesis of the target free DNA with an unprecedentedly high degree of purity, as judged by the autoradiogram of the ^{32}P 5'-monophosphates of the crude product (**Fig. 5.59.**).

The synthetic route is efficient, but an even more significant feature is the capability of the direct preparation of solid-anchored DNA oligomers. The applicability in molecular biology and diagnostics is considered to be extensive.

These findings clearly indicate the unlimited utility of organometallic reagents and transition metal catalysis. This is valid not only in traditional solution chemistry of nonpolar or weakly polar compounds, but also in solid-phase chemistry involving highly polar or even ionized substances.

d(^{5}TATGGGCCTTTTGATAGGATGCTCACCGAGCAAAACCAAGAACAACCAGGAGATTTTATT^{3})

Fig. 5.59. The solid-phase synthesis of a 60mer and the bio-image chromatogram of the crude products. Lane A: the allyl-AOC method. Lane B: the conventional method.

5.10 Organometallic asymmetric alkylation (1983 – 1985)

I had kept my interest in asymmetric organolithium or Grignard reactions ever since my time in Kyoto, when (–)-sparteine, a natural diamine, was used as a chiral ligand (**Fig. 2.13.**). Encouraged by the successful development of the BINAL-H reagent for asymmetric ketone reduction (**Fig. 5.19.**), I returned to this intriguing subject in 1983 through a collaboration with Masato Kitamura. At the beginning, he was reluctant to pursue such "simple chemistry" because of his research background with Toshio Goto, Minoru Isobe (Nagoya), and Gilbert Stork (Columbia), who were all experts in complex natural products synthesis. He eventually accepted to challenge this fundamental organic reaction and explore a new chemistry frontier.

In the 1970s, the organic chemistry community did not correctly recognize the importance of molecular aggregates of organometallic reagents. Most researchers thus avoided the use of protic chiral auxiliaries that would destroy the reactivity of reagents by elimination of hydrocarbons. However, Teruaki Mukaiyama, Dieter Seebach, and Manfred

Reetz among others showed that some protic chiral auxiliaries act effectively in the asymmetric alkylation with organolithium, Grignard, and organotitanium reagents.

In this regard, we tried to generate a Li/Mg binary organometallic reagent, in which the two different metals would play different roles as an alkyl donor and carbonyl activator. Aiming at creation of a single reactive species possessing excellent chiral recognition ability, we took our favorite, chiral 2,2'-binaphthol which brought us the successful BINAL-H asymmetric reduction (**Fig. 5.19.**). Our optimistic assumption urged us to prepare the bimetallic ethylating agent of **Fig. 5.60.** (R = Et,

Fig. 5.60. Enantioselective reaction of binaphthol-modified Li/Mg binary organometallic reagent with carbonyl compounds. S = solvent.

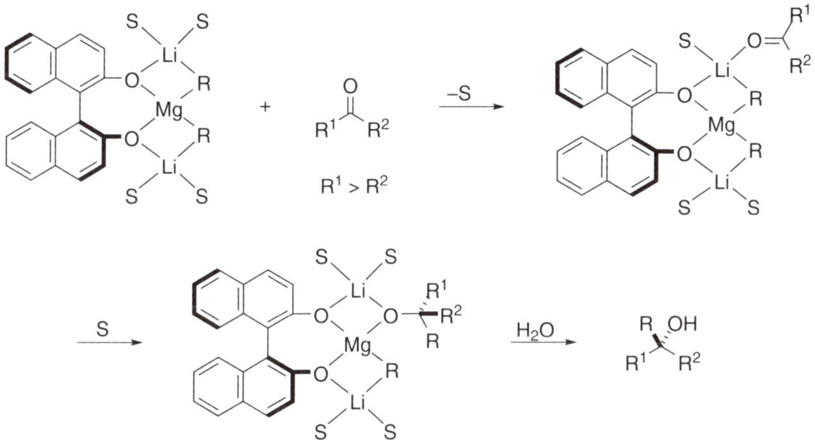

empirical formula) by treating (*S*)-binaphthol in THF with 2 equivalents of *n*-butyllithium and then 1 equivalent of diethylmagnesium. Reaction of the resulting gold-yellow solution with one equivalent of benzaldehyde (R^1 = Ph, R^2 = H) afforded (*S*)-phenyl-1-propanol in 92 % *ee* and 93 % yield. In a similar manner, various aromatic and aliphatic aldehydes could be used for the enantioselective reaction.

This asymmetric reaction, although meeting the standard nearly 40 years ago, is now classified as a stoichiometric chirality transfer from the reagent to the product. Actually, this event was our turning point to our long-term endeavors on catalytic asymmetric alkylation using main-group organometallic compounds [205]. **Fig. 5.61.** illustrates a possible

Fig. 5.61. Ligand acceleration scenario with a protic chiral molecule HX^*. M = main-group metal.

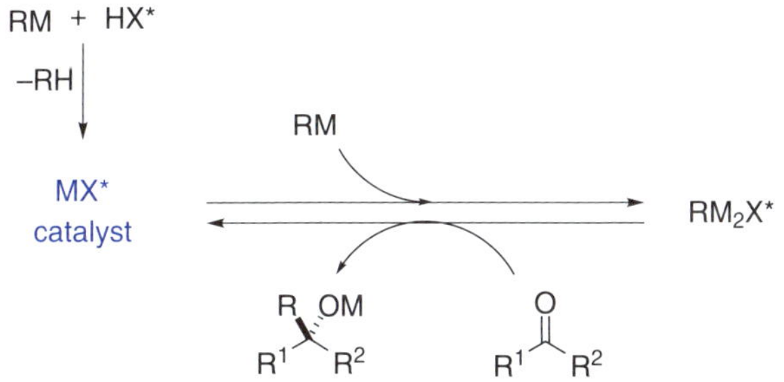

scheme of ligand-accelerated catalysis caused by a tiny amount of a protic chiral molecule. At the time, we were already confident that two metallic atoms participate in stabilizing the transition state in the alkyl transfer reaction. However, I had no idea at all how to achieve this ideal process; the intermediate RM_2X^* (or more precisely $RM^1M^2X^*$) must have some special reactivity, since the conventional, unmodified organometallic RM has inherent reactivity towards the C=O function. At the time, no one knew which type of M and X* could play such a unique role.

5.11 Nonclassical chemistry with the oldest organometallics: Chirality transfer, multiplication, and amplification (1986–2000)

Catalytic organozinc chemistry

Because of the fundamental significance of molecular chirality, chemists had pursued various ways to access enantiomerically pure substances. Fig. 5.62. illustrates the relation between chiral source and products, where the flags refer to right- or left-handed molecules. Optical resolution of racemates represents the most classical and still useful approach (Eq. 1). Chiral molecules can be transformed to other chiral molecules (Eq. 2). The intramolecular and intermolecular chirality transfer approaches (Eq. 3 and 4, respectively) are widely used, but these are stoichiometric in terms of chirality generation, requiring at least one equivalent of a chiral source in order to create a new chiral center.

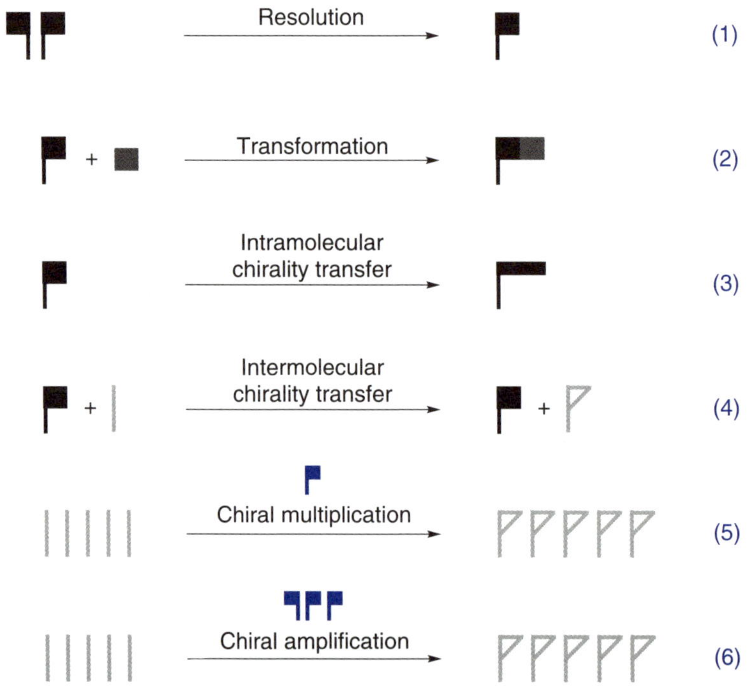

Fig. 5.62. Access to enantiomeric molecules. The black to white color change refers to intermolecular chirality transfer.

More desirably, certain chiral metal complexes catalytically *multiply* chirality via intermolecular asymmetric induction [206, 207]. as visualized in Eq. 5 (**Fig. 2.9.**). No universal way exists; these all are utilized complementally in modern organic synthesis. We now see a striking phenomenon to *amplify* chirality shown by Eq. 6, in which enantiomeric purity of the product exceeds that of the catalyst.

Fig. 5.63. Edward Frankland (1825–1899).

History creates the future. In fact, a breakthrough realizing the scheme in **Fig. 5.61.** came from the oldest organometallic compound discovered by Edward Frankland in 1849. Unlike conventional organolithium and Grignard reagents, diorganozincs are inert to carbonyl compounds. However, in 1984, Nobuki Oguni reported that diethylzinc (DEZ) reacts with benzaldehyde in the presence of 2 mol% of (*S*)-leucinol giving (*R*)-1-phenyl-1-propanol in 48.8 % *ee* and 96 % yield [Q]. Subsequently in 1986, our research group including Masato Kitamura found a highly enantioselective allylation catalyzed by (–)-3-*exo*-(dimethylamino)isoborneol [(2*S*)-DAIB, (*S*)-30], a camphor derived, skeletally congested chiral β-amino alcohol (**Fig. 5.66. a**) [208]. Thus, in the presence of 2 mol % of (*S*)-30, the reaction of benzaldehyde and DEZ proceeds rapidly in toluene at 0 °C, giving the S alcohol in 98 % *ee* and 97 % yield. This method is practical [209]. Reaction of dimethylzinc (DMZ) was slower but gave the chiral alcohol in 95 % *ee*. A range of alkylating agents and aldehydes could be used as exemplified in **Fig. 5.66. b**. A systematic study showed that bulky β-dialkylamino alcohols act much better than the normal N-monoalkyl or non-alkylated compounds and that the hydroxyl-bearing stereogenic center directs the *S*/*R* enantioselection [210]. Overall, 30 was proven to be one of the best chiral auxiliaries. The alkylation occurs better in hydrocarbons than THF. Wolfgang Oppolzer cleverly applied this chemistry for his asymmetric synthesis of (*R*)-muscone by the use of (*R*)-30 [R].

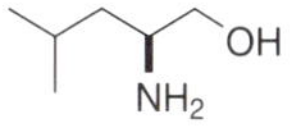

(*S*)-leucinol

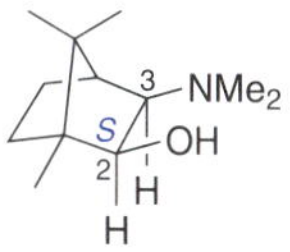

(2*S*)-DAIB, (*S*)-**30**

Fig. 5.64. Top: A linear primary amino alcohol; bottom: a sterically congested secondary dialkylamino alcohol.

In correlation with the scenario in **Fig. 5.61.**, alkylation occurs with a small amount of auxiliary (*S*)-30. Most notably, R_2Zn does not add to aldehydes in the presence of 100 % molar amount of 30. This means that RZn alkoxide (*S*)-31 formed from an equimolar mixture of R_2Zn and (*S*)-30 (**Fig. 5.66. a**) is unable to alkylate an aldehyde, and instead acts as an efficient catalyst. The Zn complex 31 with a highly polar Zn–O bond serves as a unique *bifunctional* catalyst; it provides a Lewis-acidic Zn site that accepts an aldehyde substrate, while the adjacent oxygen is a basic center that interacts with the R_2Zn reagent. Thus, alkyl transfer, as seen in **Fig. 5.67.**, intervenes with the product-forming dinuclear Zn complex 32, where both R_2Zn and aldehyde are activated cooperatively [211]. Consequently, th e alkyl group migrates from Zn to the aldehyde carbon via *anti*-5/4/5 tricyclic TS 33 producing *S*-enriched Zn alkoxide 34 with a high enantiofacial selectivity [212]. The central Zn to O interaction in TS 33 is highly electrostatic in nature.

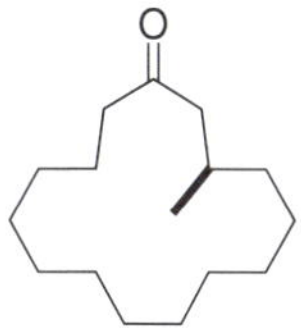

Fig. 5.65. (*R*)-muscone.

Fig. 5.66. Enantioselective alkylation with diethylzinc or dimethylzinc aided by (2*S*)-DAIB. (a) Reactivity and selectivity. (b) Selected examples of asymmetric alkylation.

(a)

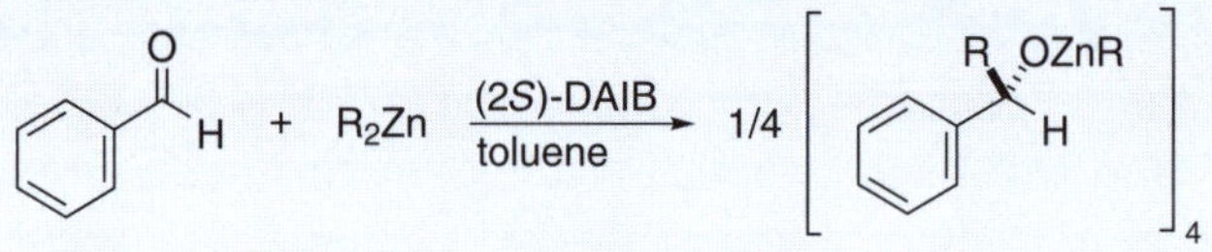

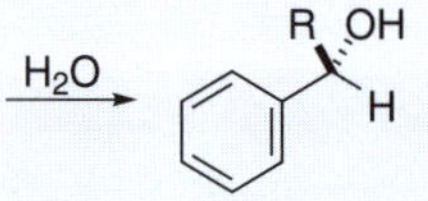

(2*S*)-DAIB, mol %	% yield	% *ee*
0	0	–
2	97	95–98
100	0	–

(2*S*)-DAIB [(*S*)-**30**] (*S*)-catalyst [(*S*)-**31**]

(b)

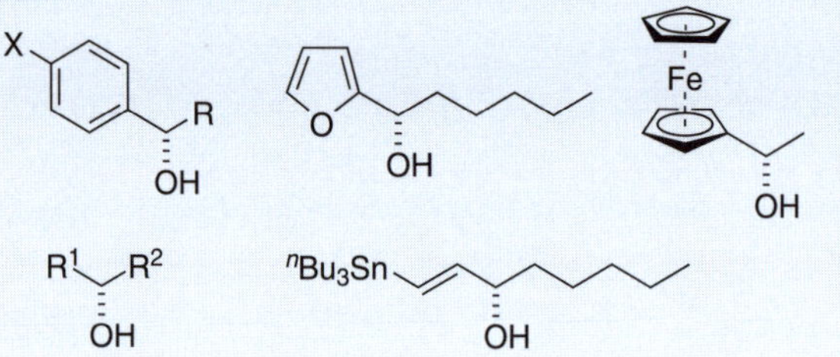

R^1 = alkenyl, alkyl
R^2 = alkyl

Fig. 5.67. Bifunctional mechanism of the alkyl transfer reaction with DMZ. (a) Reaction scheme. (b) Calculated enantioface-differentiating TS 33.

(a)

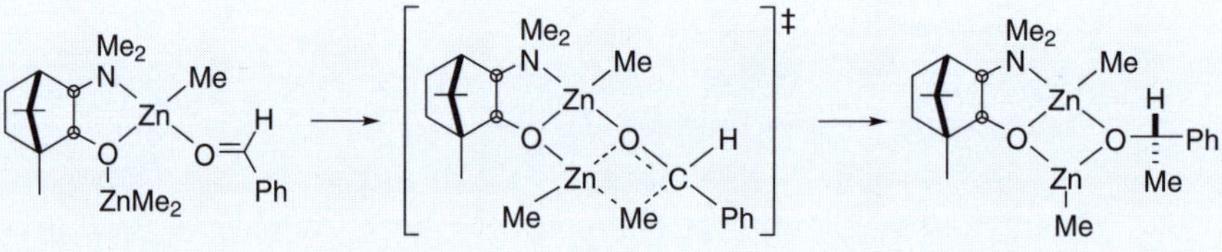

32 **33** **34**

(b)

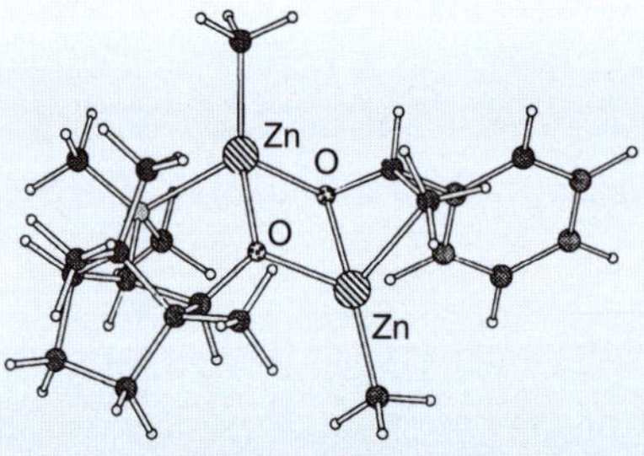

structure of TS **33**

A detailed theoretical study on the simplest model suggests the operation of the catalytic cycle of **Fig. 5.68.** [211]. Each of the steps must proceed via a low energy barrier and the overall process is to be thermodynamically downhill. The conversion of 32 to 34, via TS 33 in the real system (**Fig. 5.67.**), is the turnover-limiting and stereo-determining step, which is modeled by the conversion of F to G in **Fig. 5.68.** The catalyst / product adduct G, upon reaction with R_2Zn or aldehyde, affords tetrameric RZn alkoxide H and D or E, completing the catalytic cycle. The stability of cubic Zn tetramer H in hydrocarbon is essential to drive the catalytic reaction by preventing the product inhibition at the stage of G. The stability of monomer I or dimer J is not enough to turnover the catalytic cycle. The steric origins of enantioselectivity are: (1) the high stereo-differentiating ability of the bornane skeleton of (*S*)- 30 and (2) the nonbonded repulsion between the unreactive Zn-alkyl spectator and the aldehyde phenyl group (**Fig. 5.67. b**) [212].

Fig. 5.68. The catalytic cycle of the amino-alcohol-promoted reaction of dialkylzinc with an aldehyde.

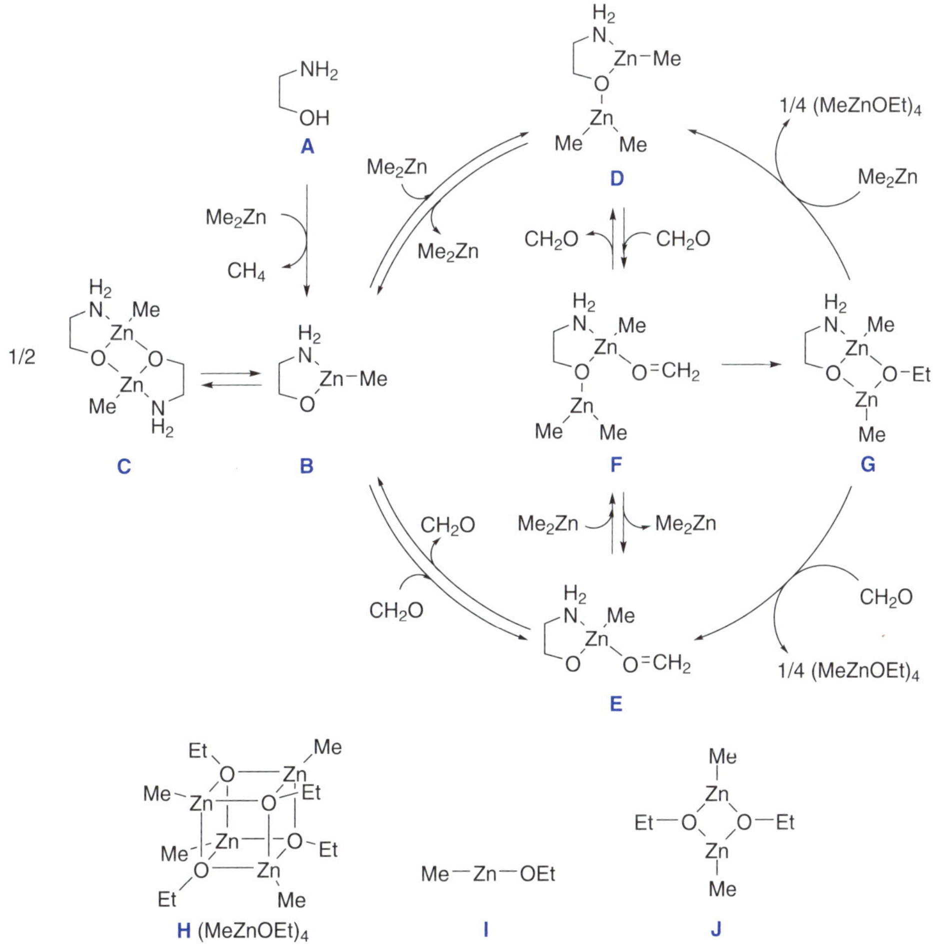

Nonlinear effects in asymmetric alkylation

What was most striking to us was the chiral amplification phenomenon. In ordinary asymmetric catalysis using a partially resolved non-enantiopure catalyst, the enantioselectivity is predicted to vary linearly as a function of the enantiomeric purity of the catalyst. In addition, the initial reaction rate depends on the catalyst concentration but is unaffected by the ratio of the coexisting *S* and *R* catalysts. However, this is not valid in this particular organozinc chemistry. In 1988, Oguni [S] discovered an unusual positive nonlinear phenomenon and subsequently we were able to elucidate its origin through detailed investigations [213 – 215].

Fig. 5.69. Nonlinear phenomena in an asymmetric reaction of benzaldehyde and DEZ catalyzed by (*S*)-31 and (*R*)-31 (R = Et). [DEZ] = [PhCHO] = 420 mM, [(*S*)-31] + [(*R*)-31] = 34 mM, toluene, 0 °C.

As illustrated in **Fig. 5.69.**, the reaction rate and stereoselectivity are greatly influenced by the enantiomeric purity of 31 [215]. Typically, the reaction of DEZ and benzaldehyde in toluene containing (*S*)-31 in mere 15 % *ee* (S/C = 12) forms the *S*-ethylation product in 95 % *ee*, which is close to the 98 – 99 % *ee* obtained with the enantiomerically pure *S* catalyst. Furthermore, the reaction of DEZ and benzaldehyde with racemic 31 is 13 times slower than with the pure *S* or *R* catalyst under identical standard conditions.

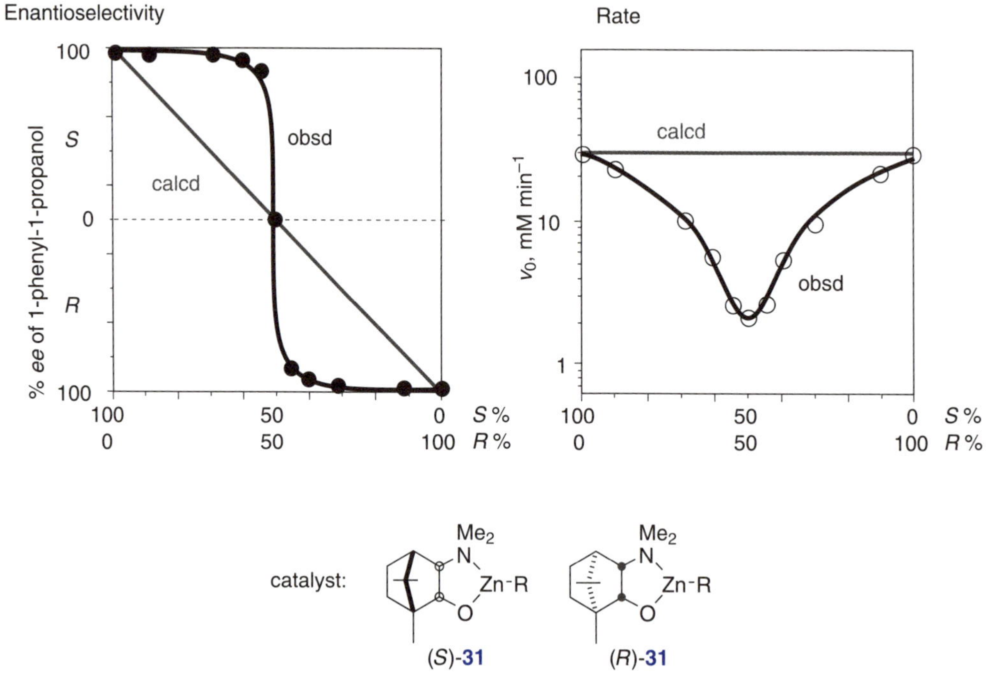

This salient nonlinear effect was proven to arise from molecular association of the chiral catalysts. The self and non-self recognition of the enantiomers is well known in crystalline state, liquid crystals, and solution phase and occurs in asymmetric catalysis (**Fig. 5.70.**) [213, 215 – 218].

The chiral RZn dimethylamino-alkoxides 31 (*S* and *R*) act as catalyst for the reaction of R_2Zn with the aldehyde. Notably, in toluene, these tricoordinate Zn monomers associate reversibly to form the homochiral dimers, *S–S* and *R–R*, and the heterochiral dimer, *S–R*, establishing a five-component equilibrium. Here, the stabilities of these diastereomeric dimers are significantly different; the heterochiral dimer is much more stable than the homochiral dimer [219]. As such, homochiral *S–S* and *R–R* dimers with C_2 chirality (see **Fig. 5.70. b**) act as a highly efficient catalyst precursor, while coexisting heterochiral *S–R* having a C_i structure is much less reactive. The detailed single-crystal X-ray analysis and NMR study on the methylzinc complexes, *S* and *R* (R = Me), clarified the origin of the distinct differences of the diastereoisomers in solution and crystalline state as well. The conclusion was consistent with the *ab initio* MO calculations for the model system [211, 212, 216].

The cryoscopic molecular weight measurement and the vapor-pressure osmometry revealed that the heterodimer is overwhelmingly more stable than the homodimers. Thus, in a 10 mM toluene solution at 40 °C, 60 % of *S–S* (R = Me) is dissolved into monomeric *S*, while

Fig. 5.70. Homochiral and heterochiral recognition of asymmetric catalysts. (a) Five-component equilibrium. (b) Crystallographic structures of the homochiral and heterochiral dimers.

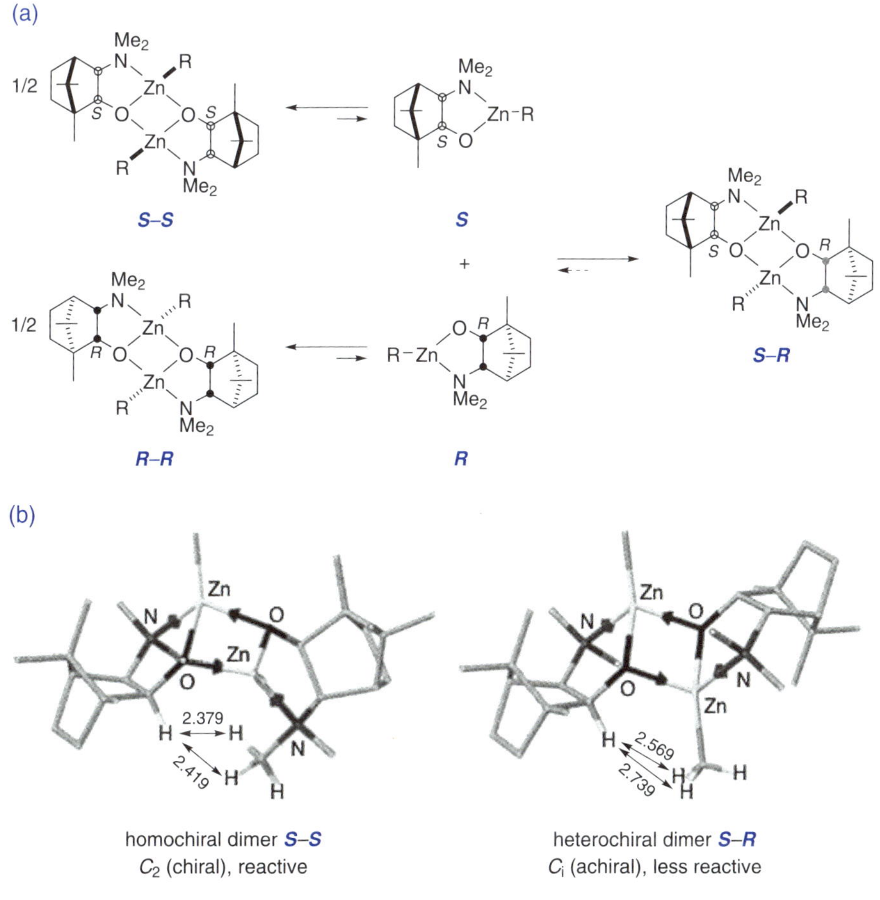

the degree of dissociation of heterochiral *S–R* is only 3 %. Moreover, the vapor-pressure osmometry analysis indicated that the dissociation constant of homochiral *S–S* (R = Me) is 3,000 times larger than that of heterochiral *S–R* in a toluene solution at 40 °C.

For these reasons, the degree of nonlinear effects is affected not only by the catalyst's *ee* but also by various reaction parameters such as the concentration of the catalyst, reagent, and substrates as well as the extent of conversion **[217]**. Further detailed computer-aided quantitative analysis confirmed that this nonlinear effect results from the competition of two *enantiomorphic* catalytic cycles involving the monomeric chiral Zn catalysts *S and R* rather than the *diastereomorphic* catalytic cycles with dinuclear Zn catalysts [215]. Moreover, the self and non-self recognition of asymmetric catalysts was proven very general; reactivity and stereoselectivity of the catalytic system are highly influenced by coexisting enantiomeric, diastereomeric, or even achiral amino alcohol auxiliaries [220].

[217] M. Kitamura, S. Suga, H. Oka, and R. Noyori: Quantitative Analysis of the Chiral Amplification in the Amino Alcohol-Promoted Asymmetric Alkylation of Aldehydes with Dialkylzincs. *J. Am. Chem. Soc.* **1998**, *120*, 9800–9809.

In principle, the sense and extent of stereoselectivity are determined by the relative stabilities of the diastereomeric transition states in the first irreversible step in the steady-state catalytic cycle. The situation, however, is more complicated when coexisting asymmetric catalysts interact with themselves or with other chiral or achiral molecules. The relative significance of self and non-self recognition was proven to strongly affect the overall profile of asymmetric catalysis, resulting in unique nonlinear phenomena.

Molecular interaction thus occurs reversibly in a sterically and electronically complementary fashion. The catalysts and reagents with a bifunctional M–X moiety (M = metal; X = halogen, OR, NR_2, etc.) tend to associate via interdependent X–M dative bonds. This metallo-organic way to form higher order structures contrasts with biochemical mechanisms in nature, where nucleic acids and peptides primarily use multiple hydrogen bonding interaction for strict molecular recognition. I came a long way to understand this intriguing phenomenon. Such significant chemical function was totally unforeseen when I first became interested in the (–)-sparteine-aided asymmetric alkylation in the late 1960s (**CHAPTER 2**).

I would like to point out that departure from the linear relationship could occur via various mechanisms and that the origins are diverse depending on the catalytic systems. Our Zn chemistry is in contrast with Henri Kagan's seminal work on the Ti tartrate-catalyzed oxidation of allylic alcohols and sulfides, where homochiral and heterochiral Ti catalysts are involved in the stereo-determining transition states **[T]**. In these cases, metal modification by a partially resolved chiral auxiliary is believed to occur in an irreversible, nonstatistical manner, forming diasterereomeric Ti complexes with two chiral auxiliary molecules of types ML^*_2 or $M_2 L^*_2$(M = metallic species; L^* = chiral ligand). The stabilities and reactivities of these catalysts must be different, resulting in a marked nonlinear effect.

[T] C. Puchot, O. Samuel, E. Dunach, S. Zhao, C. Agami, and H. B. Kagan: Nonlinear Effects in Asymmetric Synthesis. Examples in Asymmetric Oxidations and Aldolization Reactions. *J. Am. Chem. Soc.* **1986**, *108*, 2353–2357.

Notably, this organozinc chiral amplification has further been extended by Kenso Soai to catalytic asymmetric auto-induction [U]. Moreover, the clever use of this general principle has led to various metal promoted significant asymmetric catalysis by other research groups. This study at the molecular structure level may shed light on the mechanism of the propagation of chirality in nature (review: [214]).

Two great masters who guided a wandering chemist

Sir Derek Barton, a paternal presence

Professor Derek H. R. Barton (1918–1998, Nobel Prize in Chemistry 1969), who of course had a very different background and is 20 years apart in age from myself, has remained to be a fatherly figure to me. I knew this giant merely as the director of the CNRS Natural Products Chemistry Laboratory in Gif-sur-Yvette, France, from 1978, and then as a distinguished professor at Texas A & M University in College Station in the US from 1986 onwards. He was a proud British chemist and never acted like a French or an American. I, on the other hand, am a traditional Japanese, even though my country has become considerably westernized, and I have learned many things from him.

The first time I saw this legend that I admire was on stage at the opening ceremony of the IUPAC Conference on Natural Products Chemistry in Kyoto in 1964, 60 years ago. The first time I shook his hand was in 1980 at the IUPAC Conference on Organic Synthesis organized by Barry M. Trost at the University of Wisconsin in Madison. He warmly encouraged me for my presentation on "New Methodologies Related to Prostaglandins". I met him again in 1982 at the EUCHEM Conference in Louvain-La-Neuve (chaired by Leon Ghosez) and in 1983 at the Hoechst Workshop Conference at the Reisensburg Castle in Germany (organized by B. M. Trost and W. Bartmann), where he listened to the progress of my research.

We were both good travelers and met each other again and again, e.g. in 1984 at the Asian Symposium on Medicinal Plants and Spices in Seoul. — While drinking gin and tonic with him, he gave me a lecture that the bitter taste of tonic water is quinine, which was used in the British colonies to prevent malaria. Then I told him that I was actually going to France in two months for a lecture tour. He remembered this conversation and, during my stay in Paris, he suddenly invited me to his Natural Products Chemistry Laboratory in Gif. After giving a lecture, I was invited to dinner at the Barton home with his colleague Hugh Felkin (known for the Felkin-Anh transition state model for stereoselectivity) and Keith U. Ingold (son of Christopher K. Ingold) and his wife. Barton was a well-known gourmet and wine lover, and many fine wines stored at the chateau were sent to his home on special occasions such as this one.

Together Derek Barton and I were keynote speakers at more than 15 international conferences around the world. Barton's particular role at every meeting was to be in the front row and "ask the first and best questions", I believe that he heard almost every presentation I gave. His questions were short and tough; in my view he didn't like long, convoluted explanations; rather, he wanted people to be clear about their point. He appeared to be (and maybe he was) a very determined, responsive and influential person, who had long led the world's organic chemistry community, and he clearly behaved differently from the other intelligent chemists. I have heard that many people were immediately told off. Fortunately, he was always supportive of my chemistry, perhaps because it was simple enough and therefore easy to understand. Barton casually exchanged letters with me and often encouraged me with short letters at every milestone of my work.

In 1988, we were together at the IUPAC International Conference on Organic Synthesis held at the University of Nancy in France. On the last day of the conference, he asked me which university in France I would be visiting this time. I told him that I would be immediately returning home from Charles de Gaulle Airport tomorrow and promised that I would meet him again. However, the Bartons were very kind; they called me late at night and said, "It would not be good if you missed the train in Paris, so let's go back together tomorrow morning." He must have wanted to tell something to me, an

Fig. 5.71.
D. H. R. Barton (left) and the author in Seoul in 1984.

J. I. Seeman: Working with Sir Derek H. R. Barton. Chemistry, Through Chemistry and For Chemistry. *Tetrahedron* **2019**, *75*, 57–69.

oriental who was not familiar with the history of chemistry. In the cabin with just the three of us, he told me about the old days of British organic chemistry, such as the behavior of the "emperor" Robert Robinson and the perfect lectures by Christopher K. Ingold. I should have listened more attentively than I did at the time. I often met with his friendly wife, Christiane, who always brought an album of the 1969 Nobel Prize in Chemistry in her handbag, which she proudly explained to the younger generation at dinner tables. Her husband was just smiling as she did so. She did the same when I visited their home in Paris in 1984, and later in the US when I was a visiting professor at Texas A & M University in College Station in 1991 at the request of Ian Scott.

My acquaintance with Derek Barton led to the IUPAC Conference on Natural Products Chemistry in Karachi in 1994; the organizing committee was headed by Atta-ur-Rahman, the most influential person in Pakistan. It was there that the Tetrahedron Prize for Creativity of Organic Synthesis was jointly awarded to me and K. Barry Sharpless (he was absent at the event), and the jury was chaired by Barton.

In 1995, I was invited to the W. S. Johnson Symposium at Stanford University (organized by Paul A. Wender, Carl Djerassi, and Barry Trost), which was my last meeting with him. Here, Barton

Fig. 5.72. Author receiving the Sir Derek Barton Gold Medal (**Fig. 5.73.**) from RSC President David Phillips (left) at the Burlington House, London, in 2010.

gave a lecture titled "Oxygen and I" and told us that "In Chemical Sciences, the distant past, the near past and the present join together in continuous harmony." These words strongly encouraged me to write this autobiography. He passed away in 1998, while he was still active in his research until that day. He was 79. But the story doesn't end here.

The Royal Society of Chemistry established the Sir Derek Barton Gold Medal in 2001 to honor the life of this great chemist. The Medal is awarded for outstanding work after the age of 60. In the past, the medal had been awarded to active chemists such as Gilbert Stork, Teruaki Mukaiyama, and Jack Baldwin. In 2010, at the age of 71, I received this most significant award. I was exceptionally moved by this honor because of the circumstances I mentioned above. Most importantly, at the request of this great British chemist, a gourmet and a cynic with a sense of humor, the award ceremony was to be held in London with a "special dinner" organized by the RSC. In fact, the venue for the dinner, Burlington House, was later being awarded three stars by the Guide Michelin. I had a pleasant discussion with Anthony G. M. Barrett, a former student of Barton, who holds the title of Sir Derek Barton Professor at Imperial College London, in memory of our mutual mentor of the past.

Fig. 5.74. Author (left) and his wife Hiroko, learning something important from Elizabeth and Albert Eschenmoser (right) in Vilnius in 2000.

[V] A. Eschenmoser: *Prizes are Good but Discoveries are Better*. (Lives in Chemistry.) GNT Publishing, Berlin, **2025**.

Respect for Albert Eschenmoser (1925 – 2023)

Like many of my contemporaries, I had a great deal of respect for Professor Albert Eschenmoser (1925 – 2023), who was contemplating the philosophy of chemistry. After he and Woodward had achieved the famous total synthesis of vitamin B_{12} in the early 1970s [V], I was grateful to have the opportunity from time to time, to listen to his dignified lectures on his life-long research on the "Chemical Origin of Life." He was interested in the relationship between the stability of the double helix structure of DNA and the function of genetic information transfer, and the topic of his talks was often: "Why Pentose- and not Hexose-Nucleic Acids?". The unparalleled accuracy, voluminous experimental facts, objective explanations without exaggeration, and conclusions were understandable. However, as stated in the famous Chinese saying, "Ask not the sparrow how the eagle sours," we cannot grasp the full picture of his magnificent philosophy of chemical evolution. Chemical evolution is a process that precedes the evolution of life. There is no doubt that he was the chemist with the deepest understanding of the molecular science of life, and his talks always made me humble. His lectures at international conferences would leave the audience in the aftermath of admiration and I often felt sorry for the speaker who had to speak after him (fortunately, I have never had to do this). Therefore, he was often assigned as the final speaker of the session.

When I congratulated him on receiving the American Chemical Society's Roger Adams Award in the summer of 2002, his response was, "Let's not forget: prizes are good, discoveries are better". This is completely true.

I often met Eschenmoser. The last time was in 2000 at the Balticum Organicum Syntheticum conference at Vilnius University in Lithuania, which was organized by Victor A. Snieckus of Queens University of Canada. I was honored to have him listen to my overview lecture titled, "Lessons from the Oldest Organometallic Compounds" and to participate in a panel discussion alongside him. The media were also interested in the scientific event, but the headline was, "For chemists there is no vacation time," as if they did not understand the behavior of researchers. Hiroko, my wife with a very Japanese character, accompanied Eschenmoser's clever wife Elizabeth, on a sightseeing tour of the beautiful city, and asked her how she had supported her great husband.

References for chapter 5

[A] (a) J. Ciabattoni and G. A. Berchtold: Cycloaddition Reaction of Enamines with Diphenylcyclopropenone. *J. Am. Chem. Soc.* **1965**, *87*, 1404 – 1405; (b) H. M. R. Hoffmann and D. R. Joy: Cycloadditions of Allylic Cations to Conjugated Dienes: Stereochemistry. *J. Chem. Soc. B* **1968**, 1182 – 1186.

[B] J. I. Seeman: The Nozoe Autograph Books: Stories behind the Stories. *Chem. Rec.* **2013**, *13*, 483 – 514.

[C] (a) H. M. R. Hoffmann, K. E. Clemens, and R. H. Smithers: Stereospecific Generation of 2-Oxyallylic Cation from Two New Sources. Steric Course of Cycloaddition to Conjugated Dienes. *J. Am. Chem. Soc.* **1972**, *94*, 3940 – 3946; (b) H. M. R. Hoffmann, K. E. Clemens, E. A. Schmidt, and R. H. Smithers: Debromination of α,α'-Dibromo Ketones with a Zinc-copper Couple in Dimethylformamide and Dimethylacetamide. New Reaction Yielding 2-Dimethylamino-4-methylene-1,3-dioxolanes. *J. Am. Chem. Soc.* **1972**, *94*, 3201 – 3204.

[D] E. Dolgin: The Tangled History of mRNA Vaccines. *Nature* **2021**, *597*, 318 – 324.

[E] R. B. Woodward and R. Hoffmann: The Conservation of Orbital Symmetry. *Angew. Chem., Int. Ed.* **1969**, *8*, 781 – 853.

[F] P. G. Gassman, K. T. Mansfield, and T. J. Murphy: Bent sigma bonds. XI. The Reaction of Olefins with Bicyclo[2.1.0]pentane. The Formation of a Diradical intermediate. *J. Am. Chem. Soc.* **1969**, *91*, 1684 – 1689.

[G] L. Cassar, P. E. Eaton, and J. Halpern: Catalysis of Symmetry-restricted Reactions by Transition Metal Compounds. Valence Isomerization of Cubane. *J. Am. Chem. Soc.* **1970**, *92*, 3515 – 3518.

[H] F. D. Mango: Molecular Orbital Symmetry Conservation in Transition Metal Catalysis. *Adv. Catal.* **1969**, *20*, 291 – 325.

[I] (a) W. G. Dauben, M. G. Bezzolini, C. H. Schallhorn, D. L. Whalen, and K. J. Palmer: Thermal and Silver Ion Catalyzed Isomerization of the 1,1'-Bishomocubane System: Preparation of a New $C_{10}H_{10}$ Isomer. *Tetrahedron Lett.* **1970**, *11*, 787 – 790; (b) L. A. Paquette and J. C. Stowell: Silver Ion Catalyzed Rearrangements of Strained sigma bonds. Application to the Homocubyl and 1,1'-Bishomocubyl Systems. *J. Am. Chem. Soc.* **1970**, *92*, 2584 – 2586.

[J] *Chem. Eng. News* **1967**, *45*, 51.

[K] Y. Okamoto, K. Suzuki, K. Ohta, K. Hatada, and H. Yuki: Optically Active Poly(triphenylmethyl methacrylate) with One-Handed Helical Conformation. *J. Am. Chem. Soc.* **1979**, *101*, 4763 – 4765.

[L] G. Stork and M. Isobe: Simple Total Synthesis of Prostaglandins from 4-Cumyloxy-2-cyclopentenone. *J. Am. Chem. Soc.* **1975**, *97*, 6260 – 6261.

[M] Y.-R. Luo: *Handbook of Bond Dissociation Energies in Organic Compounds.* CRC Press, Boca Raton, 2002.

[N] K. N. Houk, M. N. Paddon-Row, N. G. Rondan, Y.-D. Wu, F. K. Brown, D. C. Spellmeyer, J. T. Metz, Y. Li, and R. J. Loncharich: Theory and Modeling of Stereoselective Organic Reactions. *Science* **1986**, *231*, 1108 – 1117.

[O] H. Vorbrüggen and C. Ruh-Pohlenz: Synthesis of Nucleosides. In *Organic Reactions, vol. 55.* John Wiley & Sons, New York, **1999**.
[P] A comprehensive review: B. M. Trost and T. R. Verhoeven: *Comprehensive Organometallic Chemistry, vol. 8* (eds.: G. Wilkinson, F. G. A. Stone, and E. W. Abel). Pergamon, Oxford, **1982**, 799 – 938.
[Q] N. Oguni and T. Omi: Enantioselective Addition of Diethylzinc to Benzaldehyde Catalyzed by a Small Amount of Chiral 2-Amino-1-alcohols. *Tetrahedron Lett.* **1984**, *25*, 2823 – 2824.
[R] W. Oppolzer and R. N. Radinov: Synthesis of (*R*)-(–)-Muscone by an Asymmetrically Catalyzed Macrocyclization of an ω-alkynal. *J. Am. Chem. Soc.* **1993**, *115*, 1593 – 1594.
[S] N. Oguni, Y. Matsuda, and T. Kaneko: Asymmetric Amplifying Phenomena in Enantioselective Addition of Diethylzinc to Benzaldehyde. *J. Am. Chem. Soc.* **1988**, *110*, 7877 – 7878.
[T] C. Puchot, O. Samuel, E. Dunach, S. Zhao, C. Agami, and H. B. Kagan: Nonlinear Effects in Asymmetric Synthesis. Examples in Asymmetric Oxidations and Aldolization Reactions. *J. Am. Chem. Soc.* **1986**, *108*, 2353 – 2357.
[U] K. Soai, T. Shibata, H. Morioka, and K. Choji: Asymmetric Autocatalysis and Amplification of Enantiomeric Excess of a Chiral Molecule. *Nature* **1995**, *378*, 767 – 768.
[V] A. Eschenmoser: *Prizes are Good but Discoveries are Better.* (Lives in Chemistry.) GNT Publishing, Berlin, **2025**.

BINAP FOR ASYMMETRIC CATALYSIS

6

 DOI.ORG/10.47261/1564-6

Scientific discovery can sometimes be made by good fortune, whereas meaningful invention needs persistent intellectual endeavor. Cyclic and iterative processes are crucial, but scientific knowledge can be very limited. Thus, I strongly believe that both discovery and invention are driven largely by individual intuition, rather than logical thinking.

6.1 Axially dissymmetric BINAP – why and how? (1974 – 1977)

Catalytic hydrogenation has been and will remain a core reaction. Research on this reaction was initiated at the end of the 19th century by Paul Sabatier (1912 Nobel Laureate in Chemistry) who used fine metal particles as a heterogeneous catalyst. Before starting research in this area, some of the notable achievements that attracted me included the activation of H_2 by a transition metal complex in the late 1930s (Melvin Calvin, 1961 Nobel Laureate in Chemistry, and Michael Polanyi) [A], and the homogeneous hydrogenation of olefins with $RuCl_3$ in 1961 [B], and $RhCl(PPh_3)_3$ in 1965 (Geoffrey Wilkinson, 1973 Nobel Laureate in Chemistry) [C]. Most importantly, my first memory of asymmetric hydrogenation (AH) is Shiro Akabori at Osaka University reporting that metallic Pd drawn on silk catalyzes asymmetric (heterogeneous) hydrogenation of oximes and oxazolones in 1956 [D]. In 1968, two years after our finding of the asymmetric carbene reaction (**Fig. 2.9.**), William S. Knowles (fellow Nobel Laureate in chemistry in 2001)[E] and Leopold Horner [F] independently reported the first homogeneously catalyzed AH of olefins with chiral monodentate tertiary phosphine-Ru(I) complexes, albeit in only a 3 – 15 % optical yield.

Fig. 6.1.
Paul Sabatier (1854 – 1941).

Fig. 6.2.
Melvin Calvin (1911 – 1997).

Fig. 6.3.
Geoffrey Wilkinson (1921 – 1996).

Fig. 6.4.
Shiro Akabori (1900 – 1992).

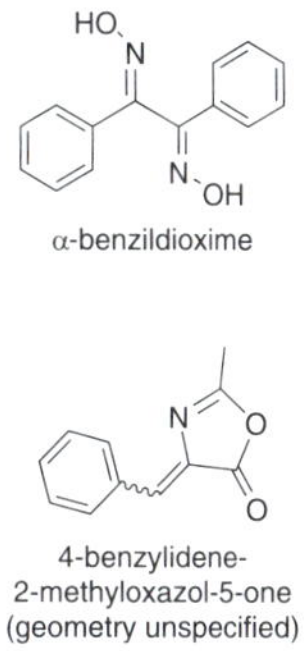

Fig. 6.5.
An oxime and an oxazolone were asymmetrically hydrogenated by S. Akabori in 1956.

My fruitful experience at Harvard (**CHAPTER 4**), coupled with my earlier encounter with asymmetric cyclopropanation at Kyoto (**CHAPTER 2**), naturally guided my research group at Nagoya to study AH. We were prepared to move forward in this area. However, we were still occupied with our other successful research in other subjects, which greatly delayed the full-scale launch of our AH project.

Fig. 6.6.
Henri Kagan's DIOP ligand that he used to synthesize phenylalanine with 80 % *ee*.

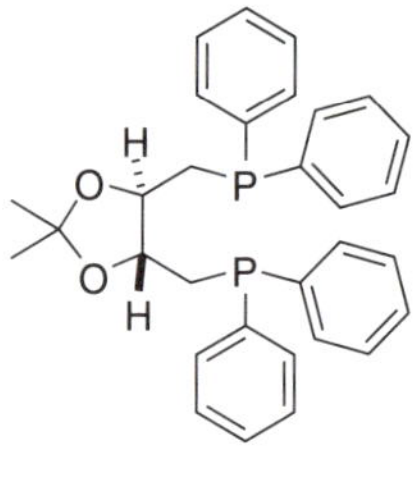

(*R*,*R*)-DIOP

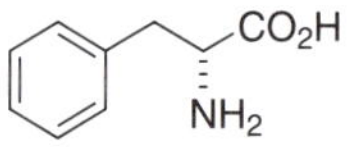

D-phenylalanine

Fig. 6.7.
William S. Knowles' DIPAMP ligand that he used to synthesize L-DOPA.

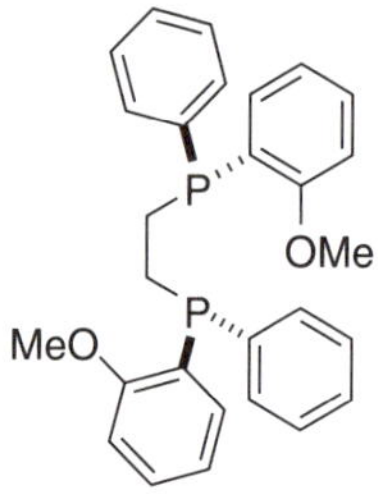

(*R*,*R*)-DIPAMP

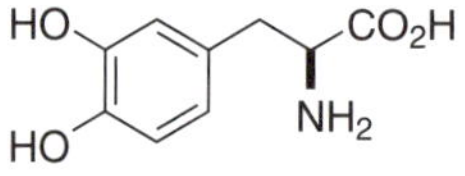

L-DOPA

The Rh center is the origin of catalytic activity, accelerating the reaction repeatedly, while the surrounding chiral ligand controls stereoselectivity in the absolute sense. If the chiral molecular catalyst is well designed, the reaction is not only accelerated but produces a chiral compound favoring either a right- or left-handed enantiomer. This cycle is repeated many times until the starting materials are fully consumed. The key to success was the design of the chiral phosphine ligands. When I entered this exciting field, Henri Kagan had already succeeded in AH of a protected dehydro amino acid catalyzed by a DIOP-Rh complex (**Fig. 6.6.**) forming phenylalanine with ca. 80 % *ee* (enantiomeric excess), then recorded as 72 % *ee* [G]. Furthermore, a Monsanto group led by Knowles, established the industrial synthesis of the drug L-DOPA (**Fig. 6.7.**) (commonly prescribed to treat Parkinson's) using a similar

Fig. 6.8.
The AH of certain dehydroamino acids was established but the AH of other olefins, ketones, and imines was unknown.

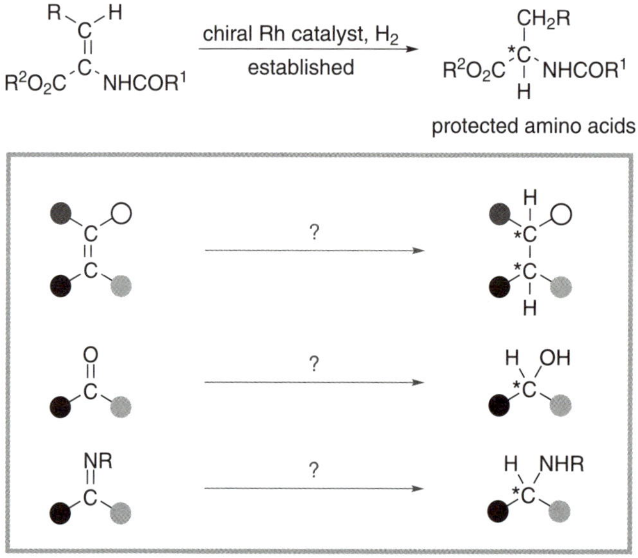

AH with a DIPAMP-Rh catalyst with a turnover number of 20,000 [H, I]. Consequently, AH chemistry was considered already complete, with no significant problems remaining to be solved. Obviously, this was not true at all. The real fact was that certain amino acids could be synthesized by chiral Rh-catalyzed AH, and this was just a starting point for a significant scientific and technological endeavor. For instance, the AH of other olefins, ketones and imines remained totally unexplored (**Fig. 6.8.**).

Our aim at Nagoya was to develop a more generally applicable catalytic system. Working towards this ambitious goal, we were eager to invent a BINAP-Rh(I) (BINAP = 2,2'-bis(diphenylphosphino)-1,1'-binaphthyl) catalyst with a fully aromatic atropisomeric C_2 chiral architecture. So what was it that attracted us to BINAP? As indicated in the molecular model of **Fig. 6.9.**, the chiral distinction in the square-planar Rh(I) complexes or the octahedral Rh(III) intermediates is provided by the presence of four *P*-phenyl rings in an alternating edge-face manner. The chirality originally generated by the binaphthyl skeleton is transmitted to the coordination sites through the *P*-phenyl rings. The in-plane sites (indicated as □ in **Fig. 6.9.**) are significantly affected by the "equatorial" phenyl substituents, while the out-of-plane sites (indicated as ■) are largely influenced by the "axial" phenyl rings. Thus, the two sets of quadrants of the complexes (first and third versus second and fourth) are clearly differentiated. These spatial characteristics are attained by combined effects of the flexible rotation about the phosphorus-phenyl (ax and eq) axes and the uniqueness of the binaphthyl-based chelate skeleton (**Fig. 5.17.**). In this regard, the seven-membered chelate structure is much more advantageous than the five-membered rings. The superiority of BINAP is thus very clear.

Fig. 6.9.
(a) (*R*)-BINAP-transition metal complex. M = metallic element.
(b) Chiral environment of the complex. The side-view is simplified by omitting the binaphthyl fragment.

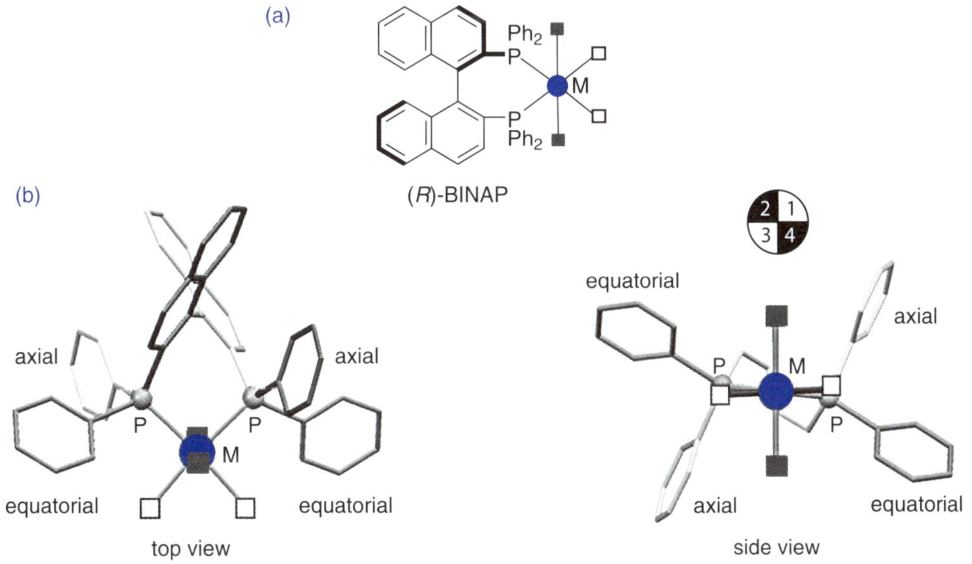

We initially planned to synthesize enantiomers starting from optically pure 2,2'-diamino-1,1'-binaphtyl (**Fig. 6.10. a**). However, a long-standing problem in the chiral biaryl synthesis lies in the difficulty in stereo-retentive conversion of the atropisomeric structures. This was also the case here even with Takaya's skill and hard work of some undergraduate students. The attempted Sandmeyer conversion of the optically pure (*R*)-diamine to the dibromide resulted in considerable racemization. Furthermore, even when the optically active dibromide (*ee* unknown due to lack of measurement methods at the time) was used, its dilithiation with *t*-butyllithium at –90 °C, followed by reaction with diphenylphosphinic chloride, led to optically inactive or only feebly active BINAP. In some cases, optically pure (*R*)-BINAP was obtained by repeated recrystallizations but in only < 10 % yield. Furthermore, the result was not reproducible. Thus, wasting a lot of time, we were urged to resolve racemic BINAP. First, the dibromide was most conveniently prepared heating 2,2'-binaphthol with dibromotriphenylphosphorane at 300 – 320 °C. Its treatment with *t*-butyllithium followed by diphenylphosphinic chloride gave the racemic diphosphine. The resolution was

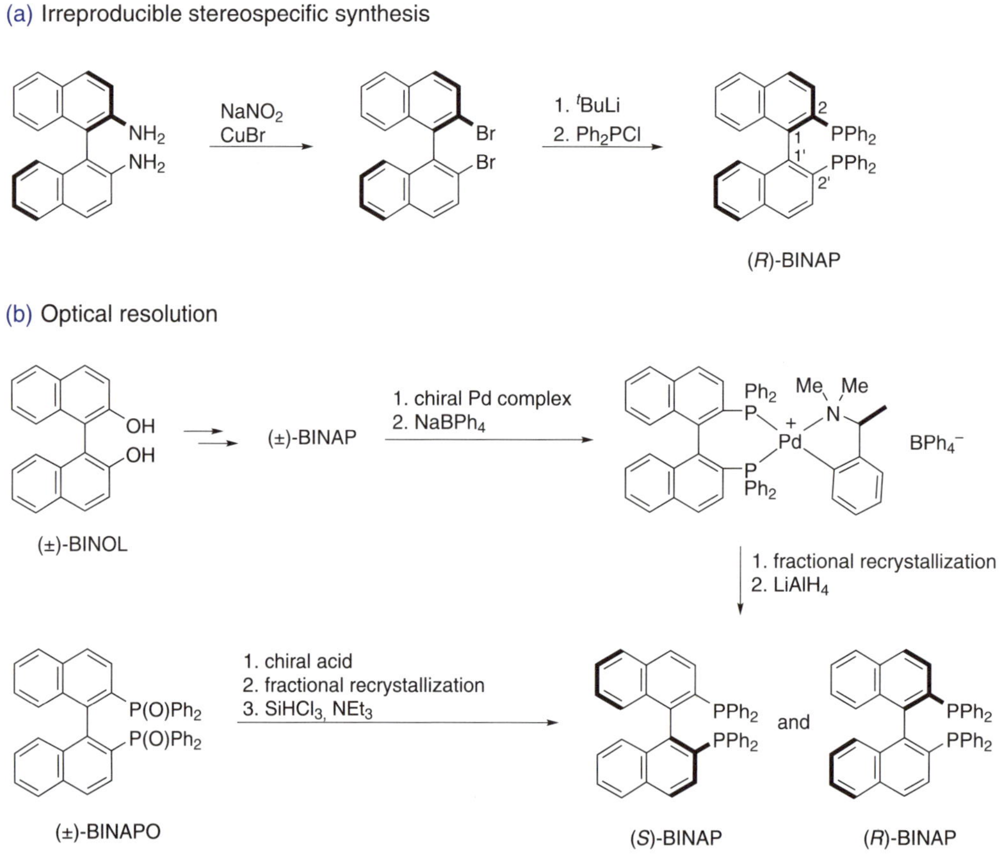

Fig. 6.10. Access to enantiomerically pure BINAP. (a) Initial attempt. (b) Optical resolution.

eventually achieved by an optimistic, patient student, Toshiaki Souchi by using optically pure dimethyl(1-phenylethyl)amine-Pd(II) complex (**Fig. 6.10. b**) [221–223] [J]. Later, our collaboration with Takasago Corp. realized the more convenient route via resolution of BINAP dioxide (BINAPO) with camphorsulfonic acid or 2,3-*O*-dibenzoyl tartaric acid (**Fig. 6.11.**) [224] [K].

Fig. 6.11.
Chiral acids.

(1*S*)-(+)-camphorsulfonic acid

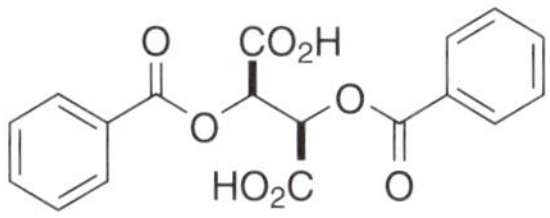

(2*R*,3*R*)-(–)-2,3-*O*-dibenzoyltartaric acid

6.2 Asymmetric synthesis of amino acids by BINAP-Rh(I) catalyzed AH (1977–1980): A momentary joy followed by a nightmare

Fortunately, we first succeeded in obtaining the long-sought optically pure BINAP in September 1977. We were able to immediately determine the crystallographic structure of the cationic (*R*)-BINAP-Rh(nbd)(ClO_4) complex, (*R*)-1 (L–L = norbornadiene) (**Fig. 6.12.**), the shape of which is beautiful. However, we were still far from achieving our final goal. Subsequently, we tried the AH of α-(acylamino)acrylic acid derivatives using the carefully prepared cationic diene-free, solvate catalyst, (*S*)- or (*R*)-1 (L = MeOH). The reaction appeared to proceed with an asymmetric sense consistent with the substrate / catalyst structures, but the enantioselectivity was highly variable and often unsatisfactory, mostly around 80 % *ee*. This was a disappointing and frustrating fact.

We worked hard to optimize the conditions and, at last, managed to establish conditions with better performance. For example, AH of the *Z*-olefinic substrate (R = Ph, R^1 = H, R^2 = Ph) using (*R*)-1 with a substrate / catalyst (S/C) molar ratio of 100 was achieved in a 0.013 M ethanol solution under an initial H_2 pressure of 4 atm and at room temperature, giving the *N*-protected phenylalanine in 97 % yield. The *S* enantioselectivity, judged from optical rotation, was almost perfect. THF was used as solvent for configurationally labile *E* substrates. α-(Benzamido)acrylic acid without β-substitution behaved like the *Z* substrates, giving the protected alanine in 98 % *ee* (**Fig. 6.12.**). At the end of 1980, we published the result on this subject for the first time [221]. We were congratulated by our colleagues and the chemistry community for our unfailing six-year effort. At this point, our former competitors seemed to have already given up on this project, and we were the only ones remaining.

Fig. 6.12.
Asymmetric hydrogenation of α-(acylamino)-acrylic acids catalyzed by BINAP-Rh complexes. L = MeOH.

After accomplishing this difficult task, we took a short break. However, I was ashamed of my ignorance of catalysis. Regretfully, this AH failed to take us to our ultimate goal for a totally unanticipated scientific reason. From the late 1970s to the early 1980s, Rh(I) was generally considered to be the best metal for AH in view of the many successful amino acid asymmetric syntheses. Furthermore, we believed that BINAP would show outstanding chiral recognition ability in AH because of its characteristic shape shown in **Fig. 6.9.** [222]. This was only partially true. In contrast to our anticipation, however, the marriage of the best metal and the best ligand did not bring about a very happy consequence. Our Rh catalyst offered no advantage over other existing candidates. Our high expectations turned into strong disappointment.

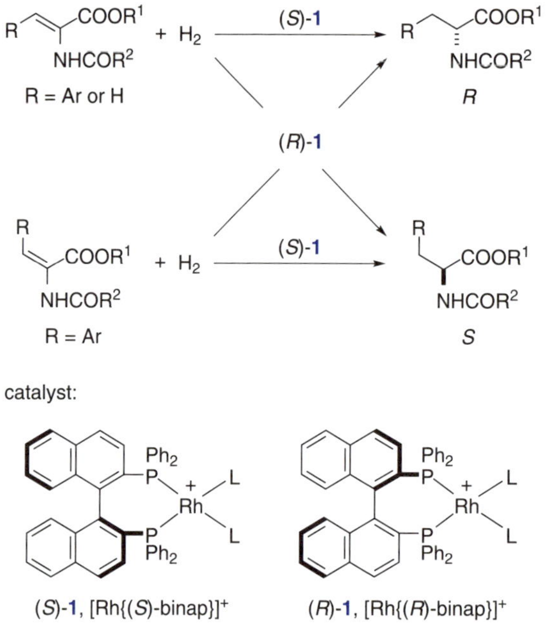

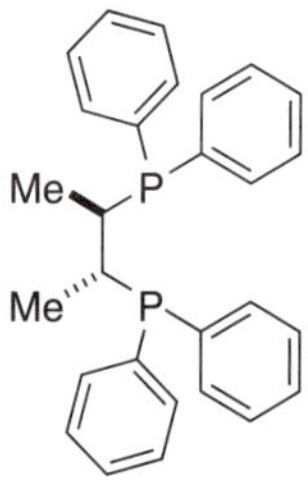

Fig. 6.13.
(*R*,*R*)-CHIRAPHOS.

Actually, the invention of the cationic BINAP-Rh complex 1 was the start of our long-lasting struggle in AH chemistry. The asymmetric synthesis of BINAP-Rh complexes (**Fig. 6.12.**) had a splendid appearance with near-perfect enantioselectivity, while the real catalytic chemistry appeared to be not so elegant from a logical point of view. As studied thoroughly by Halpern and Brown using the CHIRAPHOS-Rh catalyst (**Fig. 6.13.**), this AH proceeds via the "unsaturate/dihydride mechanism." (**Fig. 6.14.**) [L]. When a C_2-chiral diphosphine ligand is used, the Rh complex A [(*R*)-1, S = alcoholic solvent] forms a mixture of two diastereomeric enamide complexes B, depending on the *Re*/*Si* face selection at C(2), which lead to the *S* and *R* hydrogenation

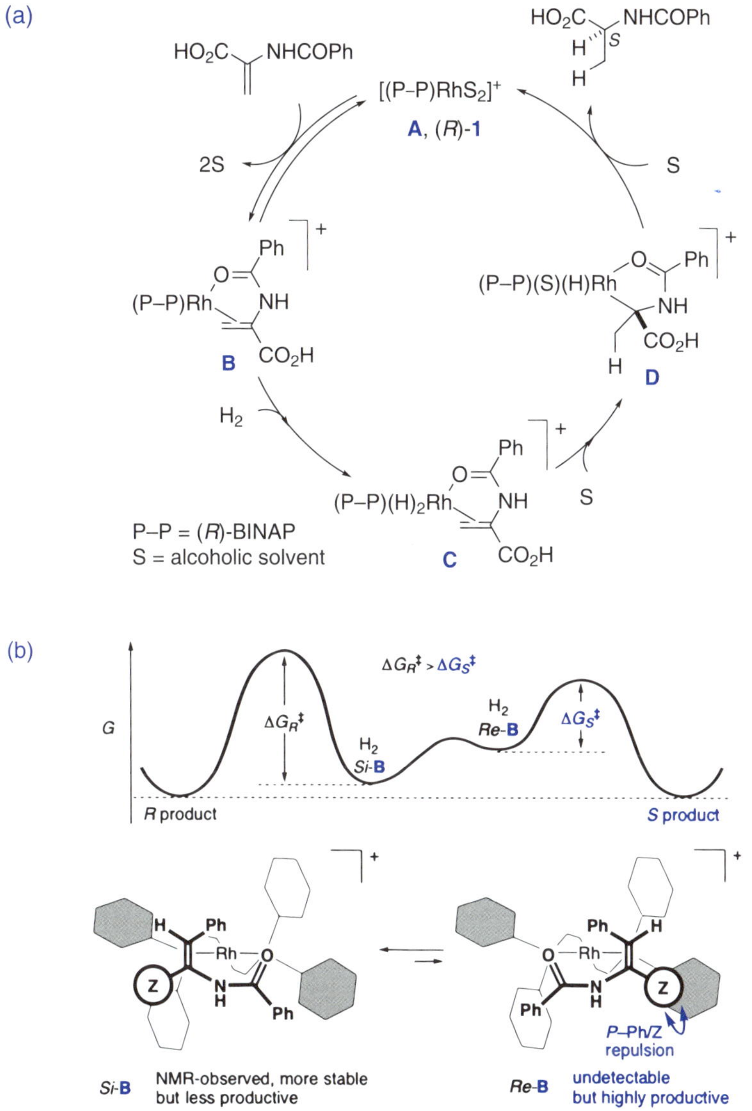

Fig. 6.14. Hydrogenation of (*Z*)-(benzamido)-cinnamic acid catalyzed by an (*R*)-BINAP-Rh complex. (a) Catalytic cycle. (b) Halpern-Brown mechanistic model of asymmetric hydrogenation. Note that the hydrogenation of B occurs in a stepwise manner. Z = CO_2H.

product, respectively. Therefore, the actual enantioselectivity is determined by the relative equilibrium ratio and reactivity of diastereomeric B (**Fig. 6.14. a**). Most importantly, under the reaction conditions (S/C = 100, 4 atm, room temperature, ethanol as a solvent), the more stable, major diastereoisomer *Si*-B is consumed much more slowly than the less stable, minor isomer *Re*-B due to the lower reactivity toward H_2, giving C, or, if this step is reversible, this leads to the more difficult migratory insertion of the Rh dihydride, C to D (**Fig. 6.14. b**). This process is a purely kinetic phenomenon based on the Curtin-Hammett principle in which the relative stability of the substrate-Rh complexes seriously contrasts with their reactivity.

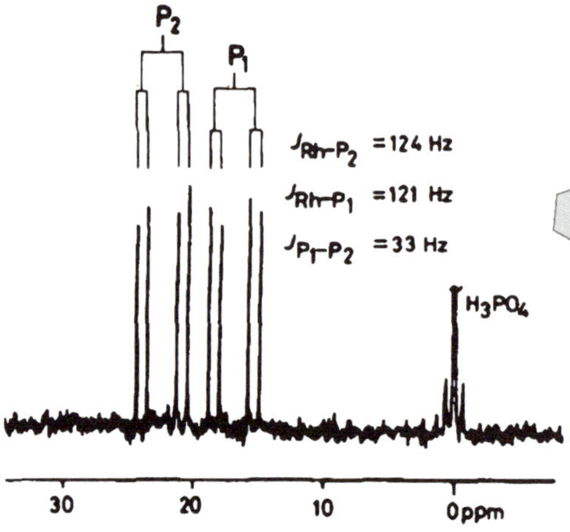

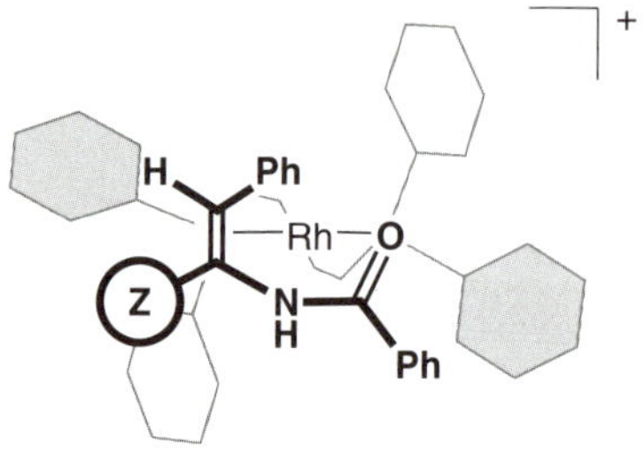

Si-B

Fig. 6.15. ^{31}P NMR spectrum of a 6 : 1 mixture of (*Z*)-α-(benzamido)cinnamic acid and [Rh{(*R*)-binap}(MeOH)$_2$]$^+$ClO$_4^-$ in methanol at 20 °C.

This reaction situation can be monitored by ^{31}P NMR, as the diastereomeric complexes B are known to give a separate eight-line signal due to Rh–P and P–P coupling. As illustrated in **Fig. 6.15.** (taken 40 years ago on a JEOL-FX 100 instrument), the spectrum of a 1 : 6 mixture of (*R*)-1 and the enamide in methanol displayed a single set of eight-line signals. Unlike for other chiral diphosphine-Rh complexes giving two separate eight-line signals, no other signals were observed. Thus, the enamide enantioface differentiation by (*R*)-1 was proven excellent, as we had expected from the BINAP 3D shape (**Fig. 6.9.**) [222]. This favored complex *Si*-B, however, must be converted to the less stable, NMR-undetectable diastereoisomer *Re*-B by de-coordination / re-coordination prior to its hydrogenation. Notably, the formation of *Si*-B is not instantaneous, occurring slowly on an NMR time scale via some intermediates.

Therefore, the observed enantioselectivity of near 100 % is a result of managing the delicate balance of the stabilities and reactivities of diastereomeric complexes B. The optimum conditions leading to near perfect enantioselectivity are obtainable only by a very careful choice of reaction parameters. I greatly appreciate the effort of my young colleagues, particularly Akira Miyashita (educated by Akio Yamamoto and Robert Grubbs and later appointed as a Professor at Saitama University in Japan), in finding this narrow window as exaggerated in (**Fig. 6.16.**) in the mechanistic black box. The reaction must be conducted under a very low substrate concentration and low hydrogen pressure so as to exclude the contaminated pathway via stable *Si*-B. As such, the BINAP-Rh-catalyzed reaction occurs slowly, because the reactive substrate-Rh complex *Re*-B is present in a very low concentration under these hydrogenation conditions.

Such an aspect of AH is distinct from formally similar enzymatic catalysis where the enzyme-substrate lock-and-key binding specificity is imperative in determining the product's overall stereoselectivity. We had

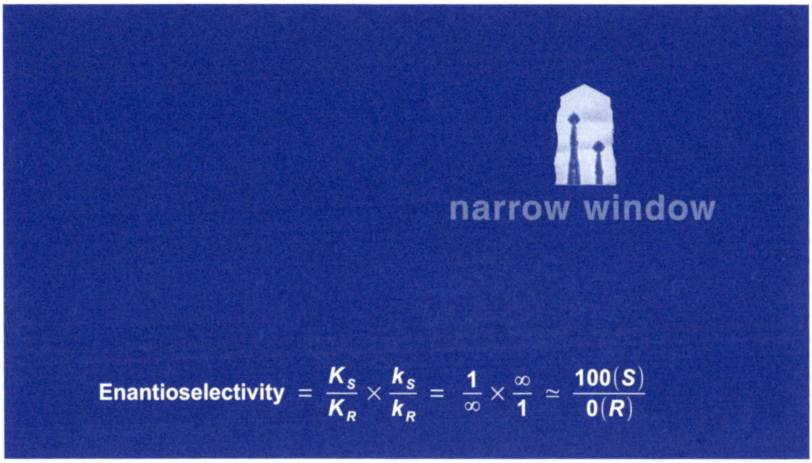

Fig. 6.16. The narrow window in the mechanistic black box.

already invented the BINAP-Rh catalyst, had applied it to the asymmetric synthesis of amino acids, and filed a patent for this research in 1978. Nevertheless, due to insufficient understanding of the reaction mechanism, our journal publication was delayed until 1980.

The study of BINAP-Rh-catalyzed AH in the 1970s was a nightmare for our research group. In retrospect, we noticed merely static, not kinetic, characteristics. Although the molecular architecture of metal complexes can be precisely analyzed by X-ray crystallography and other spectroscopic methods, their functions were not immediately apparent from their structures. Despite insurmountable difficulties, its molecular beauty motivated me to continue this project. We learned a lot from this painful experience, and our imaginative optimism led the BINAP AH chemists to the next advancement in 1986 (**CHAPTER 6.4**).

Henri Kagan — A descendant of Louis Pasteur

The significance of Pasteur's discovery of molecular chirality is unmeasurable. France has produced countless scientists in this field, and among them was Henri B. Kagan (* 1930), who has been an outstanding figure of our time. His invention of the C_2-chiral diphosphine ligand DIOP in 1971 paved the way for the development of Rh-catalyzed asymmetric hydrogenation. I met him for the first time in 1975, when he visited us in Nagoya with his wife, Claude Kagan, and this was also the time when I was still struggling with the synthesis of the BINAP ligand. Since then, we have met more than 20 times in different countries around the world and have developed a long-lasting friendship. Unlike other lively French colleagues of mine, Kagan is a quiet and sincere man whose interest centers only on academic matters. I have been impressed by his manner; he never lost his curiosity and always faced the basic issues of molecular chirality. He once said, "to foresee and then to realize is exciting, but even more rewarding is to discover the unforeseeable." His focus was more on conceptual issues than on technical applications. Although he was a major speaker at many conferences, he actually preferred dialogues and discussions at the table with me and our students based on detailed experimental data rather than debates with a large audience. I can well understand the feeling of this scholar.

Kagan and I have tackled the same or similar problems, including the asymmetric hydrogenation and

Fig. 6.17.
The 2001 Wolf Prize recipients. Front row, from the left: K. Barry Sharpless, center: the author, fourth: Henri B. Kagan. Back row: the fourth from left, Avram Hershko (2004 Nobel laureate in chemistry). At the Chagall Hall of Knesset, Jerusalem, May 2001.

non-linear phenomena in catalysis, and I would like to thank him for the useful advice he had always given to me like an elder brother. When Kagan was elected as a member of the prestigious French Academy in 1992 and received a beautiful Academic Sword, I delivered a congratulatory lecture at Paris-Sud University with my peers, K. Barry Sharpless, John M. Brown, Jean-Pierre Sauvage, and Lionel Salem. I also participated in the symposium at the ACS meeting in Washington, D.C., where he was awarded the Tetrahedron Prize in 2000. One of the happiest moments for me, however, was when the three of us, Kagan, Sharpless, and myself were jointly awarded the Wolf Prize in 2001.

Fig. 6.18.
From the left: author and Hiroko Noyori, Claude and Henri Kagan. Jerusalem, May 2001.

6.3 A gift in return: Industrial asymmetric menthol synthesis (1982)

During the study of BINAP-Rh-catalyzed amino acids syntheses that we started in 1974, I was taught the difference in values held between researchers in academia and industry. Young academics would venture into the unknown and reach for the unattainable, while corporate researchers, even those working at chemical laboratories conducting fundamental research, focused on achieving targeted economic or societal benefits by applying accumulated knowledge. The BINAP ligand seemed promising from the beginning and, in fact, did grow to become one of the pivotal players in the field of asymmetric catalysis. However, its development was not straightforward.

In 1978, two years before our first scientific publication describing the amino acid synthesis, the Ajinomoto Company agreed to submit a patent application for chiral BINAP-Rh catalyst 1 on our behalf. They had widely commercialized mono-sodium (*S*)-glutamate (**Fig. 6.19.**), an umami taste enhancer, and other amino acids, and was backed with a long tradition of excellent fermentation technology. However, its researchers were all faithful followers of Louis Pasteur and were not convinced of the practicality of artificial asymmetric catalysts rivaling or sometimes even outperforming microorganisms using enzymes. I felt they undervalued the future potential of chemical asymmetric synthesis. However, science is destined to make progress, leading to technological developments that even the best professionals cannot correctly predict. This is the very nature of science and technology.

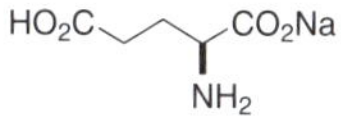

Fig. 6.19. Mono-sodium (*S*)-glutamate.

Inventors: R. Noyori, H. Takaya. Applicant: Ajinomoto Co. Inc. Catalysts for asymmetric hydrogenation. Jpn. Pat. Appl. Sho-53-136605, November 6, 1978; Jpn. Kokai Tokkyo Koho Sho-55-61937, May 10, 1980.

Everything has context. Our chiral BINAP-Rh complexes became famous immediately after their appearance in 1980, unanticipatedly, not as AH catalysts but for other purposes. Although our Rh chemistry was not beneficial for the initial purpose to the extent that we expected [221], the cationic (*S*)-BINAP-Rh complex, (*S*)-1, was found to serve as an excellent catalyst for the asymmetric 1,3-hydrogen shift reaction of diethylgeranylamine (**Fig. 6.20.**) with an *E*-configuration, giving the (*E*,*R*)-citronellal enamine with an excellent *ee* and chemical yield (**Fig. 6.21.**). This process has acted as a key step in the production of (–)-menthol at Takasago International Co. [225, 226]. In a similar manner, the *Z*-olefinic nerylamine in the presence of (*S*)-1 affords the enantiomeric (*E*,*S*)-citronellal enamine with an equally high *ee*. Thus, the pro-*S* hydrogen (H_S) at C(1) in allylic amines moves to C(3) in a suprafacial manner.

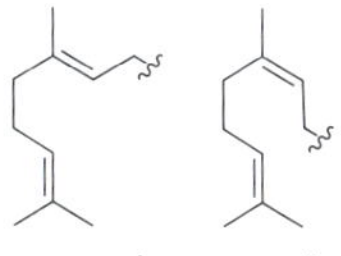

Fig. 6.20. Geranyl and neryl group.

Various mechanisms are possible for double-bond shift reactions. We postulated a nitrogen-triggered mechanism leaving the olefinic bond intact (**Fig. 6.22. a**), which differs from conventional metal hydride addition / elimination or a π-allylmetal hydride pathway. Molecular orbital (MO) calculations on a simplified model (**Fig. 6.22. a**) indicated

Fig. 6.21. Asymmetric synthesis of (–)-menthol.

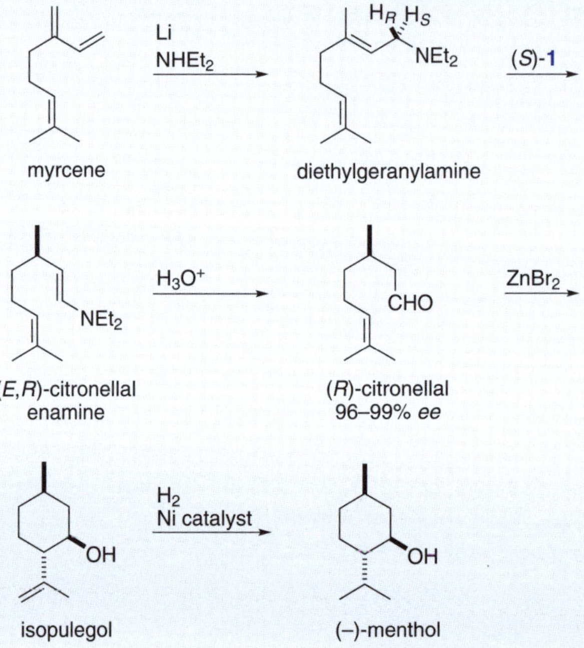

that the 1,3-hydrogen shift occurs via intramolecular oxidative addition of the C(1)–H bond to the Rh(I) center, giving a distorted octahedral Rh(III) hydride, followed by reductive elimination accompanied by allylic transposition [227]. With the geranylamine, the (*S*)-BINAP-Rh catalyst, (*S*)-1, clearly differentiates between the pro-*S* and pro-*R* hydrogens on the flexible skeleton. The NMR study indicated the steady-state catalytic cycle starts with a nitrogen-coordinated geranylamine-Rh complex, formed from the chain-carrying enamine-Rh complex (the only detectable species) and the allylic amine substrate. The enantioselection is made at the cyclometallation (intramolecular oxidative addition) step. MO calculation on the real transition state (TS) structures shows clearly the preference of the C(1)–H_S activation resulting in the (*R*)-citronellal enamine as illustrated in **Fig. 6.22. b**.

The asymmetric (–)-menthol production was realized through a fruitful academic-industrial collaboration (**Fig. 6.23.**). First, the research group of Juntaro Tanaka and Kunihiko Takabe at Shizuoka University (Japan) found a regioselective method for synthesizing diethylgeranylamine by $LiNEt_2$-catalyzed addition of diethylamine ($NHEt_2$) to myrcene (**Fig. 6.21.**) [M]. Kazuhide Tani at the Otsuka Laboratory of Osaka University, who had been a classmate of mine at Kyoto, happened to be struggling to find a catalyst for the asymmetric hydrogen-shift reaction of the allylic amine. The reactivity of Ti and Co catalyst systems was not satisfactory. He identified the high reactivity of cationic Rh(I) complexes but was unable to find suitable chiral ligands. Fortunately, the problem

Fig. 6.22. (next page) Mechanism of allylamine to enamine isomerization catalyzed by an (*S*)-BINAP-Rh complex.
(a) Calculations on a model system.
(b) Diastereomeric TSs in the C(1) hydrogen activation.
(Color figure available in the supporting material. ›Links [6].)

(a) Model reaction

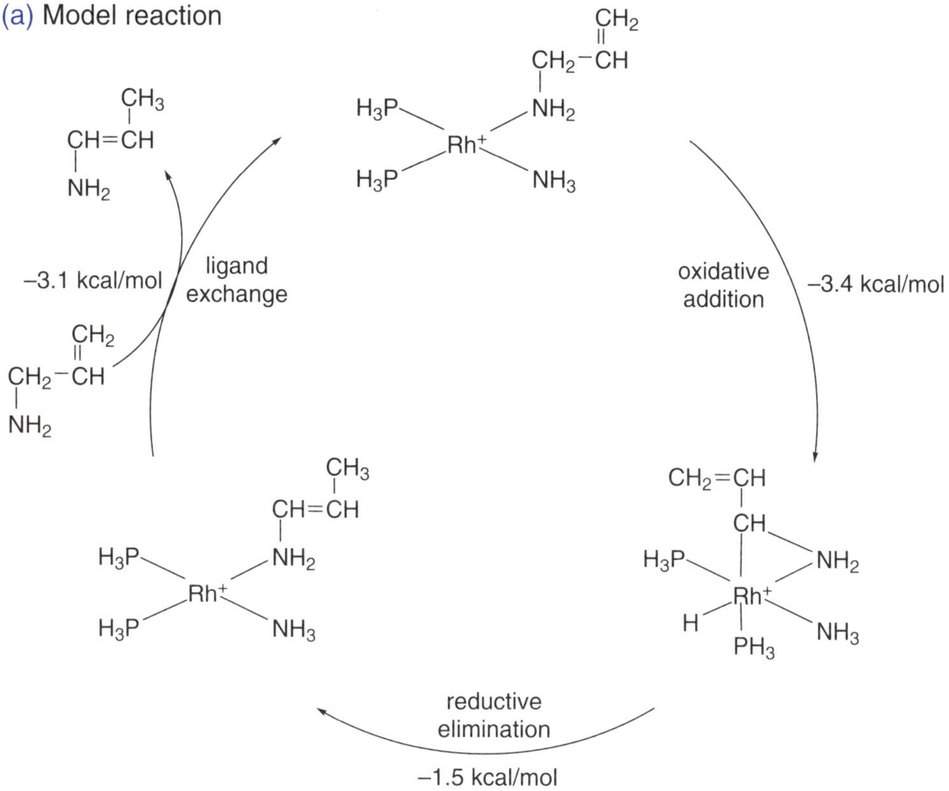

(b) Diastereomeric transition states in C–H activation

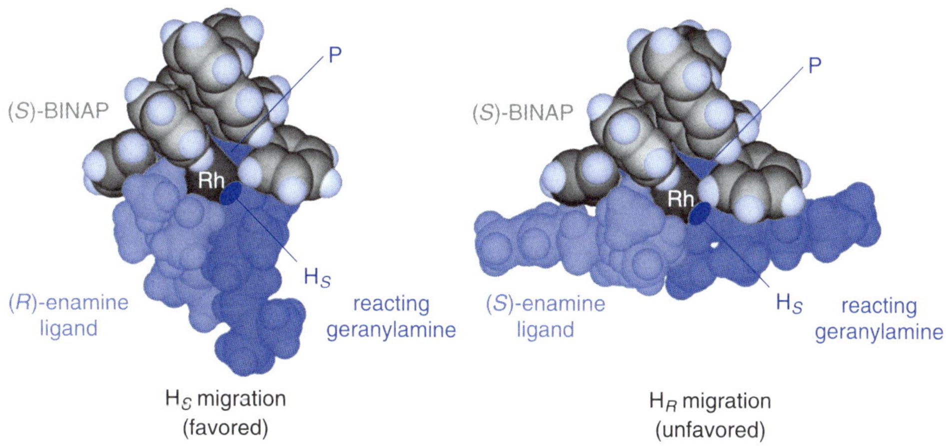

Fig. 6.23. Photo of major collaborators for (–)-menthol synthesis at the Takasago plant (February 1984). From the left, K. Tani, H. Takaya, R. Noyori, S. Otsuka, S. Akutagawa, and H. Kumobayashi.

was instantly solved by the use of our newly developed BINAP-Rh complex (*S*)-1 coordinated with 1,5-cyclooctadiene, THF, or acetone [N]. Among the various candidates, only the BINAP ligand worked well, displaying an outstanding reactivity and enantioselectivity to give (*R*)-citronellal enamine in 96 – 99 % *ee*.

In 1983, technical refinements at the Takasago Company enabled a full-scale industrial terpene production from myrcene [O]. This catalytic method, which was under total quality control allows a reaction using a 9-ton quantity of the geranylamine and 9.8 kg of $[Rh\{(S)\text{-}Tolbinap\}_2]^+ClO_4^-$ with 2.7 m^3 of THF per batch, giving a turnover number (TON) of 6,800. The crystalline BINAP-Rh catalyst could be

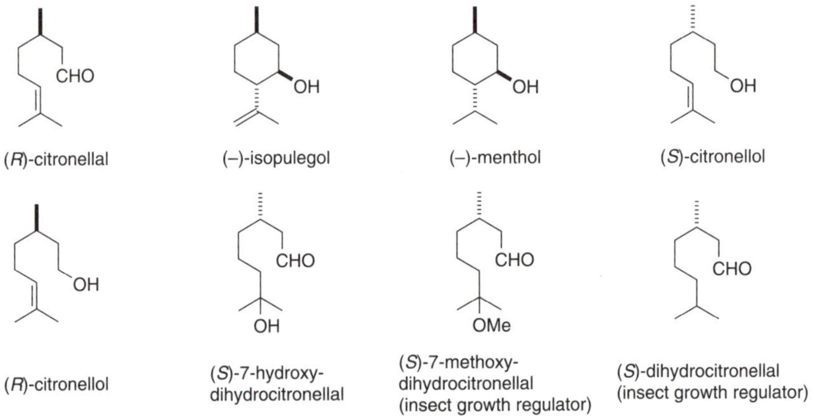

Fig. 6.24. Industrially synthesized asymmetric terpenes.

recovered, with only a very slight loss, and was effectively recycled. This chemical method enables the production of both *R*- and *S*-configurated aroma ingredients. I had learned that natural (*S*)-citronellol in an *ee* up to 92 % is obtainable from rose oil but in a limited quantity. Later, a research group at Hoffmann-La Roche tried many different chiral Rh catalysts for this important reaction, but no one could find a complex that could beat our catalysts. Thus, our catalytic system was able to synthesize citronellol with almost perfect enantioselectivity (99.5 % *ee*). The major optically active terpenoids produced by our method are listed in **Fig. 6.24.** [95].

Later, the production procedure was further improved. In the mid-2000s, the market was running out of natural mints grown in India and China, but this compact chemical process allows a more stable supply of this living necessity, compared to the climate-variable natural production from peppermint, which requires large land area. Although BASF and Symrise developed an efficient production scheme from petroleum-based materials, "naturalism" customers are comfortable to use the Takasago menthol derived from bio-based starting material myrcene. Currently, the Takasago Company is producing over 2,000 metric tons of (–)-menthol per year, enriching our daily life in the civilized society.

6.4 Chiral molecular catalysts: Beyond their shape (1986 – 2003)

The accumulated scientific knowledge of our predecessors has unraveled many unknowns and has transformed many impossible to possible. Yet, researchers must humbly admit that benchmarks in science are always accompanied by Socratic ignorance. Honestly, in the late 1970s, I understood little about asymmetric catalysis. Aiming at achieving a high degree of chiral recognition, we focused merely on the molecular architecture but ignored the functional engineering in the real catalytic environment. I remained too naive in this regard. Asymmetric catalysis is *four-dimensional chemistry* [228]. High efficiency can be achieved only by using a combination of both an ideal three-dimensional structure (*x*, *y*, *z*) and suitable kinetics (*t*) [229]. We must venture to consistently create new knowledge to open new worlds of the unknown.

Fig. 6.25. illustrates a typical, though not general, catalytic scheme that utilizes a molecular catalyst composed of a metallic element and chiral organic ligand(s). The process primarily consists of an induction process and a catalytic cycle. Under the established reaction conditions, the initially applied chiral precatalyst A is converted to the true catalyst B, but this induction step is often not straightforward or even unpredictable. Then, in the steady-state catalytic cycle, B activates achiral molecules X and Y and transforms them to the chiral product X–Y. Although the ligand must create a distinct enantio-differentiating

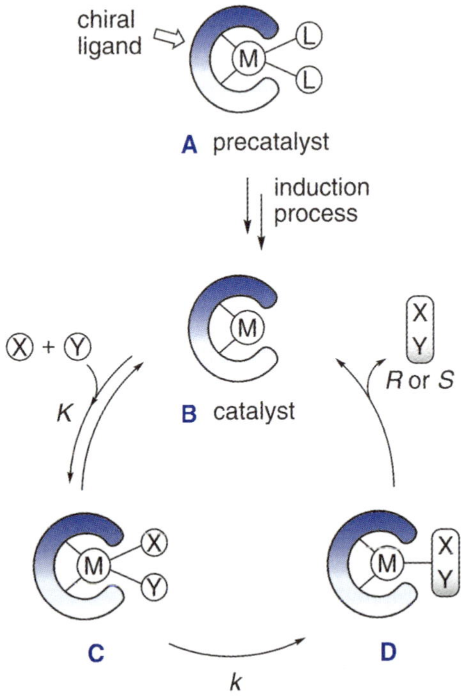

Fig. 6.25. The principle of asymmetric catalysis with chiral organometallic catalysts. M = metal; X, Y = reactant and substrate; L = neutral or anionic ligand.

environment in the key transition metal-based complexes, such architectural design does not suffice to achieve effective asymmetric catalysis. Some of the steps in the multi-stage transformation are reversible, while the first irreversible step, for example, C to D, kinetically determines the absolute configuration of X–Y. Furthermore, efficient asymmetric catalysis requires a high turnover number (TON) and high turnover frequency (TOF), but the best way to generate high catalytic activity is not immediately apparent. All the transition metal-based entities B – D in this cycle must be neither very unstable nor very stable, avoiding substrate and / or product inhibition. Instead, these conversions must take place smoothly via a low kinetic barrier, and without any destructive side-reactions. The reaction conditions strongly influence the stability and reactivity of A – D. In general, both suitable architectural and functional engineering are crucial for obtaining sufficient TON and TOF. Overall, however, "molecular catalysis" can cope with such requirements, because any molecule, in theory, can be designed and synthesized at will.

Looking back at my long experience in this exciting field, I now view that asymmetric catalysis lies totally outside the realm of traditional organic synthesis. It is a pervasive, all-embracing endeavor involving synthetic organic chemistry, catalytic chemistry, structural chemistry, inorganic and coordination chemistry, physical chemistry, and theoretical chemistry, as well as chemical engineering for practical production. We should be optimistic but not innocent.

6.5 Switching from Rh to Ru catalysts in asymmetric hydrogenation (1986 –)

Dihydrogen (H_2) is the simplest molecule and its properties are fully understood. The H–H bond is readily cleaved by transition metal complexes, but truly useful asymmetric hydrogenations (AHs) were elusive. BINAP-transition metal complexes are shaped in a manner that is beneficial for chiral recognition (**Fig. 6.9.**) [230, 231] and these complexes can act as AH catalysts. However, their efficiency is highly dependent on the structures of unsaturated substrates, the properties of the central metal and the auxiliary anionic or neutral ligands, and the reaction conditions (e.g. hydrogen pressure, temperature, solvent, and additive). No universal catalysts are known, because of the diversity of unsaturated organic compounds and their different reactivities.

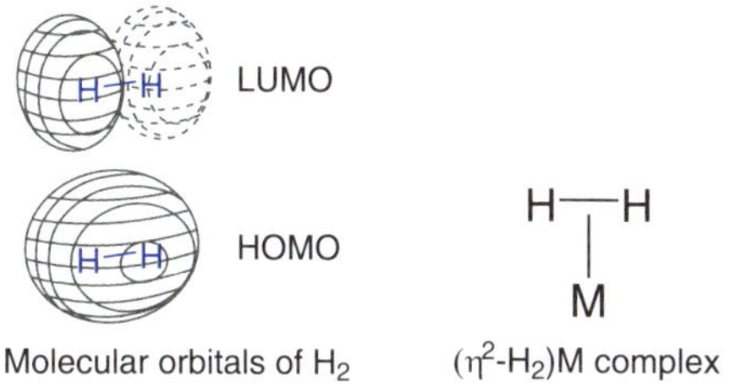

Fig. 6.26. MOs of dihydrogen and general formula of a dihydrogen metal complex.

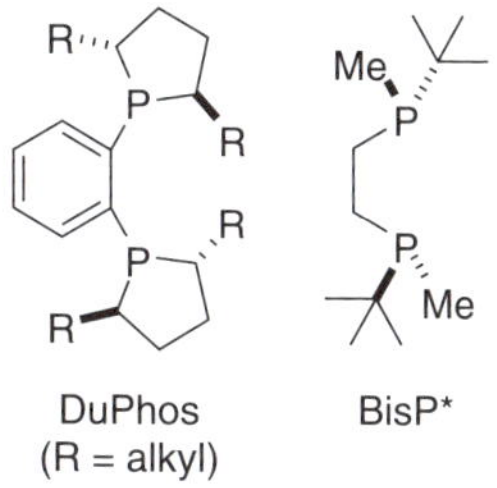

Fig. 6.27. Chiral ligands DuPhos and BisP*.

6.5.1 General asymmetric hydrogenation of functionalized olefins: Significance of metallic elements

The way to chemically manipulate H_2 remained limited despite our long and laborious research efforts. As described in **CHAPTER 6.1**, the scope of Rh-catalyzed AH was far from ideal for mechanistic reasons. It was effective but, until around 2000, excellent enantioselectivity was obtainable only with the complexes of DuPhos, certain monodentate phosphites, or phosphoramidites [P]. Notably, some unique Rh-electron-rich BisP* catalysts, found later by Tsuneo Imamoto, trigger AH via the "dihydride / unsaturate mechanism," giving α-amino acid derivatives in high *ee* [Q].

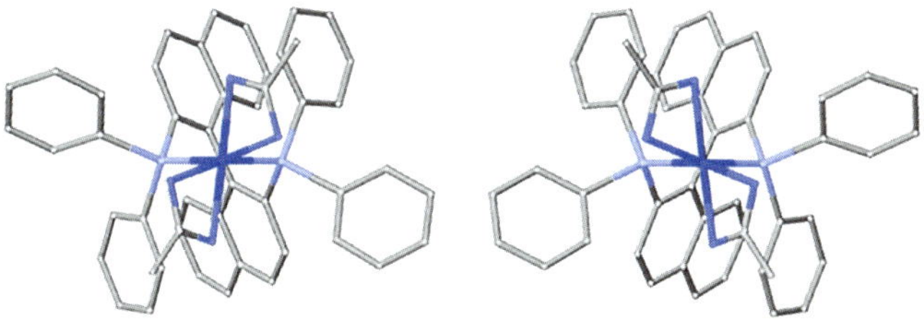

(*R*)-**2**, [Ru(OAc)$_2${(*R*)-binap}] (*S*)-**2**, [Ru(OAc)$_2${(*S*)-binap}]

Fig. 6.28. Structures of BINAP-Ru diacetate complexes. (Color figure available in the supporting material. ›Links [6].)

Earlier in 1986, we were able to overcome this long-standing problem by switching the metallic element from d^8 Rh(I) to d^6 Ru(II). The newly invented BINAP-Ru diacetate catalysts, (*R*)- and (*S*)-2, (**Fig. 6.28.**) [232, 233] opened the door to a new generation of homogeneous AHs, with a broad synthetic scope [234]. In recognition of this achievement, Hidemasa Takaya and I were awarded the Fluka Prize, Reagent of the Year in 1989.

Fig. 6.29. General asymmetric synthesis of isoquinoline alkaloids. Ru catalyst = (*R*)-2 or (*S*)-2.

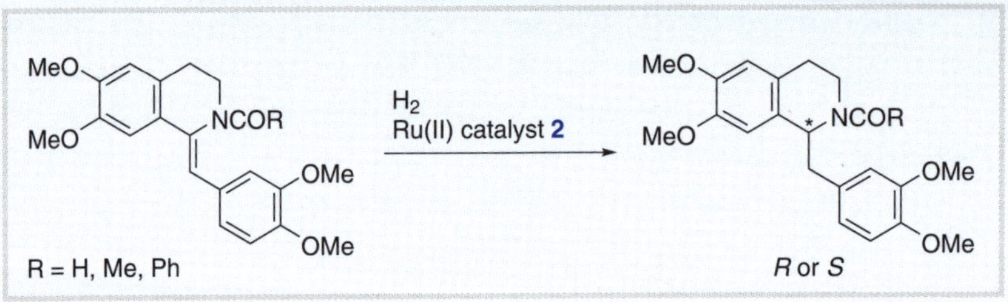

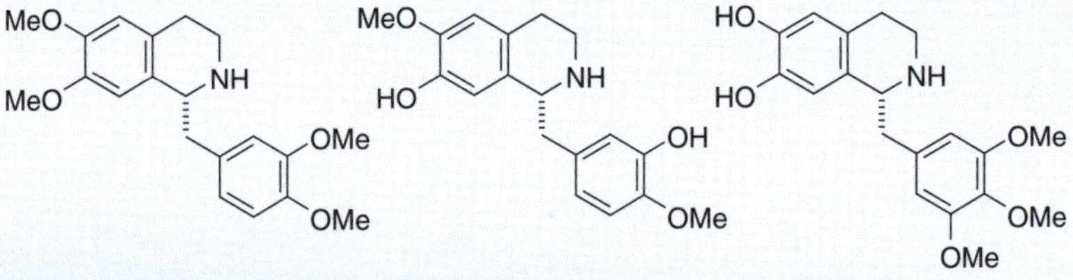

(*R*)-tetrahydropapaverine (*R*)-norreticuline (*R*)-tretoquinol

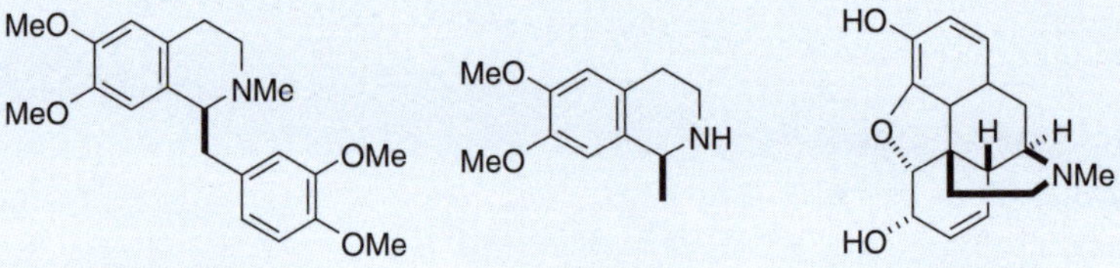

(*S*)-laudanosine (*S*)-salsolidine morphine

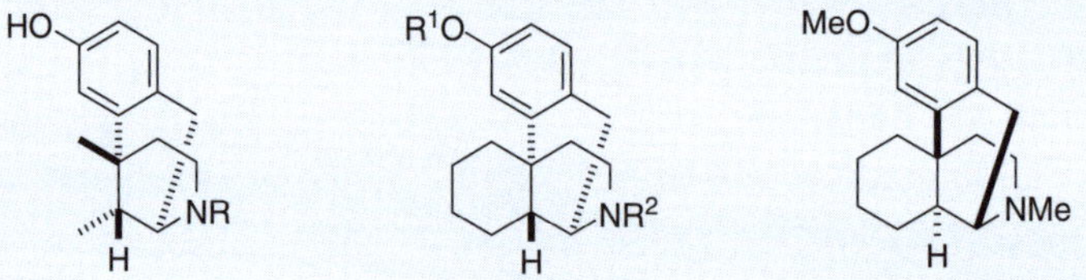

benzomorphans morphinans dextromethorphan

Fig. 6.30. Reversal of asymmetry in the hydrogenation of dehydroamino acid derivatives.

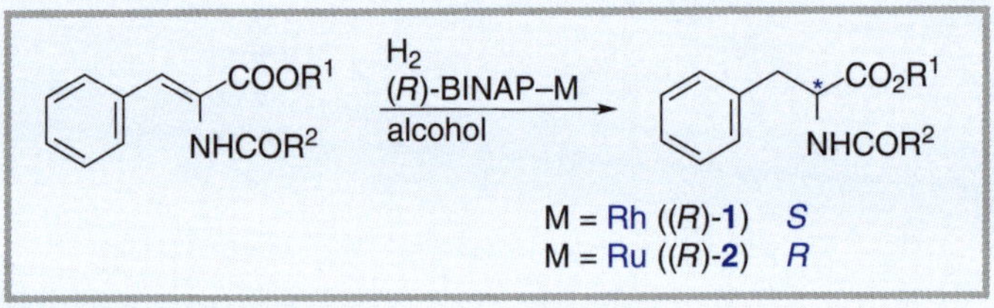

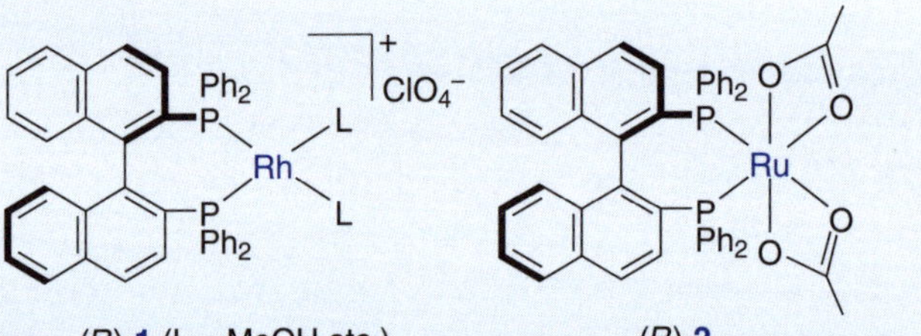

(*R*)-**1** (L = MeOH etc.) (*R*)-**2**

With the new chiral Ru catalysts in hand, we accomplished the first general asymmetric synthesis of isoquinoline alkaloids. The AH with 2, shown in **Fig. 6.29.**, proceeds smoothly, giving the N-protected tetrahydroisoquinolines in 95–100 % *ee* [235, 236]. This method is highly reliable and widely applicable to the synthesis of natural morphine and various benzomorphan analogues [237].

For a decade after the discovery of this reaction, we focused on further exploring practical AH processes utilizing these efficient catalysts. The fruitful consequences urged us to elucidate why the Ru, but not the Rh, catalyst works so efficiently. The substrates in **Fig. 6.29.** belong to an interesting class of (*Z*)-configurated enamides [238], with which the Rh chemistry gave us an unforgettable bitter experience (**Fig. 6.12.**). Therefore, we compared the behavior of Ru- and Rh-BINAP catalysts using standard α-(acylamino)cinnamic ester substrates [239]. As illustrated in **Fig. 6.30.**, when methyl (*Z*)-α-(acetamido)cinnamate is hydrogenated in the presence of the Ru catalyst (*R*)-2, the corresponding (*R*)-phenylalanine derivative is produced in 92 % *ee* and in 100 % yield. Most notable is the sense of asymmetric induction; the observed *R* to *R* catalyst / product relation is opposite to that attained with the Rh catalyst (*R*)-1, i. e. from *R* to *S* [239] [R]. This reversal of asymmetric configuration in going from Ru to Rh was also seen in the isoquinoline alkaloid synthesis.

This contrasting behavior results from the operation of a "monohydride / unsaturate mechanism" for Ru (**Fig. 6.32. a**) [239] instead of the well-studied "unsaturate / dihydride mechanism" for Rh (**Fig. 6.14.**) [222]. First, the Ru precatalyst A [(*R*)-2] causes the heterolytic cleavage of H_2 generating monohydride B, the actual catalyst. Most importantly, RuH B is generated before substrate coordination giving C. The configuration of the product is determined by the irreversible cleavage of the Ru–C bond in D by H_2. The enantiomer ratio corresponds well to the relative stability of the diastereomeric substrate / RuH complexes Si- and *Re*-C (**Fig. 6.32. b**). Here, the major *Si*-C is converted to the *R* hydrogenation product in a stereospecific manner, i. e., migratory insertion followed by mainly hydrogenolysis and some protonolysis. Notably this net hydrogenation uses two different H_2 molecules, as proven by isotope labeling experiments. This result stands in sharp contrast to the Rh-catalyzed reaction, which uses only one H_2 molecule / substrate affecting the pairwise H atom incorporation into the product.

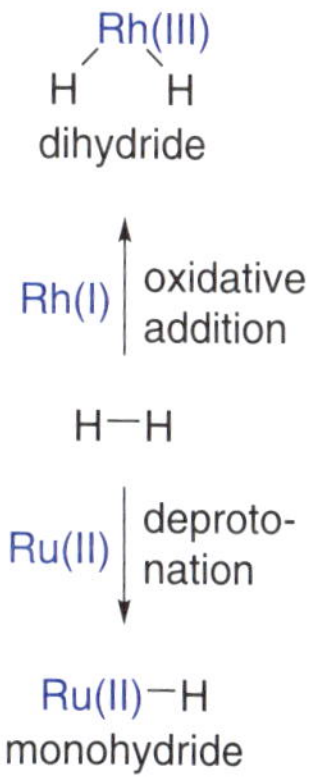

Fig. 6.31. Reaction of dihydrogen with Rh(I) and Ru(II).

This method is also useful for an asymmetric synthesis of certain β-amino acids [240] and α-amino phosphonic acids [241].

The substrate scope of this AH is broad, allowing asymmetric synthesis of a wide variety of biologically significant compounds [230, 231]. The BINAP-Ru(II) complex catalysts 2 exhibit an unprecedented efficiency in AH of prochiral allylic and homoallylic alcohols [242, 243]. Geraniol with two C=C double bonds is hydrogenated only at the allylic

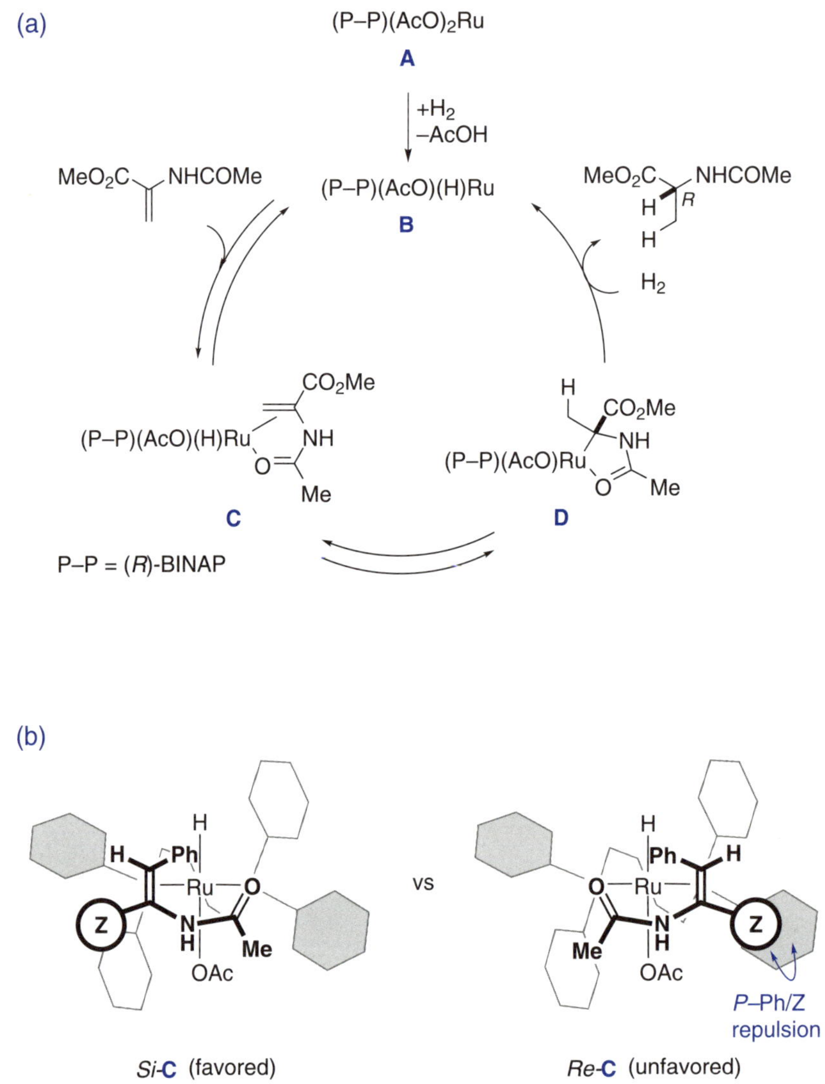

Fig. 6.32. Hydrogenation of methyl (*Z*)-α-(acetamido)-cinnamate catalyzed by an (*R*)-BINAP-Ru complex. (a) Catalytic cycle, in which the substrate structure is simplified. (b) Diastereomeric (*Z*)-enamide-Ru complexes. Z = CO_2Me.

alcohol part due to the directive hydroxyl group. This AH occurs in methanol containing (*R*)-BINAP-Ru$(OCOCF_3)_2$ with S/C = 50,000 at an initial pressure of 30 atm to provide (*S*)-citronellol quantitatively in 96–99 % *ee* (**Fig. 6.33.**). Again, the asymmetry was reversed in going from Ru to Rh, although the Rh catalyst generated citronellol only in < 60 % *ee* [244]. In the past, (*R*)-citronellol was obtained by hydrogenation of naturally occurring citronellal with < 85 % *ee*, while its optically pure compound is accessible by microbiological reduction of geraniol.

This large-scale Ru-catalyzed hydrogenation opened a straightforward way to synthesize the chiral C_{15} building block of α-tocopherol (vitamin E) and vitamin K_1. Moreover, the (*R*)-TolBINAP-Ru catalyzed reaction can be used to prepare a precursor of an important

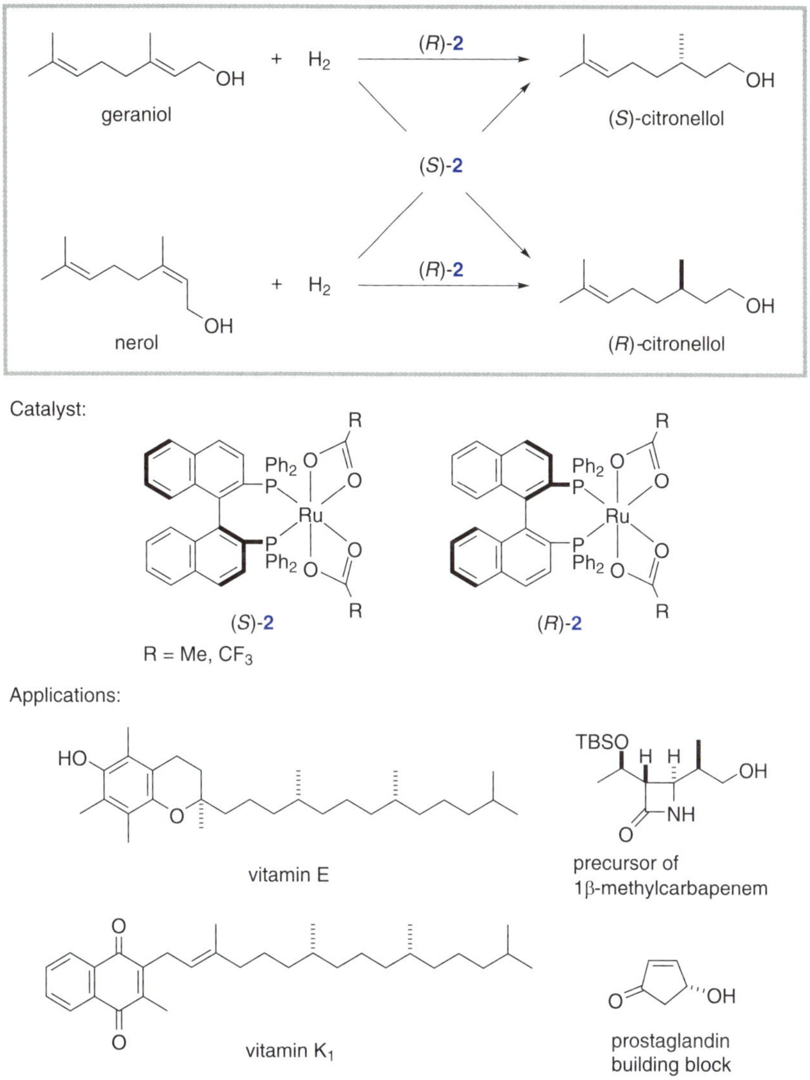

Fig. 6.33. Asymmetric hydrogenation of allylic alcohols. TBS = *tert*-butyldimethylsilyl.

1β-methylcarbapenem antibiotic [245]. Furthermore, this AH resolves certain chiral allylic secondary alcohols, allowing multi-kilogram access to (*R*)-4-hydroxycyclopentenone, a key intermediate for our prostaglandin (PG) synthesis (CHAPTER 5) [116].

Importantly, a range of α,β-unsaturated carboxylic acids are enantioselectively hydrogenated in an alcoholic medium as shown in Fig. 6.34. [246]. Some chiral lactones are conveniently prepared under these catalytic conditions (Eq. 2–4). The high pressure hydrogenation of an α-arylacrylic acid in methanol afforded (*S*)-naproxen, a very important anti-inflammatory agent, in 97 % *ee* (Eq. 5). (*S*)-2-Methylbutanoic acid obtained by this reaction is a fruit flavor (Eq. 6). Certain β,γ-unsaturated carboxylic acids can also be used as substrates in this hydrogenation reaction.

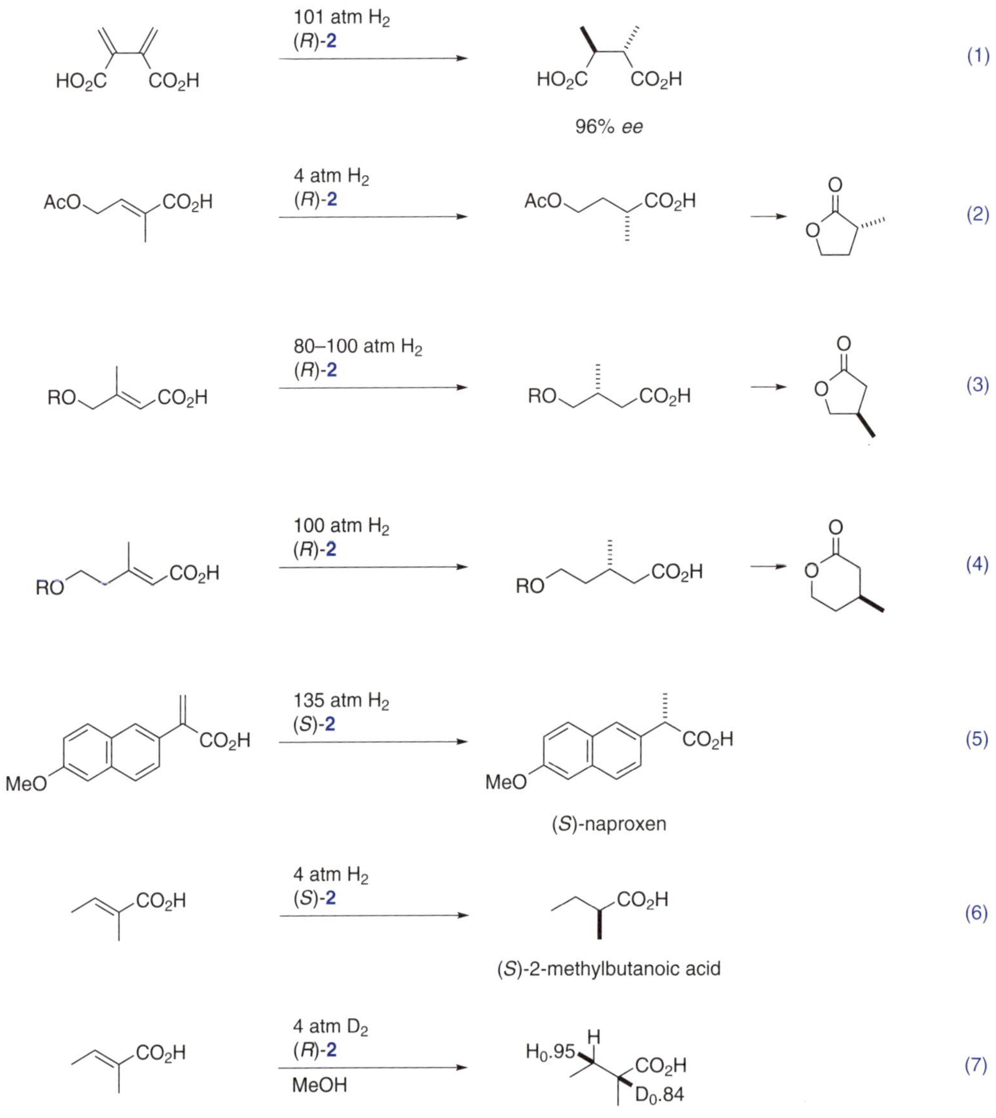

Fig. 6.34. Application of asymmetric hydrogenation of some unsaturated carboxylic acids.

The catalytic pathway is considered to start with the exchange of carboxylate ligands on Ru with the unsaturated carboxylic acids. This AH, however, appears to be not as simple as it was expected. First, the configuration and degree of enantioselectivity catalyzed by the (*S*)-BINAP-Ru complex, (*S*)-2, are markedly dependent on the structure of the olefinic substrates and the hydrogen pressure (**Fig. 6.35.**). The substrates are categorized in three classes, indicating the operation of multiple catalytic cycles. In the early mechanistic picture in **Fig. 6.32.**, the Ru–C bond in D is cleaved mainly by H_2 but now, as seen in Eq. 7 of **Fig. 6.34.**, its protonolysis becomes conspicuous in the *cis* hydrogenation of (*E*)-2-methylbutenoic acid [247]. This could be due to the enhanced solvent acidity, a subtle difference in the properties of the Ru–C bond, or even a change in the mechanism [S]. Under high-pressure conditions, the yet unknown dihydride or polyhydride mechanism could be in operation. Thus, we were unable to predict the full potential of the Ru dicarboxylate catalysts.

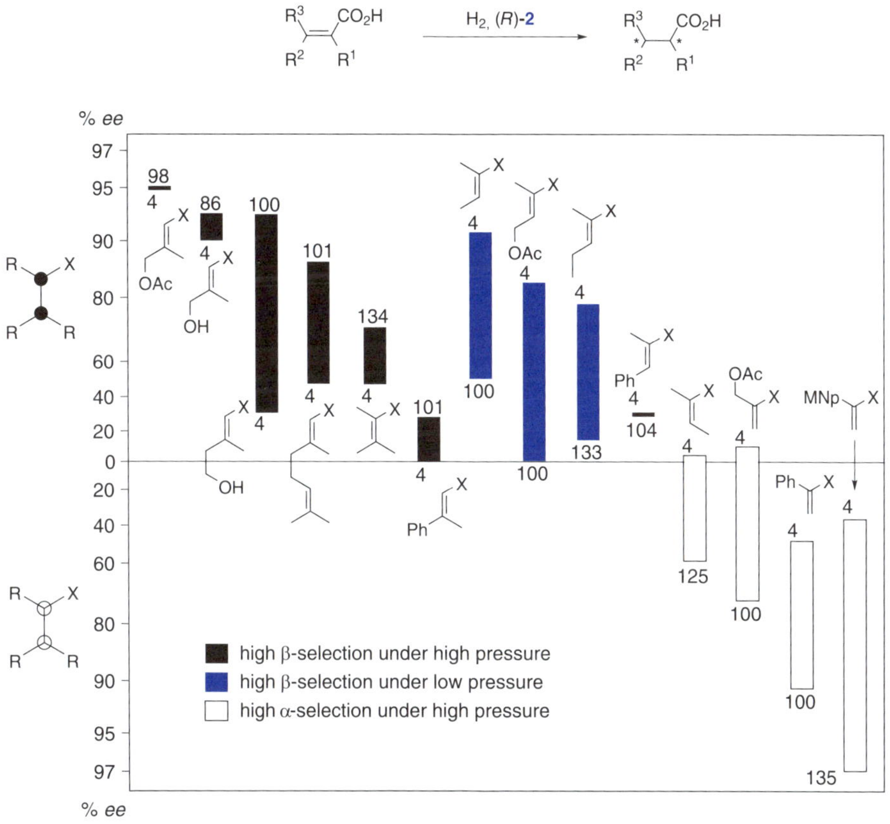

Fig. 6.35. Pressure dependence on the profile of asymmetric hydrogenation of unsaturated carboxylic acids in methanol. X = CO_2H, MNp = 6-methoxy-2-naphthyl. Conditions: H_2 pressure: 4 and 84–135 atm (Each number refers to the initial H_2 pressure in atm).

6.5.2 AH of functionalized ketones

Effect of anionic ligands on reactivity

BINAP-Ru dicarboxylates 2 (**Fig. 6.28.**) have enjoyed an enormous diversity of catalytic functions. We were therefore surprised to learn that simple anionic ligands on Ru have quite an effect on the reactivity. The Ru diacetate 2, though excellent for AH of functionalized *olefins*, is totally inactive for structurally similar *ketones* such as β-keto esters. This fact disappointed us. The problem, however, was immediately solved, when Masato Kitamura added a drop of HCl to the hydrogenation system of methyl 3-ketobutanoate with (*S*)-2 in methanol [248]. We were delighted to see the HPLC trace of the (*R*)-MTPA esters of the product showing a 99.6 : 0.4 enantiomeric ratio

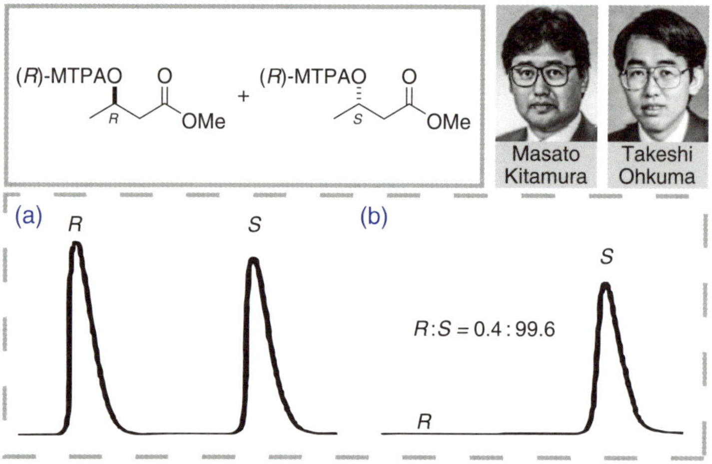

Fig. 6.36. HPLC analysis of (*R*)-MTPA esters of methyl-3-hydroxybutanoate. (a) Diastereoisomers derived from the racemic product. (b) Product obtained via (*S*)-BINAP-Ru catalyzed hydrogenation. MTPA = α-Methoxy-α-(trifluoromethyl) phenylacetyl.

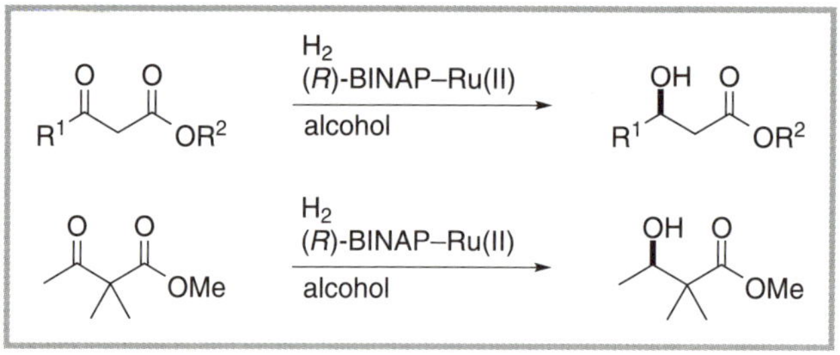

Fig. 6.37. Asymmetric hydrogenation of β-keto esters. R^1 = Me, Et, ^{n}Bu, ^{i}Pr, Ph. R^2 = Me, Et, ^{i}Pr, ^{t}Bu.

Catalysts:

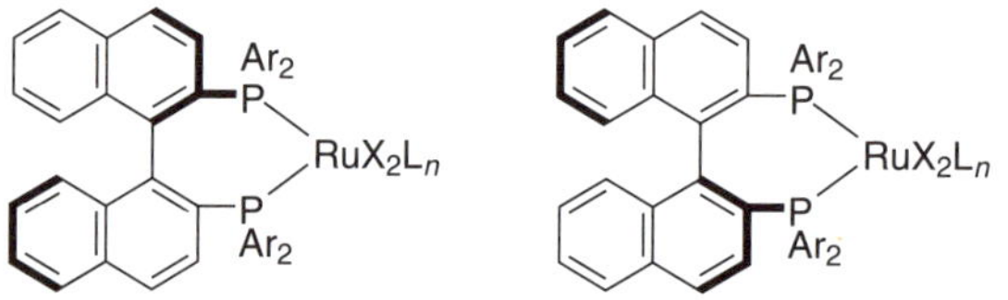

(*R*)-**3**, (*R*)-BINAP–Ru(II) (*S*)-**3**, (*S*)-BINAP–Ru(II)

X = halogen, L = neutral ligand

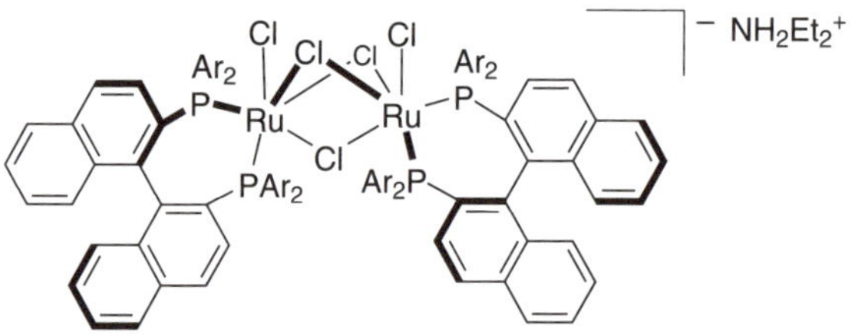

Ar = *p*-methoxyphenyl

Fig. 6.38. A Ru complex used for the AH of ß-keto esters [T].

(**Fig. 6.36.**). I would like to add that this happened in 1987, when chiral stationary phases for direct HPLC analysis were not yet available; the Mosher ester was a powerful tool in the study of asymmetric synthesis.

As illustrated in **Fig. 6.37.**, various β-keto esters are hydrogenated in alcoholic solvents in a predictable manner with a consistently high enantioselectivity of > 99 % *ee* [248, 249]. The efficiency surprised us. Reactions can be conducted at 4 – 100 atm of H_2 at room temperature with S/C of up to 10,000 on any scale even > 1,000 kg of the substrate, with up to 50 % (v/v) substrate concentration (Takasago International Corp.). This purely chemical approach is quite general and far superior to any of the biological processes, including baker's yeast reduction, the efficiency of which is often variable (**CHAPTER 3**).

For this purpose, we used Ru halide complexes with an empirical formula RuX_2(binap) 3 (X = Cl, Br, I; polymeric form). The efficient precatalysts were prepared by mixing the diacetate and HX or Me_3SiI in a 1 : 2 molar ratio [248, 250] but, most conveniently in our hands, by simply heating commercial $[RuCl_2(benzene)_2]_2$ and BINAP in DMF at 100 °C [251]. [RuX(binap)(arene)]Y (X = halogen, Y = halogen or BF_4) [252], $(NH_2Et_2)[\{RuCl(binap)\}_2(\mu\text{-}Cl)_3]$ [T] (**Fig. 6.38.**), and other *in situ*-formed halogen-containing BINAP-Ru complexes [U] can be used as well.

Our mechanistic model is shown in **Fig. 6.39. a**. The precatalyst A [(*R*)-3] is converted to the actual catalyst RuHCl B by reaction with H_2. B interacts reversibly with the β-keto ester to form the σ-type chelate complex C, in which metal-to-carbonyl hydride transfer is geometrically difficult [253]. Subsequently, protonation at the carbonyl oxygen increases the electrophilicity of the carbon and converts the geometry from σ (in-plane interaction) to π (out-of-plane interaction), thereby facilitating the hydride migration. The hydroxy ester ligand in the resulting product D is liberated by solvent molecules. The cationic Ru complex E once again cleaves H_2, thus regenerating B. This catalytic cycle turns over repeatedly until the keto esters are fully consumed.

Enantioselectivity in a ratio of > 99 : 1 is achieved in the hydride transfer step C to D. Ligation of the ester C=O to the Ru center stabilizes the TS in the hydride transfer step. Thanks to the shape of (*R*)-BINAP, diastereomeric *Si*- and *Re*-F are clearly differentiated (**Fig. 6.39. b**). Thus, the *R*-generating *Si* structure is highly favored over the *S*-generating *Re* isomer, which suffers from the *P*-Ph/R repulsive interaction. Here, the key is the carbonyl protonation in D, which is caused by HCl [254] [V] generated in the A to B induction step. When $Ru(OAc)_2$(binap) 2 is used, acetic acid is produced in place of HCl. The weak carboxylic acid is unable to generate high catalytic activity. Thus, even achiral anionic ligands in the precatalyst are not to be ignored in catalytic hydrogenation [W].

Fig. 6.39. Hydrogenation of β-keto esters catalyzed by an (*R*)-BINAP-Ru complex. (a) Catalytic cycle. (b) Diastereomeric TS.

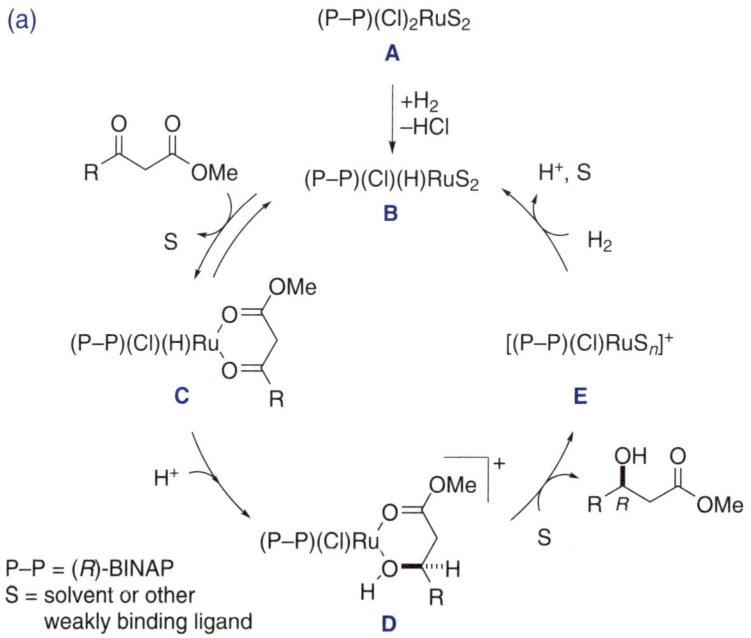

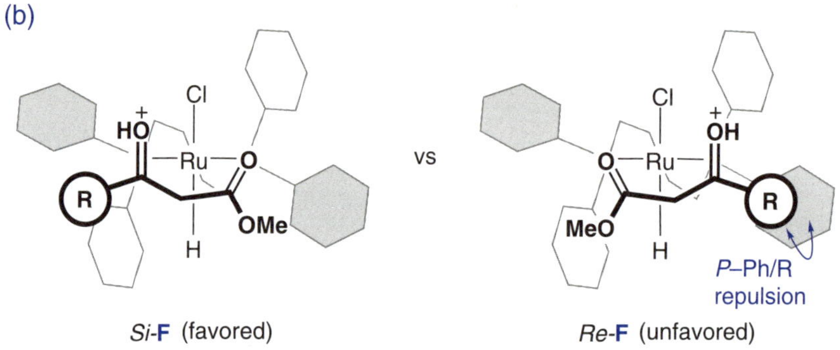

Wide scope of the substrates

This new AH is highly general as illustrated in **Fig. 6.40.** A wide array of functionalized ketones can be hydrogenated in a predictable manner and with consistently high enantioselectivity [230, 231, 255–257]. The heteroatom-containing "directive functional groups" include a dialkylamino, hydroxyl, alkoxyl, siloxyl, keto, alkoxycarbonyl, alkylthiocarbonyl, dialkylaminocarbonyl, phosphoryl, and sulfoxyl group, among other possibilities. Notably, neighboring halogen atoms also direct the stereochemical outcome. Certain β-diketones give near enantiopure *S*,*S* or *R*,*R* diols by double stereo-differentiation. AH products of γ-keto esters can be converted efficiently to chiral 4-substituted γ-lactones in high ee on any scale.

The resulting (*R*)-1,2-propandiol was used for the industrial synthesis of the antibacterial levofloxacin (Daiichi Pharmaceutical Co.)

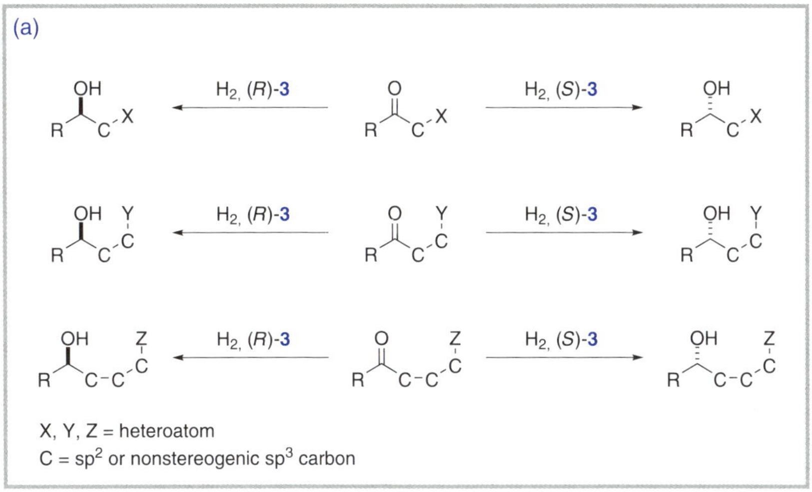

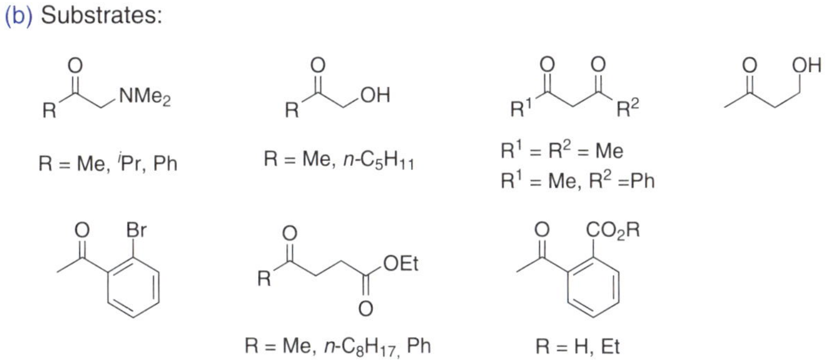

Fig. 6.40. General asymmetric hydrogenation of functionalized ketones. (a) Asymmetric induction depending on the catalyst's configuration. (b) Examples of substrates.

(**Fig. 6.41.**). In addition, carnitine (a carrier of long-chain fatty acids through membranes), γ-amino-β-hydroxybutyric acid (GABOB) and a compactin intermediate are obtainable in high enantiomeric purity [258]. The pre-existing stereogenic centers in the ketonic substrate significantly affect the steric course. Statines can also be obtained in high diastereo- and enantioselectivities [259].

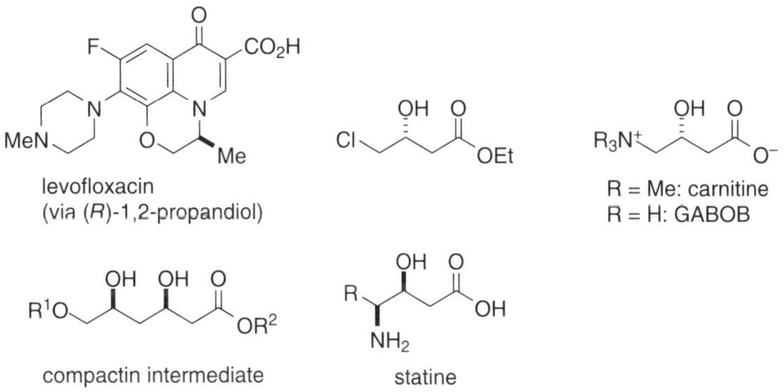

Fig. 6.41. Applications of BINAP-Ru catalyzed hydrogenation.

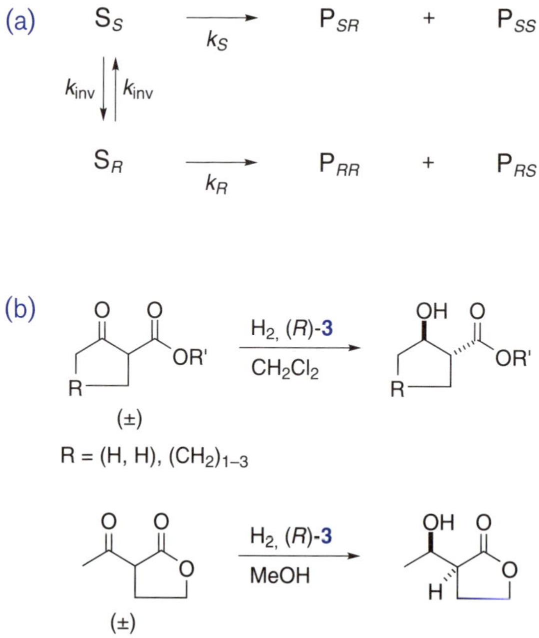

Fig. 6.42.
(a) The second-order stereoselective transformation; (b) Dynamic kinetic resolution of racemic keto esters.

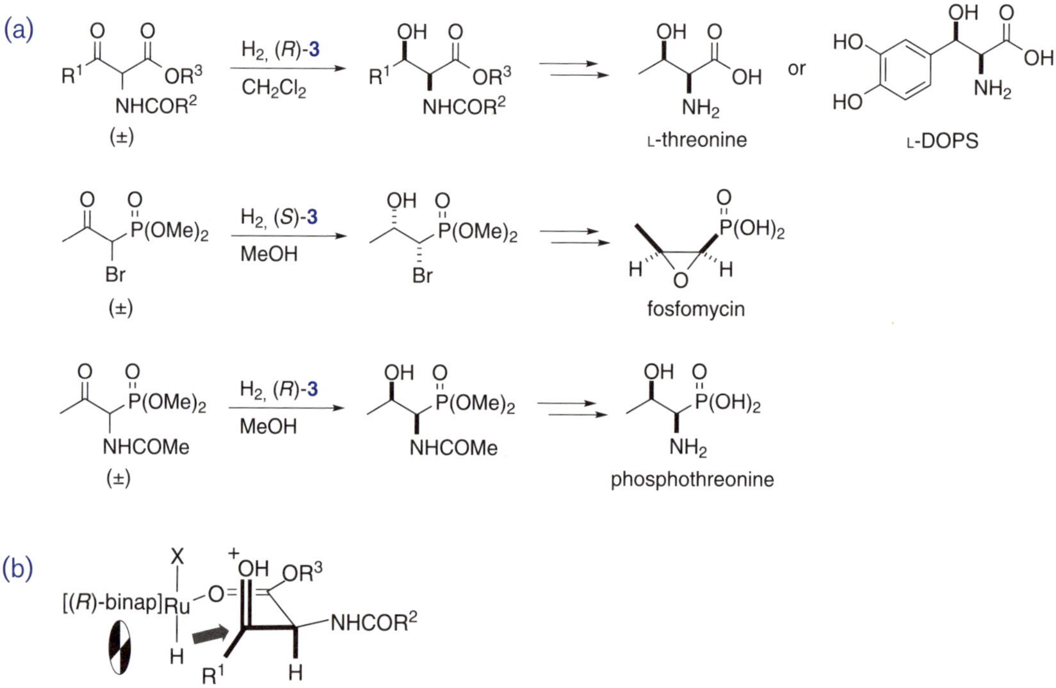

Fig. 6.43.
(a) Asymmetric synthesis of physiologically significant compounds via dynamic kinetic resolution. (b) Felkin-Anh transition state. X = halogen, solvent, etc.

Importance of solvents for selectivity: Dynamic kinetic resolution

The molecular structure of the chiral catalyst is of prime significance. In this 4D chemistry, however, the overall selectivity is optimized by choosing a suitable solvent, temperature, concentration and / or ratio of a substrate, reactant, and catalyst. Unlike ordinary kinetic resolution that gives a maximum of 50 % yield of the enantiomer, "dynamic kinetic resolution" (defined by us in 1989) [260 – 262] allows the conversion of configurationally labile α-substituted β-keto esters to a single stereoisomer out of four possible isomers, as shown in **Fig. 6.42. a**. The reaction medium strongly affects the kinetic parameters of the stereoselective reaction. The degree of C_3 *S*/*R*-face selection is consistently high regardless of the solvents, but the C_2/C_3 *syn* / *anti* preference is highly influenced by the nature of the solvent, either dichloromethane or standard alcohols, in addition to the substrate structure and functionality (**Fig. 6.42. b**).

As illustrated in **Fig. 6.43. a**, the second-order stereoselective transformation under slightly acidic conditions constitutes a practical asymmetric catalysis, leading to threonines, anti-Parkinsonian DOPS, clinically important antibiotic fosfomycin [263], and phosphothreonines [264]. The *syn* selectivity is rationalized by considering the Felkin-Anh transition state in the hydrogen transfer step shown in **Fig. 6.43. b**.

The usefulness of this reaction system was highlighted by the industrial synthesis of carbapenem antibiotic intermediate at Takasago International Co. (**Fig. 6.44.**). The (*R*)-BINAP-Ru-catalyzed hydrogenation of racemic methyl α-(benzamidomethyl)acetoacetate 4, under suitable conditions, produces the 2*S*,3*R* hydroxy ester 5 with 96 : 4 *erythro* : *threo* diastereoselectivity and 99.5 : 0.5 enantioselectivity [260]. Its cyclization to a β-lactam followed by Murahashi's Ru-catalyzed oxidative acetoxylation [X] gives the requisite chiral 4-acetoxyazetidinone 6.

Dynamic kinetic resolution is characterized by the presence of parallel reactions interrelated by the stereoinversion of the enantiomeric substrates. The efficiency is decisively affected by kinetic parameters, in particular the relative rates of the stereoinversion and reactions as well as the intrinsic stereochemical parameters of the catalyst and substrate

Fig. 6.44. Stereoselective synthesis of carbapenem antibiotics. Hydrogenation conditions: [(±)-4] = 0.2 M in CH_2Cl_2, [(*R*)-3] = 13 mM, 100 atm, 50 °C).

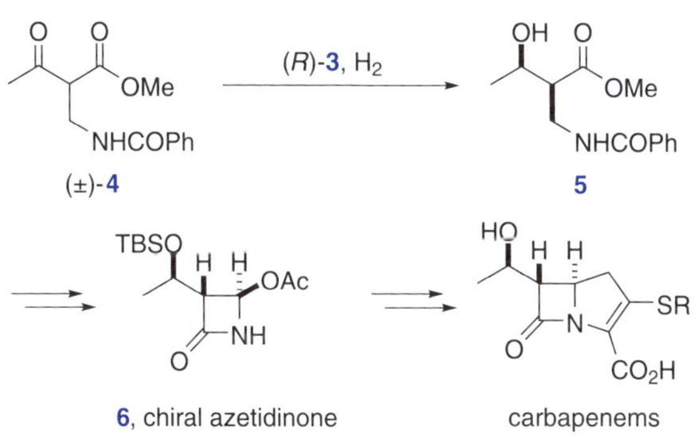

(**Fig. 6.45.**). Computer-aided quantitative analysis of the (*R*)-BINAP-Ru aided AH of (±)-4 revealed that: (1) the inherent, highest *S,R* : *S,S* : *R,R* : *R,S* selectivity at time $t = 0$ ($w:x:y:z$ partition parameter), where the starting *S* and *R* ketones are expected to exist in equal amounts, is 93.8 : 0.1 : 5.6 : 0.49; (2) (*S*)-4 is hydrogenated 15 times faster than (*R*)-4; (3) the inversion of the slow-reacting (*R*)-4 to (*S*)-4 takes place 92 times faster than hydrogenation; (4) the extent of the (*R*)-BINAP catalyst-based asymmetric induction (C_{cat}) is calculated to be 104 : 1 in favor of the 3*R* isomer; and (5) the substrate-based asymmetric induction (C_{sub}) is 9 : 1 in favor of *syn* C(2) / C(3) stereochemistry [265]. Thus, various steric and kinetic parameters can be beneficially combined to selectively produce the desired 2*S*,3*R* stereoisomer. When the reaction is conducted in polar methanol, both (*S*)- and (*R*)-4 are hydrogenated smoothly, and consequently the diastereoselectivity is drastically reduced to form a 1 : 1 mixture of (*S,R*)-5 in 93 % *ee* and (*R,R*)-5 in 97 % *ee*.

Fig. 6.45. Stereoselective hydrogenation of racemic methyl α-(benzamidomethyl) acetoacetate 4 in CH_2Cl_2.

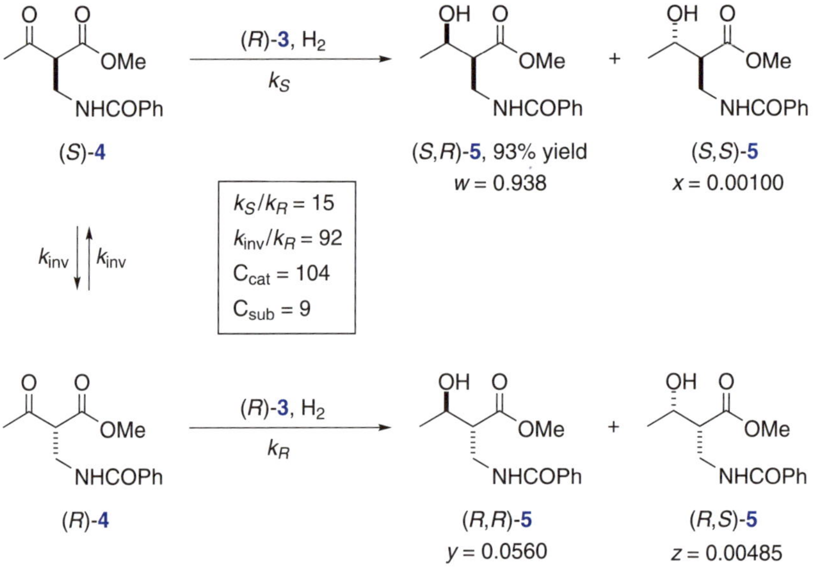

References for chapter 6

[A] (a) M. Calvin and M. Polanyi: Homogeneous Catalytic Hydrogenation. *Trans. Faraday Soc.* **1938**, *34*, 1181 – 1191; (b) M. Calvin: Homogeneous Catalytic Hydrogenation. *J. Am. Chem. Soc.* **1939**, *61*, 2230 – 2234.

[B] J. Halpern, J. Harrod, and B. R. James: Homogeneous Catalytic Hydrogenation of Olefinic Compounds. *J. Am. Chem. Soc.* **1961**, *83*, 753 – 754.

[C] J. F. Young, J. A. Osborn, F. H. Jardine, and G. Wilkinson: Hydride Intermediates in Homogeneous Hydrogenation Reactions of Olefins and Acetylenes Using Rhodium Catalysts. *Chem. Commun.* **1965**, 131 – 132.

[D] S. Akabori, S. Sakurai, Y. Izumi, and Y. Fujii: An Asymmetric Catalyst. *Nature* **1956**, *178*, 323 – 324.

[E] W. S. Knowles and M. J. Sabacky: Catalytic Asymmetric Hydrogenation Employing a Soluble, Optically Active, Rhodium Complex. *Chem. Commun.* **1968**, 1445 – 1446.

[F] L. Horner, H. Siegel, and H. Büthe: Asymmetric Catalytic Hydrogenation with an Optically Active Phosphinerhodium Complex in Homogeneous Solution. *Angew. Chem., Int. Ed.* **1968**, *7*, 942 – 942.

[G] T. P. Dang and H. B. Kagan: The Asymmetric Synthesis of Hydratropic Acid and Amino-acids by Homogeneous Catalytic Hydrogenation. *J. Chem. Soc. D* **1971**, 481 – 481.

[H] (a) W. S. Knowles: Asymmetric Hydrogenations (Nobel Lecture). *Angew. Chem., Int. Ed.* **2002**, *41*, 1998 – 2007; (b) W. S. Knowles: Asymmetric Hydrogenation. *Acc. Chem. Res.* **1983**, *16*, 106 – 112.

[I] One of the early examples of successful asymmetric catalysis was the chiral phosphine / Ni-catalyzed hydrovinylation of norbornene: B. Bogdanović, B. Henc, A. Lösler, B. Meister, H. Pauling, and G. Wilke: Asymmetric Syntheses with the Aid of Homogeneous Transition Metal Catalysts. *Angew. Chem., Int. Ed.* **1973**, *12*, 954 – 964.

[J] K. J. Brown, M. S. Berry, K. C. Waterman, D. Lingenfelter, and J. R. Murdoch: Preparation and Lithiation of Optically Active 2,2'-Dihalo-1,1'-binaphthyls. A General Strategy for Obtaining Chiral, Bidentate Ligands for Use in Asymmetric Synthesis. *J. Am. Chem. Soc.* **1984**, *106*, 4717 – 4723.

[K] For direct stereospecific synthesis, see: D. Cai, J. F. Payack, D. R. Bender, D. L. Hughes, T. R. Verhoeven, P. J. Reider: (*R*)-(+)- and (*S*)-(−)-2,2'-Bis(diphenylphosphino)-1,1'-binaphthyl (BINAP) [Phosphine, [1,1'-binaphthalene]-2,2'-diylbis[diphenyl-, (*R*)- and (*S*)]. *Org. Synth.* **1998**, *76*, 6 – 11.

[L] (a) A. S. C. Chan and J. Halpern: Interception and Characterization of a Hydridoalkylrhodium Intermediate in a Homogeneous Catalytic Hydrogenation Reaction. *J. Am. Chem. Soc.* **1980**, *102*, 838 – 840; (b) C. R. Landis and J. Halpern: Asymmetric Hydrogenation of Methyl (*Z*)-α-Acetamidocinnamate Catalyzed by {1,2-Bis((phenyl-*o*-anisoyl)phosphino)ethane}rhodium(I): Kinetics, Mechanism and Origin of Enantioselection. *J. Am. Chem. Soc.* **1987**, *109*, 1746 – 1754; (c) J. M. Brown and P. A. Chaloner: The Mechanism of Asymmetric Homogeneous Hydrogenation. Rhodium(I) Complexes of Dehydroamino Acids Containing Asymmetric Ligands Related to Bis(1,2-diphenylphosphino)ethane. *J. Am. Chem. Soc.* **1980**, *102*, 3040 – 3048.

[M] K. Takabe, T. Yamada, T. Katagiri, and J. Tanaka: Telomerization of Isoprene with Dialkylamine: *N,N*-Diethylnerylamine [2,6-Octadien-1-amine, *N,N*-diethyl-3,7-dimethyl-, (*Z*)-]. *Org. Synth.* **1989**, *67*, 48 – 51.

[N] (a) S. Otsuka and K. Tani: Catalytic Asymmetric Hydrogen Migration of Allylamines. *Synthesis* **1991**, 665 – 680; (b) S. Otsuka: Discoveries of the Catalysis of Asymmetric Isomerization of Allylamines and its Significance in Science and Industry. *Acta Chem. Scand.* **1996**, *50*, 353 – 360.

[O] (a) S. Akutagawa: *Organic Synthesis in Japan: Past, Present, and Future* (eds.: R. Noyori, T. Hiraoka, K. Mori, S. Murahashi, T. Onoda, K. Suzuki, O. Yonemitsu). Tokyo Kagaku Dozin, Tokyo, **1992**, 75; (b) Personal correspondence from Dr. Noboru Sayo, Takasago International Corporation, **2012**.

[P] (a) M. J. Burk: Modular Phospholane Ligands in Asymmetric Catalysis. *Acc. Chem. Res.* **2000**, *33*, 363 – 372; (b) M. van den Berg, A. J. Minnaard, E. P. Schudde, J. van Esch, A. H. M. de Vries, J. G. de Vries, and B. L. Feringa: Highly Enantioselective Rhodium-Catalyzed Hydrogenation with Monodentate Ligands. *J. Am. Chem. Soc.* **2000**, *122*, 11539 – 11540; (c) C. Claver, E. Fernandez, A. Gillon, K. Heslop, D. J. Hyett, A. Martorell, A. G. Orpen, and P. G. Pringle: Biarylphosphonites: a Class of Monodentate Phosphorus(III) Ligands that Outperform their Chelating Analogues in Asymmetric Hydrogenation Catalysis. *Chem. Commun.* **2000**, 961 – 962; (d) M. T. Reetz and G. Mehler: Highly Enantioselective Rh-Catalyzed Hydrogenation Reactions Based on Chiral Monophosphite Ligands. *Angew. Chem., Int. Ed.* **2000**, *39*, 3889 – 3890.

[Q] T. Imamoto: Synthesis and Applications of High-performance P-chiral Phosphine Ligands. *Proc. Jap. Acad., Ser. B* **2021**, *97*, 520 – 542.

[R] T. Ikariya, Y. Ishii, H. Kawano, T. Arai, M. Saburi, S. Yoshikawa, and S. Akutagawa: Synthesis of Novel Chiral Ruthenium Complexes of 2,2'-Bis(diphenylphosphino)-1,1'-binaphthyl and their Use as Asymmetric Catalysts. *J. Chem. Soc. Chem. Commun.* **1985**, 922 – 924.

[S] M. T. Ashby and J. Halpern: Kinetics and Mechanism of Catalysis of the Asymmetric Hydrogenation of α,β-Unsaturated Carboxylic Acids by Bis(carboxylato){2,2'-bis(diphenylphosphino)-1,1'-binaphthyl}ruthenium(II), $[Ru^{II}(BINAP)(O_2CR)_2]$. *J. Am. Chem. Soc.* **1991**, *113*, 589 – 594.

[T] (a) S. A. King and L. DiMichele: *Catalysis of Organic Reactions* (eds.: M. G. Scaros and M. L. Prunier). Dekker, New York, **1995**, 157 – 166; (b) T. Ohta, Y. Tonomura, K. Nozaki, H. Takaya, and K. Mashima: An Anionic Dinuclear BINAP–Ruthenium(II) Complex: Crystal Structure of $[NH_2Et_2][\{RuCl((R)\text{-}p\text{-MeO-BINAP})\}_2(\mu\text{-Cl})_3]$ and Its Use in Asymmetric Hydrogenation. *Organometallics* **1996**, *15*, 1521 – 1523.

[U] J. P. Genet: *Reduction in Organic Synthesis* (ed: A. F. Abdel-Magit). Am. Chem. Soc., Washington, D.C., **1996**, 31 – 54.

[V] S. A. King, A. S. Thompson, A. O. King, and T. R. Verhoeven: An Improved Procedure for the Synthesis and Use of $[RuCl_2(BINAP)]_2 \cdot NEt_3$. Dependence of the Ru(II)-BINAP Catalyzed Asymmetric Hydrogenation of β-Keto Esters on Trace Amounts of Acid. *J. Org. Chem.* **1992**, *57*, 6689 – 6691.

[W] K. Fagnou and M. Lautens: Halide Effects in Transition Metal Catalysis. *Angew. Chem., Int. Ed.* **2002**, *41*, 26 – 47.

[X] S. Murahashi, T. Naota, T. Kuwabara, T. Saito, H. Kumobayashi, and S. Akutagawa: Ruthenium-catalyzed Oxidation of Amides and Lactams with Peroxides. *J. Am. Chem. Soc.* **1990**, *112*, 7820 – 7822.

MY WAY TO GREEN CATALYSIS

7

 DOI.ORG/10.47261/1564-7

7.1 The ERATO project with generous funding (1991 – 1996)

My academic career has been generously and continuously supported by the Ministry of Education (now known as MEXT), the Japan Society for the Promotion of Science (JSPS), and various industrial corporations. In addition, from 1991 to 1996, I was appointed as the Project Leader of the Exploratory Research for Advanced Technology (ERATO) Noyori Molecular Catalysts Project of the Japan Research and Development Corporation (JRDC) (now Japan Science and Technology Agency (JST)). I was grateful to the JRDC committee's recognition towards my research capability. This new project attracted considerable attention from talented young researchers of countries in Europe and North America and also from various industries in Japan. Takao Ikariya from the NKK Corporation acted as Technical Officer and the capable industrial researchers Shohei Hashiguchi (Takeda Pharmaceutical Co.) and Tatsuya Miyatake (Sumitomo Chemical Co.) joined our project as team leaders. Every Friday, our international group met off-campus to spend wonderful times discussing the progress of this vibrant basic-science project.

The research theme of the ERATO program was totally unrestricted. The only soft request from JRDC was to develop new technologies, which could be possibly applicable to the chemical industry. We then set Green Chemistry as the overall goal and set several research themes aiming toward developing practical catalytic processes, for which high selectivity, reactivity, and productivity were our major interest. A useful catalysis must show a high turnover number (TON), defined as mols of product per mol of catalyst, and a high turnover frequency (TOF), defined as TON per hour or second. In addition, reactions are expected to be operationally simple, safe, and environmentally friendly. These important attributes could be obtained only by designing suitable molecular catalysts and reaction conditions through a deep understanding of, or unique insight into, the catalytic cycle. Although combinatorial approaches coupled with high throughput screening had already been prevailing at that time, it was, and still remains my belief that to a large extent, the discovery of efficient catalytic reactions relies on serendipity and originates from sound, comprehensive scientific knowledge. Discovery is the most joyful aspect above anything else.

Synthesis of polyacetylenes

Asymmetric hydrogenation (AH) was the major topic of this project. We also studied polymerization and green reaction media, some areas of which were totally new to me. An outgrowth of my favorite Rh chemistry, coupled with my friendship with Yoshio Okamoto at Nagoya, an inventor of man-made single-handed polymers, I attempted the polymerization of acetylenes with the aim of preparing new helical molecular structures.

Fig. 7.1.
Rh(I)-initiated stereospecific living polymerization of phenylacetylenes.

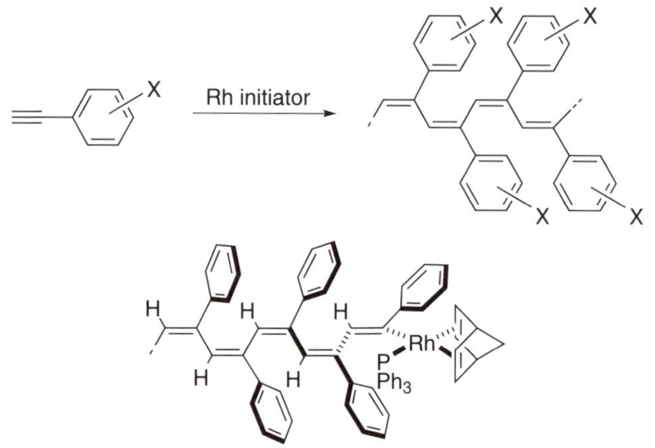

nbd = norbornadiene.

Polyacetylenes possessing alternating olefinic bonds along the main chain are of great interest because of their unique physical properties. We found that a tetracoordinate Rh complex, generated from a structurally defined Rh(C≡CPh)(nbd)$(PPh_3)_3$, with the aid of 4-(dimethylamino)pyridine, acts as an excellent initiator of the stereospecific living polymerization of phenylacetylene and its *m*- and *p*-derivatives [266–269]. The resulting polyacetylenes have a narrow molecular-weight distribution, M_w/M_n = 1.06–1.2. An active Rh(I) complex bearing a growing polymer chain, NBD, and PPh_3 was isolated from the reaction mixture and characterized (**Fig. 7.1.**). The living nature of the Rh-catalyzed reaction allows the synthesis of an AB type block copolymer from different phenylacetylenes. Detailed analysis indicated that polymerization proceeds via a 2,1-insertion mechanism, instead of via the metathesis pathway, to provide stereoregular poly(phenylacetylene)s with a cis-transoidal backbone structure. Although I had planned the synthesis of single-handed polymers by the use of a chiral diene or a chiral norbornadiene derivate as ligands, we were unable to find time to pursue this project further.

7.2 Asymmetric transfer hydrogenation (1992–1996)

We set a new theme to establish asymmetric transfer hydrogenation (ATH). Transfer hydrogenation is an intriguing natural and artificial process. In nature, ethanol is oxidized to acetaldehyde using alcohol dehydrogenase and nicotinamide adenine dinucleotide (NAD) as a cofactor (**Fig. 7.2.**). Acetaldehyde is further oxidized to acetic acid. The alcohol to aldehyde conversion is reversible; acetaldehyde is reduced by using the same enzyme and NADH, a reduced form of NAD, as a cofactor. Such enzymatic reactions are very selective and stereospecific in an

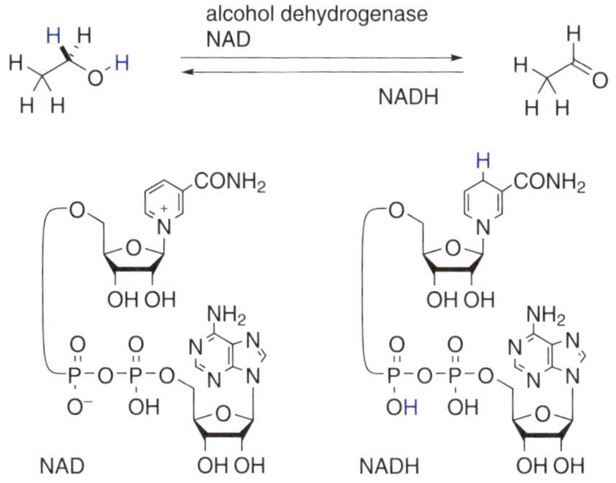

Fig. 7.2.
In vivo redox process of ethanol.

absolute sense. I recalled a beautiful lecture by Frank H. Westheimer on this topic [A], when I participated in a stereochemistry symposium in Geneva in March 1988, before starting this study. I was puzzled, however, why this enzymatic reduction uses only two hydrogen atoms in a large NADH molecule with a molecular weight of 665. Obviously, the chemical conversion, "transfer hydrogenation", needs cheaper, readily available cofactors and catalysts.

2-Propanol is among the most convenient hydrogen donors, because it is stable, easy to handle (b. p. 82 °C), nontoxic, inexpensive, and dissolves many organic substrates. It is a small alcohol with a molecular weight of only 60, and two hydrogen atoms among a total of eight hydrogen atoms in the molecule can be used. As such, numerous attempts had been made to bring about an asymmetric Meerwein-Ponndorf-Verley reaction of ketones occurring via aluminum 2-propoxide, but without success. Some pioneering trials with transition metals and lanthanides were also known but needed much improvement to be practical, because of low catalytic activity, insufficient enantioselectivity, low S/C molar ratio, or narrow scope.

Aiming for a breakthrough on this subject, we conducted an extensive study with numerous experiments based on bold, speculative brain storming, resulting in an entirely new catalytic asymmetric transfer hydrogenation [270, 271]. In 1995, we first published a report on non-phosphine Ru(II) complex 1 created from [$RuCl_2(\eta^6$-arene$)_2$ (arene = *p*-cymene, mesitylene, etc.)] and (1*S*,2*S*)- or (1*R*,2*R*)-*N*-(*p*-toluenesulfonyl)-1,2-diphenylethylenediamine (TsDPEN); 1 acts as an excellent catalyst for the ATH of acetophenone derivatives giving 1-phenylethanols in up to 97 % *ee* (**Fig. 7.3.**, Eq. 1) [272]. The reaction occurs in 2-propanol containing a small amount of strong base. We were lucky enough to achieve this reaction but were unable to answer the question of why we used the arene ligand in this reaction. Arenes serve as a *fac* six-electron donor

Fig. 7.3.
ATH of carbonyl compounds.
Ru catalyst = [RuCl{(*S*,*S*)-tsdpen}(η^6-arene)]

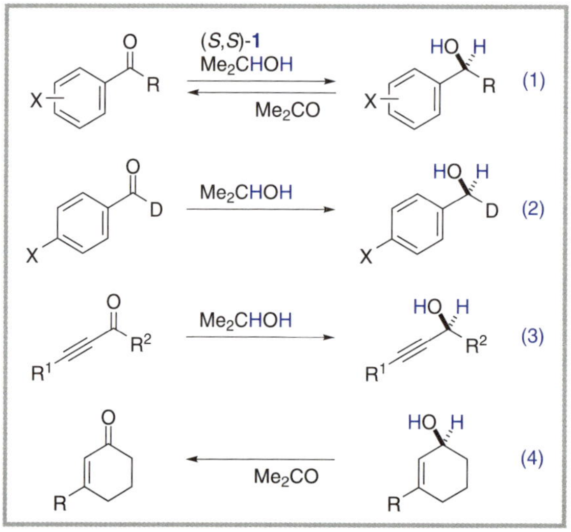

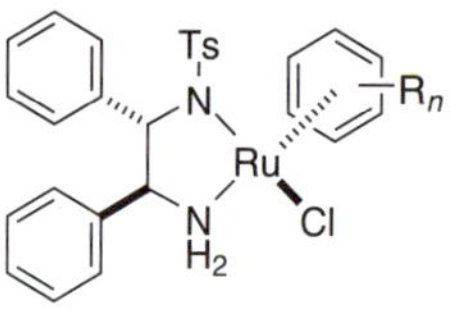

(*S*,*S*)-**1**

Kanizsa's triangle

in an octahedral complex and are synthetically flexible (the Kanizsa's triangle in **Fig. 7.3.**). We realized the importance of their convenient properties but were left without an appropriate scientific explanation; we were indifferent to their steric and electronic effects. In any event, the arene ligands remain on the Ru center throughout the reaction.

7.2.1 The discovery of an unimagined catalytic system

TsDPEN is a readily available, reliable chiral ligand. Likewise, the presence of chiral ethanolamines markedly accelerates ATH of aromatic ketones [273]. Then what is their kinetic function? Soon after this discovery, we found that well-shaped C_2-symmetric $P_2(NH)_2$–Ru(II)Cl_2 complex 2 (**Fig. 7.4.**) also serves as a similar catalyst for ATH [274]. Notably, dehydro P_2N_2–Ru complex 3 was almost inactive. Ru complexes 2 and 3 are spatially very similar (identified by X-ray analysis) but functionally distinct.

Fig. 7.4.
Active and inactive Ru catalysts.

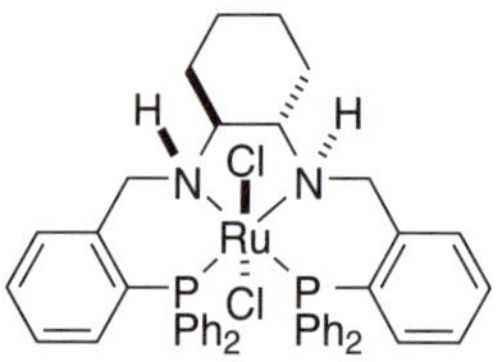

(*S*,*S*)-**2**, active

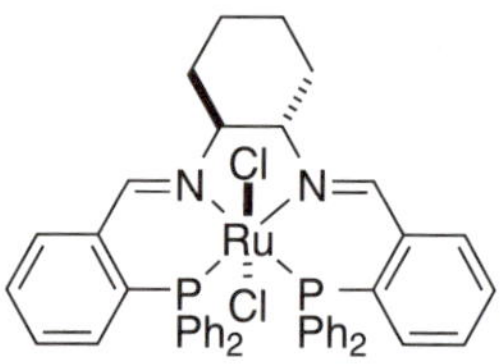

(*S*,*S*)-**3**, inactive

The NH effect

Despite having no concrete evidence in hand, I immediately recognized the significance of the NH function in this catalytic system. The neutral amine ligand is not a mere spectator but plays an active role. Separately, I learned of the origin of the anticancer activity of cisplatin, *cis*-$[PtCl_2(NH_3)_2]$, at the Fifth International Conference on the Chemistry of the Platinum Group Metals, of the Royal Society of Chemistry – Dalton Division held at St. Andrews in July 1993. I was invited by the chair of the Inorganic Chemistry division at that time, David J. Cole-Hamilton, to give a keynote lecture, "Asymmetric Catalysis: Science and Opportunities." Other keynote lecturers were Thomas J. Meyer (University of North Carolina), Jan Reedijk (Leiden University), Neville V. Richardson (University of Liverpool), Peter Maitlis (University of Sheffield), and Brice Bosnich (University of Chicago). In particular, I was impressed by Reedijk's lecture, "New Insights in the Structure-Activity Relationships and Mechanism of Action of Platinum Amine Compounds with Anti-tumour Properties" [B]. A free NH proton is not a strong hydrogen bond donor; however, when the nitrogen atom interacts with the Pt center, the NH group has increased acidity allowing for interaction with a phosphate moiety in DNA. As a consequence, the DNA double-stranded structure is disrupted, and the Pt center forms a complex with a guanine moiety. Thus, the Pt metal and protic neutral ligand cooperate in developing antitumor activity. It was believed that if the strength of hydrogen bonding is greater in the transition state than in the ground state molecules, then hydrogen bonding must contribute to the acceleration of chemical reactions. This intuitive mechanistic hypothesis was later proven true. Although this conference was in a different field to the conferences I usually attended, it was actually Reedijk's talk that gave me the insight and decisive conviction for the "NH Effect in Asymmetric Hydrogenation" that I discovered later after hearing his lecture.

The NH moiety is decisively responsible for reactivity in our system. We also devised a new diamine-Ir(I) complex catalyst **[275]**. Likewise, (pentamethylcyclopentadienyl)-Ir(III) and -Rh(III) complexes act as excellent chiral catalysts [276].

As illustrated in Eq. 2 in **Fig. 7.3.**, the Ru complexes of type 1 convert various substituted benzaldehydes to benzyl-1-*d* alcohols in 95 – 99 % *ee* with > 99 % isotopic purity [277]. α,β-Acetylenic ketones are reduced to chiral propargylic alcohols of high enantiomeric purity without saturation of the triple bond (Eq. 3) [278]. The chiral products are useful building blocks for the synthesis of various biologically active and structurally interesting compounds.

[275] S. Inoue, K. Nomura, S. Hashiguchi, R. Noyori, and Y. Izawa: Asymmetric Transfer Hydrogenation Catalyzed by Diamine-Iridium(I) Complexes. *Chem. Lett.* **1997**, *26*, 957 – 958.

Catalytic mechanism

Our question was: is substrate / metal complexation essential for hydrogenative saturation of unsaturated compounds? The answer we found:

no, it is not always necessary. In asymmetric catalysis using transition metal / chiral ligand complexes, chemists had simply considered that: (1) the metallic element generates reactivity, while the attached chiral ligand differentiates three-dimensional spaces; (2) reactants and substrates undergo inter-ligand reactions in the metal complex template. In truly efficient asymmetric catalysis, however, the metal and ligand interplay more strongly with each other to result in high reactivity and selectivity. In other words, all electronic and steric factors cooperate in generating sufficient chemical functions.

We elucidated that this transfer hydrogenation takes place on the chiral molecular surface of the coordinatively saturated metal hydride entity via the "metal-ligand bifunctional mechanism" [271]. A detailed mechanistic study taught us a series of non-classical aspects: (1) this ATH requires the NH function in the ligand to secure catalytic activity; (2) steric repulsion is not a sole factor for chiral discrimination and needs to be balanced with electronic attraction. In addition, the role of strong bases is unique; they do not facilitate Ru alkoxide formation.

Fortunately, we could identify both the actual catalyst and reactive intermediate by isolating Ru amide complex 5 and Ru hydride 6 in a crystalline state (**Fig. 7.5.**) [279]. Treatment of orange-colored precatalyst 4 with one equivalent of KOH affords a purple-colored complex 5 by elimination of HCl. X-ray analysis indicate that 5 has a formal 16e configuration with a square-planar structure, in which the amido-Ru bond, 1.897 Å, is much shorter than the normal amino-Ru dative bonds, 2.1 Å, due to their partial double-bond character. This purple complex reverts

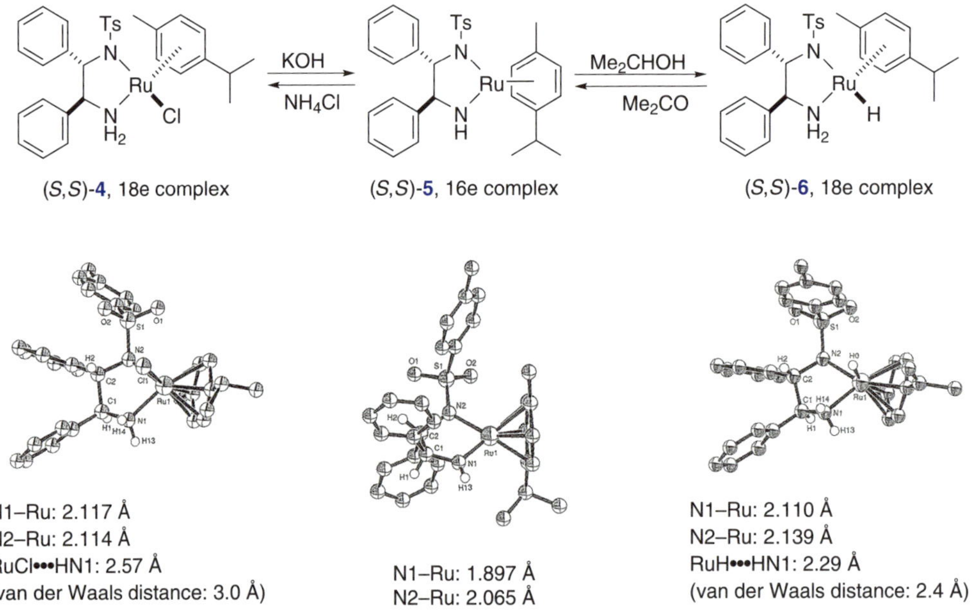

Fig. 7.5. Identification of precatalyst (*S*,*S*)-4 and the 16e and 18e actual catalysts, (*S*,*S*)-5 and (*S*,*S*)-6, and their respective X-ray crystal structures.

to the orange RuCl complex by the addition of ammonium chloride. Thus, a strong base is necessary to remove HCl from amino RuCl 4. Because of its characteristic bonding scheme, Ru complex 5 dehydrogenates 2-propanol to give yellow-orange Ru hydride 6. Notably, 6 forms cleanly *without any alkaline base* and returns to 5 by the action of acetone.

This mechanism is remarkably simple. Isolated 5 and 6 equally act as catalysts for the ATH reaction as illustrated in **Fig. 7.5.** The 16-electron Ru amide 5 dehydrogenates a secondary alcohol to form the 18e RuH species 6. This delivers two hydrogen atoms on Ru and nitrogen to a ketone directly to provide an alcohol and the 16e Ru amide catalyst 5, which is a resting state in the catalytic cycle. No other complexes that limit the turnover rate are involved. Dehydrogenation of 2-propanol is the turnover-limiting step and reduction of acetone is more feasible.

The mechanism is totally different from those presumed for the classical Meerwein-Ponndorf-Verley type reaction. A detailed *ab initio* MO and DFT (Density Functional Theory) study on a simple model, conducted by Masashi Yamakawa, a computation expert at Kinjo Gakuin University in Nagoya, indicated that transfer hydrogenation takes place via a six-membered pericyclic mechanism (**Fig. 7.6. a**). The reducing RuH species 7 having an NH ligand acts as a 1,4-dipole, which fits well with the carbonyl dipole. Thus, the hydride on Ru and the proton on nitrogen are simultaneously delivered to the C=O function, as we had long conceived. Most significantly, neither the oxygen on the alcohol nor on the ketone touches the Ru center of the catalyst during the reaction. This is also the transition state of dehydrogenation of an alcohol by the Ru amide complex.

Fig. 7.6. Calculated transition state of transfer hydrogenation. (a) The simplest formaldehyde to methanol conversion (MP2; Second order Møller-Plesset perturubation theory). Ru catalyst = [RuH(N-$H_2CH_2CH_2O$)(η^6-benzene)]. NPA (natural population analysis) charges (B3LYP) are given in blue. (b) The *S*-alcohol producing *Re* TS in ATH of acetophenone catalyzed by (*S*,*S*)-6. Generated by a global reaction route mapping (GRRM) method.

(a)

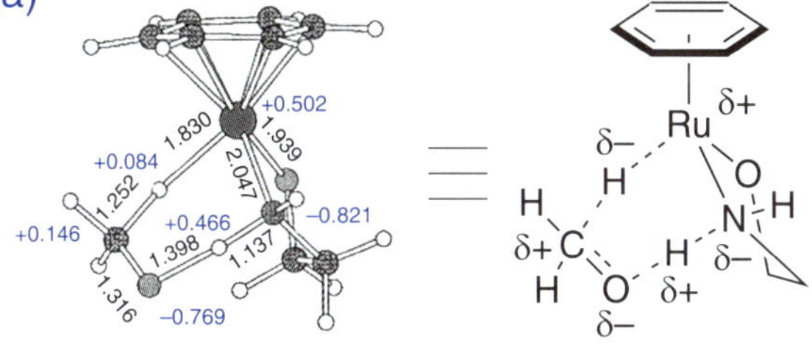

7

(b)

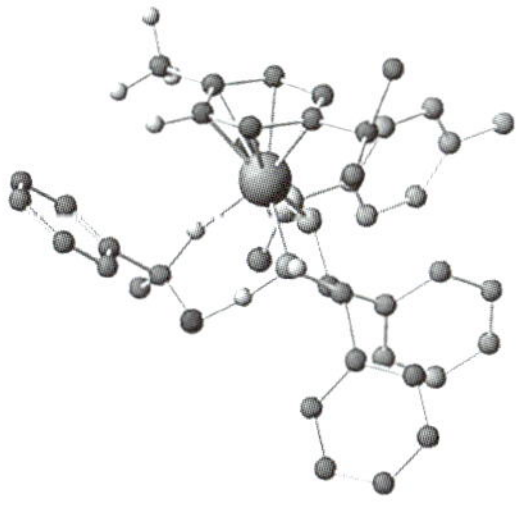

Re transition state

The forward and reverse reactions take place via low energy barriers. The unique nature of the NH function is predicted from the crystal structures of the 18e complexes (*S*,*S*)-4 and (*S*,*S*)-6 (**Fig. 7.5.**). The RuCl / HN and RuH / HN distances are shorter than the van der Waals distance, suggesting the presence of certain electrostatic interactions involving the $H^{\delta+}–N^{\delta-}$ dipole. This catalytic process is in sharp contrast to many other metal-centered reactions and explains the C=O *vs* C=C chemoselectivity.

The origin of the enantioselectivity achieved with Ru hydride (*S*,*S*)-6 was proven far beyond our imagination. Obviously, the chirality originates from the *S*,*S* configuration of the TsDPEN ligand (**Fig. 7.5.**). The two phenyl groups, oriented in the equatorial direction, form a skewed five-membered chelate ring in a δ-conformation, which in turn defines the *R* configuration of the newly formed Ru stereogenic center, identical to that of RuCl precatalyst (*S*,*S*)-4. *p*-Cymene-RuH catalyst (*S*,*S*)-6 hydrogenates acetophenone in 2-propanol to give (*S*)-1-phenylethanol in > 95 % *ee*.

At this point, I was surprised to learn that the η^6-*p*-cymene ligand plays a pivotal role in determining both the sense and degree of enantioselectivity. First, the observed asymmetric correlation between the *S*,*S* catalyst 6 and *S* alcoholic product indicates that the sterically more congested *Re* transition state is favored over the less crowded *Si* transition state (**Fig. 7.6. b**). Secondly, this disubstituted benzene ligand behaves better than the parent benzene. This is due to the presence of multiple CH/π attractions between the cymene CH and the phenyl ring of acetophenone. These were proven by detailed theoretical calculations [271, 280, 281]. The *Re* transition state is favored over the most stable *Si* transition state by 11.8 kJ mol^{-1}. The electron flow from the cymene ligand to the ketone phenyl ring via the transient Ru–H–C linkage results in an increase in acidity of the cymene $C(sp^2)$–H and also the electron density of the *ortho* and *meta* carbons in the reacting acetophenone. As such, the *p*-cymene or mesitylene Ru catalysts perform better than simpler benzene or fully alkylated hexamethylbenzene complexes. Consistent with this view, the η^6-mesitylene / (*S*,*S*)-TsDPEN Ru complex catalyzes the ATH of skeletally symmetrical 4-cyano-4'-methoxybenzophenone to give the corresponding (*S*)-diphenylmethanol in 66 % *ee*! [282] The overall lesson we learned is that ligand substituents are present not only for enhancing non-bonding repulsions but also for exerting electrostatic attractions to stabilize transition state structures.

7.2.2 An ideal asymmetric transfer hydrogenation of ketones and imines

Similar to the *in vivo* system (**Fig. 7.2.**), this chemical redox process is reversible. 2-Propanol is a much better *in vitro* cofactor than NADH for carbonyl reduction, while acetone in turn acts as a much less expensive cofactor (**Fig. 7.3.**, Eq. 1 and 4) than NAD, which is essential for the transformation with oxidoreductases, such as horse liver alcohol dehydrogenase. This property can be utilized for kinetic resolution (a convenient resolution method without derivatization) of certain alcohols with a high reduction potential simply by the use of acetone as a hydrogen acceptor [283]. This procedure is much more appealing when applied to the meso substrates, because the theoretical yield of the chiral desymmetrized product is 100 % rather than 50 %, as demonstrated by access to the intermediate compound for pharmaceuticals ML-236A and FK-506.

This reversibility, however, causes a serious problem as well. Hydrogen-donating 2-propanol has a similar structure to secondary alcohol products. Therefore, unlike in ordinary one-direction catalysis, the *ee* value tends to decrease as the conversion increases due to contamination from the reverse process. The problem was clearly solved by using formic acid as a hydrogen donor, which is viewed as an adduct of H_2 and very stable CO_2 [282]. Since oxidation of alcohols with CO_2 is impossible, the reaction proceeds irreversibly under truly kinetic control. Acetophenones with an electron-releasing substituent or benzo-fused cyclic ketones, both known as difficult substrates, provide alcohols in > 97 % *ee* consistently. The reaction can be conducted in a high substrate concentration.

This process is highly practical and applicable on an industrial scale. **Fig. 7.7.** illustrates some applications. This method with the *R*,*R* catalyst allows for the selective synthesis of (*R*)-8 or (*R*)-9 and (*R*)-10, which serves as intermediates of MK-0417 (carbonic anhydrase inhibitor) and L-699,392 (LTD_4 antagonist), respectively. This ATH is applied for the reduction of *m*-trifluoromethylacetophenone to (*S*)-11 using an S/C ratio of 5,000 : 1 and at a scale, which is industrially feasible. The *S* alcohol is an important intermediate for (*S*)-MA-20565 ((*S*)-12), an agrochemical fungicide with a wide spectrum. The double reduction of benzils, coupled with dynamic kinetic resolution of the intermediary benzoins, provides a practical way to synthesize enantiomerically pure hydrobenzoins [284, 285].

Furthermore, the invention of the chiral TsDPEN-Ru(II) catalyst of type 4 realized, for the first time, a highly efficient asymmetric reduction of imines with an S/C ratio of up to 1,000 : 1 (**Fig. 7.8.**) [286]. A variety of natural and unnatural isoquinoline and indole alkaloids were synthesized. This reaction is applicable to the synthesis of an intermediate for MK-0417 and certain chiral indoles. Interestingly, the reaction in aprotic polar solvents such as DMF, DMSO, MeCN, or CH_2Cl_2, exhibits a remarkable C=N / C=O selectivity as high as 1000 : 1.

The saturation occurs selectively at a keto or imino group without affecting olefinic linkages or many functional groups such as esters, sulfides, sulfones, nitro groups, aryl chloride, cyanide, and many heteroaromatics.

Fig. 7.7. ATH of ketones with a 5:2 HCO_2H/NEt_3 mixture. Ru catalyst = [RuCl{(*S*,*S* or *R*,*R*)-tsdpen}(η^6-mesitylene)].

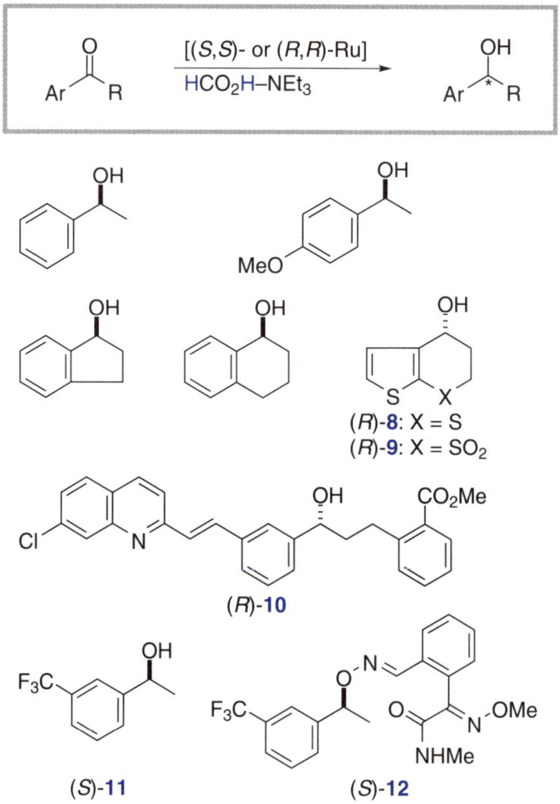

Fig. 7.8. ATH of imines in a 5:2 HCO_2H/NEt_3 mixture. Rucatalyst = [RuCl{(*R*,*R*)-tsdpen}(η^6-benzene or *p*-cymene)].

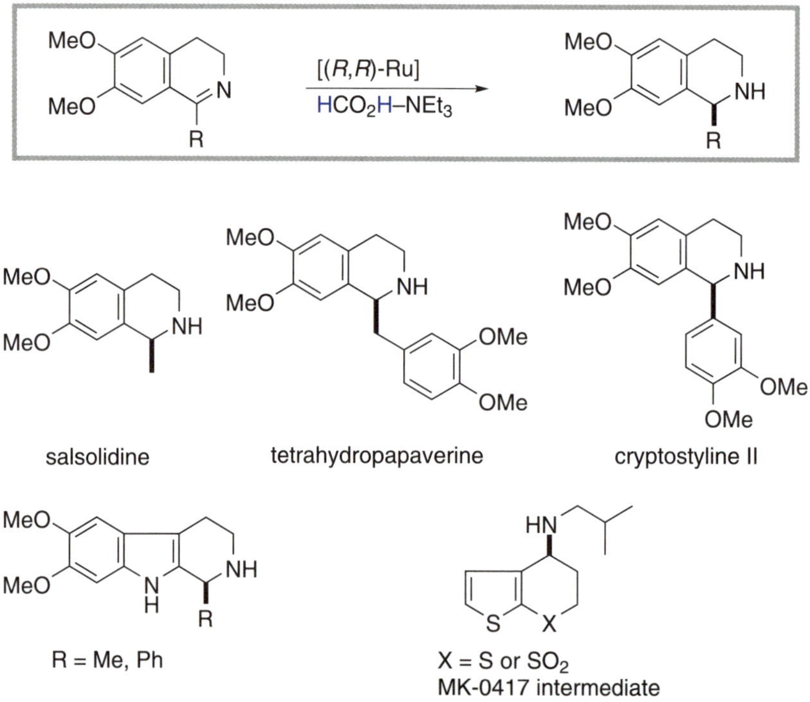

7.2.3 Transfer hydrogenation / hydrogenation network

This story continues further. The "atom economy" of transfer hydrogenation is greatly improved in going from the use of NADH to 2-propanol and now to formic acid. However, the simplest molecule H_2 is the ideal reducing agent in view of achieving 100 % atom economy. Unfortunately, RuCl complex 4 was not applicable as a precatalyst for this purpose. Still, however, I imagined that transfer hydrogenation and hydrogenation are mechanistically linked, because both reactions involve a common Ru hydride intermediate.

Going back to the original process in 2-propanol, shown simply by black arrows in **Fig. 7.9.**, a strong base eliminates HCl from amino RuCl complex A (X = Cl) to form Ru amide B, which leads to transfer hydrogenation in 2-propanol via Ru hydride C. Unfortunately, H_2 gas cannot be used as a reducing agent under such conditions. This scheme, however, suggests the possibility of a reaction using the same catalyst A and H_2 simply by switching the conditions from basic to perhaps neutral or slightly acidic [287, 288]. As shown in blue arrows, the ionization of the Ru–X bond in A forms the 16e cationic Ru species D (solvate), which reversibly accommodates H_2 forming the η^2-H_2 Ru complex E. Subsequent deprotonation gives the same RuH C which hydrogenates a ketonic substrate. The resulting Ru amide B cannot be protonated by pure alcohols but is protonated under slightly acidic conditions to go back to amino complex D, completing the hydrogenation cycle.

The key steps in this mechanism-based scenario are the efficient generation of cationic Ru species D from A and the protonation from B to generate D. Obviously, alcoholic solvents with a high dielectric constant

Fig. 7.9. Transfer hydrogenation / hydrogenation network. Basic (black) and acidic (blue) conditions. X = Cl, OTf.

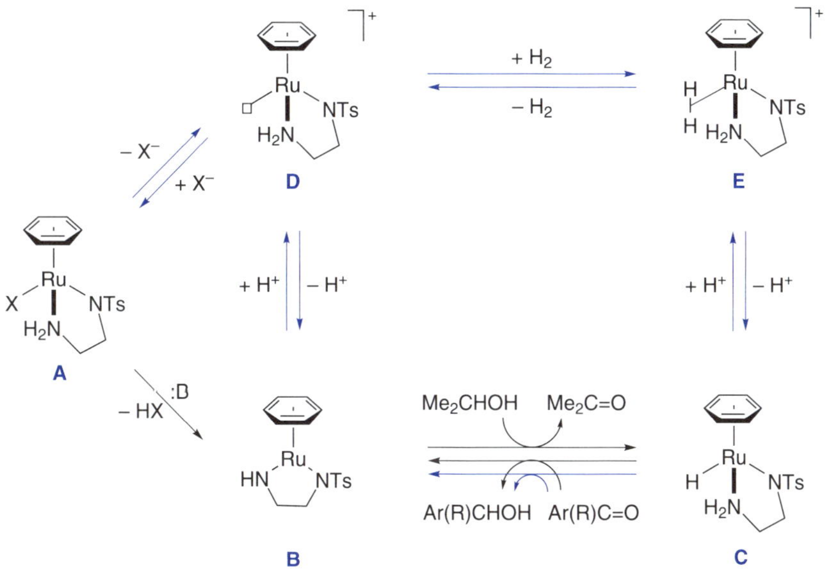

are preferred and the chloride is to be replaced by a triflate. We tested this attractive possibility by using 4-chromanone whose AH had remained difficult because of the cyclic planar structure and the relatively high acidity of $C(3)H_2$ caused by the oxygen atom. In fact, as shown in **Fig. 7.10.**, when a 2.4 kg scale hydrogenation of 4-chromanone was conducted in methanol containing a chiral, highly ionizable Ru triflate catalyst (S/C up to 7000) and a trace amount of triflic acid, the *S* alcohol was obtained in 98 % *ee*. Various base-sensitive ketones including α-chloro aromatic ketones [289] can be reduced smoothly by this new AH method.

Enantioselective Ru chemistry has displayed a range of aspects that are not recognized in conventional synthetic chemistry. Designing a practical asymmetric catalyst needs an integrated molecular approach. The desired molecular function emerges by combining various steric and electronic properties together with appropriate reaction parameters. Most importantly, we first observed the presence of an effective hydrogenation / transfer hydrogenation network. This notion not only provided a new mechanistic paradigm but also greatly expanded the scope of hydrogenation. We realized a range of catalytic asymmetric transfer hydrogenations (ATH) using inexpensive, stable organic donors. After completion of the ERATO Noyori project in 1996, this chemistry was further extended by others, including Takao Ikariya, who was then appointed Professor of the Tokyo Institute of Technology.

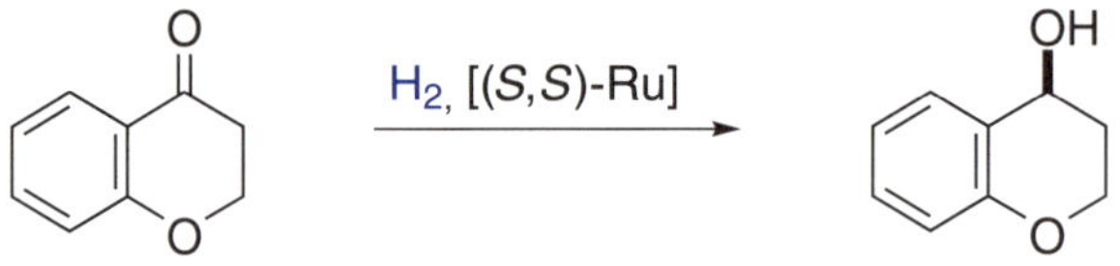

Fig. 7.10. AH of 4-chromanone. Ru catalyst = [RuOTf{(*S*,*S*)-tsdpen}(η^6-*p*-cymene)] (9.6 g). Reaction conditions: methanol (8 L), 60 °C, 17 atm, 8 h.

7.3 Advent of Ru-BINAP / diamine catalysts for AH (1995 –)

Despite the high utility of ATH in laboratories and medium-sized manufacturing, we had been inclined to develop asymmetric hydrogenation (AH) for full use by extending the capability of BINAP-Ru catalysts [290].

The simplest way is not always the easiest. In view of the superb efficiency of the BINAP-$RuCl_2$ catalyst in AH of β-keto esters (CHAPTER 6), one might expect that the same catalyst is effective for AH of simple ketones as well. But this is not true at all. This standard BINAP-Ru catalyst smoothly hydrogenates methyl 3-oxobutyrate in *aqueous acetone*, giving methyl 3-hydroxybutyrate in 99 % *ee*. The simplest ketone used as a solvent is almost inert to hydrogenation, while the structurally more elaborate keto substrate is easily hydrogenated.

This difference is distinct and whipped up my scientific passion. A hard-working collaborator in this regard was Takeshi Ohkuma, (now Professor at Hokkaido University) who joined our project after finishing his postdoctoral work with Paul Wender at Stanford University. Although our endeavors clearly solved this problem in 1995 [291] by applying our original "metal-ligand bifunctional mechanism," I would like to first describe a long-term struggle that I experienced in the 1980s [253].

7.3.1 My personal view on mechanistic complexity

Without doubt, Ru(II) complexes readily form metal hydrides (MH) upon interaction with H_2 (**Fig. 6.31.**). I conceived that the real problem is the strong Lewis acidity of the d^6 Ru center, a carbocation analogue in my view, which tends to form a σ complex with the C=O function rather than a π complex. In this Lewis acid complex, the relative locations of the nucleophilic H and electrophilic carbon center are inappropriate. The requisite interaction between the Ru–H bond and π-face of the C=O group is achieved only through drastic change of the ground-state σ structure. The desired ground-state π complexation is feasible only with more electropositive transition metals with a low-lying LUMO, which facilitates the metal to substrate backbonding. Furthermore, the resulting metal alkoxide intermediate may undergo reverse β-elimination unless some subsequent process rapidly cleaves the M–O bond. Although most textbooks state that the reaction between C=O compounds and metal alkyls or hydrides occur in a stoichiometry of 1 : 1, the simple [$\pi2 + \sigma2$] process may be kinetically unfavorable. Instead, many organometallic reactions are considered to involve two metallic centers in the transition state (TS), in which one metal carries a nucleophile and the other interacts with the carbonyl oxygen atom, enhancing the nucleophilicity of the carbon atom, as is seen in our organozinc chemistry (**CHAPTER 5**).

Thus, we thought that protonation or hydrogen bonding to the carbonyl oxygen atom would also facilitate π-face attack of a nucleophilic on the metal center. Unfortunately, unlike for the reaction of functionalized ketones such as β-keto esters (**Fig. 6.42.**), addition of strong acids does not assist hydrogenation of acetophenone with $[RuCl_2(binap)(dmf)_n]$, because both the protonated ketone and the RuH species are present in very low concentrations in the catalytic system. Then we envisaged that, if the anionic ligand on a metal, for example X in $[RuHX(binap)(solv)_2]$, is protic, then it might facilitate the hydride delivery through the assistance of ligand / substrate hydrogen bonding. Furthermore, unlike the pathway involving metal alkoxide intermediates, such a process would produce alcohol directly without any M–O bond cleavage reactions. The candidates for X include a conjugate base of hydroxy carboxylic acids, dicarboxylic acids, phosphoric acid, phosphonic acids, etc. Young Makoto Tokunaga , now a Professor at Kyushu University, conducted more than

200 experiments along this line, and the best result was obtained with a catalyst prepared *in situ* from [Ru(OCOMe)$_2${(*R*)-xylbinap}] and phenylphosphonic acid in a 1:2 ratio (xyl = 3,5-xylyl). This elaborate system catalyzed AH of acetophenone in CH_2Cl_2 (S/C = 700, 100 atm, 100 °C, 18 h) to give (*S*)-1-phenylethanol in > 99 % yield and 78 % *ee* (M. Tokunaga, Dissertation, Nagoya University, 1995). Regretfully, the result did not meet the standard of modern organic synthesis.

7.3.2 Discovery of high reactivity in 1995

Research is not a routine business. Deep passion is needed because failure often provides an opportunity for a new discovery. The key factors for the reactivity of β-keto esters originate from the presence of a functional group and acidic conditions, while we now know that hydrogenation of simple, unfunctionalized ketones needs basic conditions.

During our search for a reactive catalyst system for hydrogenation and transfer hydrogenation using [$RuCl_2(PPh_3)_3$], 2-propanol, and KOH, we found an interesting additive effect (**Fig. 7.11.**). Acetophenone is reduced with a turnover frequency (TOF) of 70 h^{-1} even in the absence of hydrogen gas. The addition of one equivalent of ethylenediamine retarded this transfer hydrogenation. This effect surprised us because structurally similar ethanolamine and *N*-tosylethylenediamine significantly facilitate the transfer hydrogenation in the presence of (η^6-arene)Ru(II) chloride complexes (**Fig. 7.3.**). However, we had to be

Fig. 7.11. Rate of reduction of acetophenone catalyzed by [$RuCl_2(PPh_3)_3$] in 2-propanol.

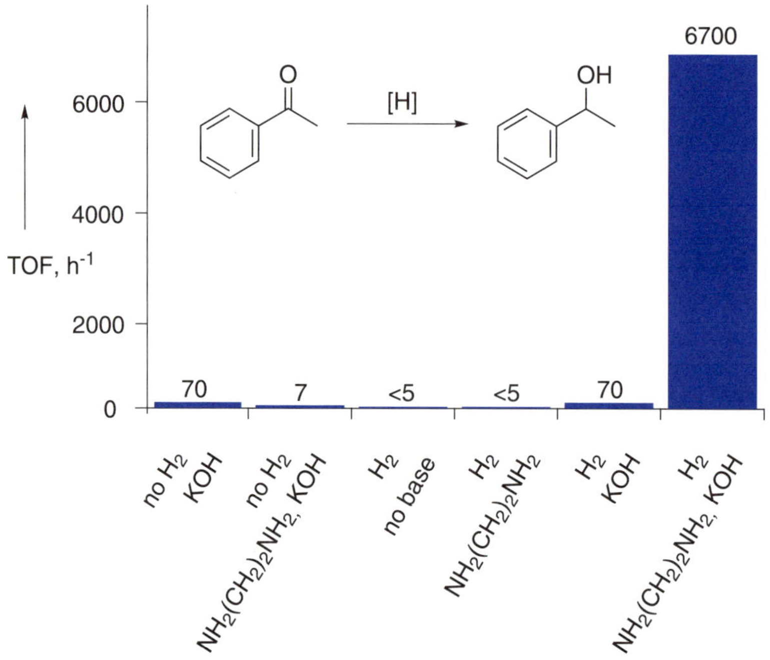

careful. Most significantly, ethylenediamine together with KOH greatly accelerates hydrogenation with molecular hydrogen at 3 atm, giving a TOF number of 6,700 h^{-1}. This is a net hydrogenation as proved by the reaction using Me_2DCOH. What is crucial is the presence of both the inorganic base and NH-bearing organic base, instead of conventional *N,N,N′,N′*-tetramethylethylenediamine (TMEDA). This new catalyst system does not catalyze hydrogenation of enolizable β-keto esters.

The NH effect is to be utilized for asymmetric hydrogenation (AH). As illustrated in **Fig. 7.12.**, we devised various chiral [$RuCl_2$(diphosphine)(1,2-diamine)] complexes which serve as excellent precatalysts for AH of simple ketones [253]. The use of BINAP and a suitably shaped chiral 1,2-diamine allows for practical asymmetric hydrogenation of various ketones in 2-propanol containing alkaline bases such as KOH, KO^iPr, and KO^tBu. The chiral environment is highly tunable, and the asymmetric catalysts can be immobilized by binding them to a polystyrene resin [292].

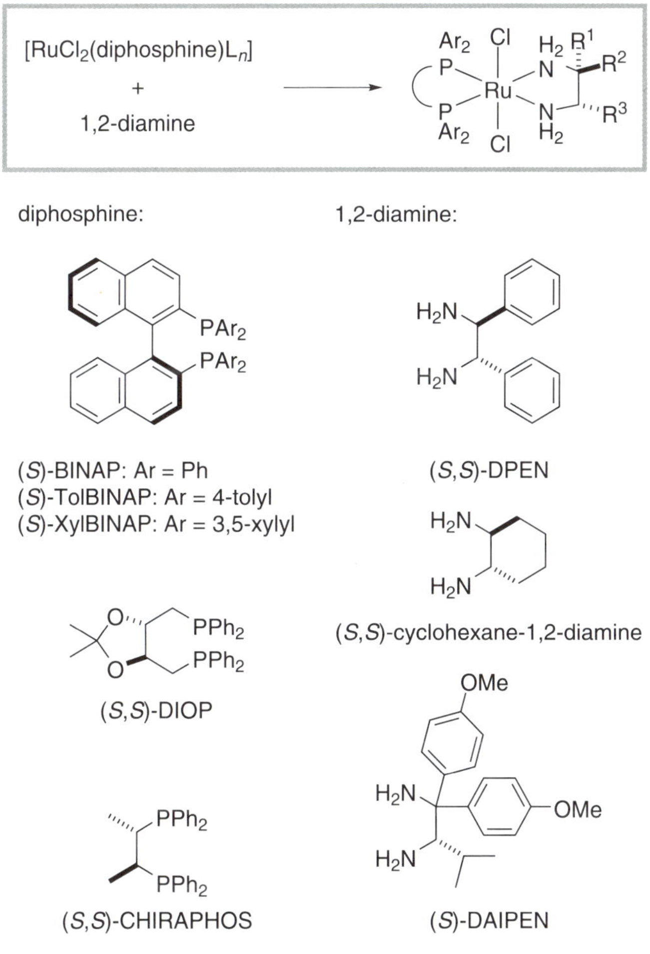

Fig. 7.12. Synthesis of shelf-stable asymmetric catalysts.

7.3.3 General AH of simple ketones and its mechanism

AH of aromatic and hetero-aromatic ketones

A catalyst system consisting of [$RuCl_2$((*S*)-binap)(dmf)$_n$], (*S*,*S*)-DPEN, and KOH in 2-propanol enables AH of various alkyl aryl ketones with high enantioselectivity (**Fig. 7.13.**). The high efficiency is a result of the synergetic effects of chiral diphosphine and diamine. Under selected conditions, AH of acetophenone (601 g) with the Ru catalyst (2.2 mg) proceeds in 80 % enantioselectivity and with a TON of at least 2,400,000 and a TOF at 30 % conversion of 228,000 h^{-1} or 63 s^{-1}. This high rate appears to outperform many enzymes. The enantioselectivity is greatly improved to 99 % by combining sterically more demanding (*S*)-Xyl-BINAP and (*S*)-DAIPEN (or *R* and *R*) with little decrease in reactivity. This simple AH was conducted with an S/C ratio of 100,000 : 1 to give various ring-substituted 1-phenylethanols with high *ee* values.

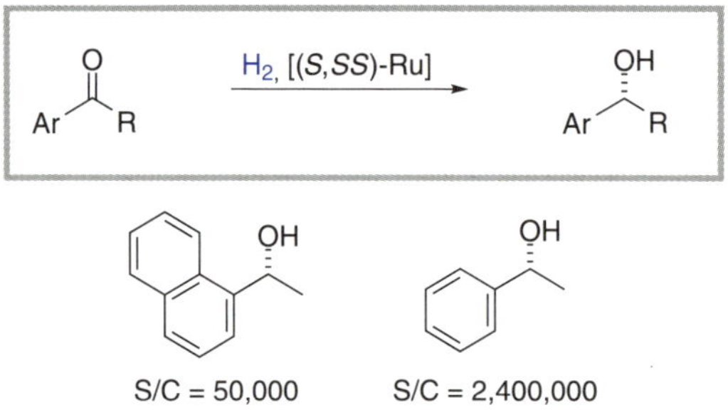

Fig. 7.13. Rapid, productive, and enantioselective hydrogenation of aromatic ketones. Ru catalyst = [$RuCl_2${(*S*)-tolbinap}{(*S*,*S*)-dpen}]. Reaction conditions: ^{t}BuOK in 2-propanol.

The (*R*)-1-(1-naphthyl)ethanol product was used for desymmetrization of prochiral 3-substituted glutaric anhydrides giving chiral building blocks for the synthesis of (+)-compactin, an HMG-CoA-reductase inhibitor, and 1233A, a hypocholesterolemic agent.

This AH tolerates many substituents, including halogens, CF_3, alkoxyls, esters, amides, NO_2, and NH_2 [293]. The utility is shown in **Fig. 7.14.** AH of *ortho*-substituted benzophenones proceeded with an

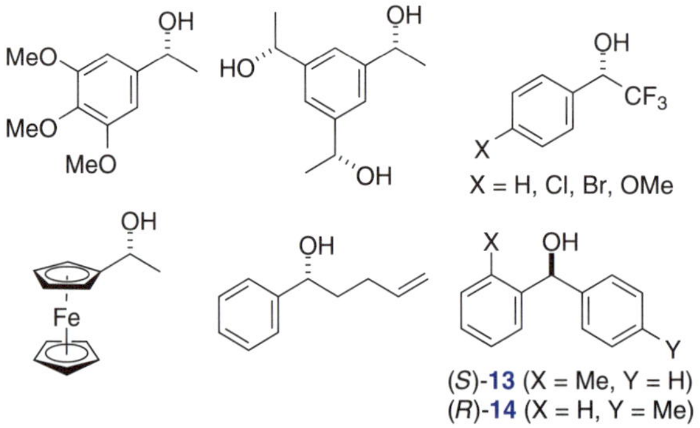

Fig. 7.14. Some chiral alcohols obtained by AH with (*S*,*SS*)-(diphosphine)-(diamine)Ru catalysts. Conditions: ^{t}BuOK in 2-propanol.

S/C ratio of up to 20,000:1 and in high enantioselectivity, allowing for the synthesis of (*S*)-13 and (*R*)-14, which are precursors of orphenadrin (antihistaminic and anticholinergic) and neobenodine (antihistaminic), respectively [294]. In this AH, *ortho* halogens are regarded as specially enantio-directive, protective substituents which are easily removable from the chiral products.

Cyclopropyl phenyl ketones provide the chiral alcohol in high *ee* with an opposite sense of asymmetric induction, implying the presence of a unique electronic effect of the three-membered ring in the transition state [293].

Under this new standard condition, various ketones possessing an electron-rich or -deficient heterocyclic substituent can be hydrogenated in very high enantioselectivity (**Fig. 7.15.**) [295]. The asymmetric nature is identical with normal aromatic ketones. Sulfur-containing heterocycles such as thiophene and thiazole do not hinder the catalytic activity.

Fig. 7.15. AH of heteroaromatic ketones, Ru catalyst = *trans*-[$RuCl_2${(*S*)-xylbinap}{(*S*)-daipen}]. Reaction conditions: ^{t}BuOK in 2-propanol.

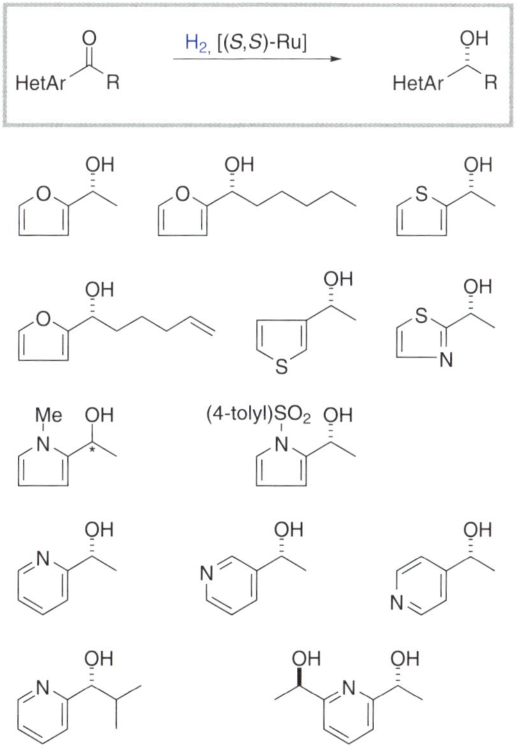

AH of amino ketones

Importantly, the XylBINAP / DAIPEN-Ru complex appears to act as an efficient catalyst for AH of α-amino ketones (**Fig. 7.16.**) [296]. Since the substrates behave like simple ketones, in going from methyl ketones to phenyl ketones (Eq. 1 vs. Eq. 2), asymmetric induction is reversed

Fig. 7.16.
AH of α-amino ketones.
Ru catalyst = trans-[$RuCl_2$\{(*R*)-xylbinap\}\{(*R*)-daipen\}]. Conditions: ${}^tBuOK$ in 2-propanol.

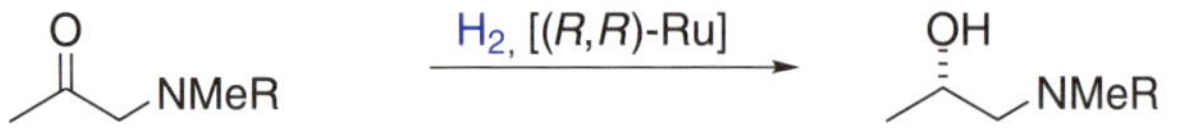

(1)

R	*ee*, %
Me	92
Ph	81

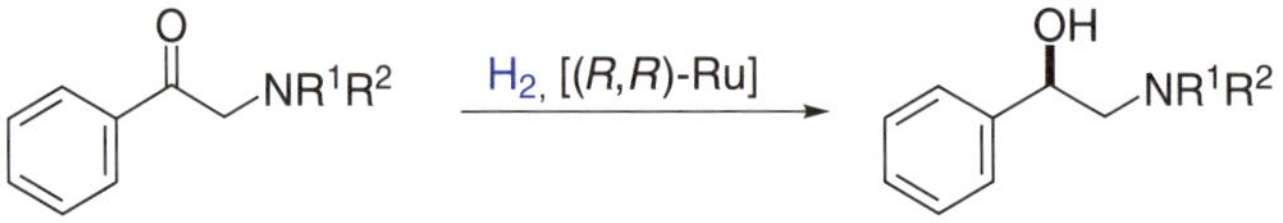

(2)

R^1	R^2	*ee*, %	R^1	R^2	*ee*, %
Me	Me	93	COPh	Me	99.8
COMe	Me	99	$CO_2{}^tBu$	Me	99
COPh	H	95			

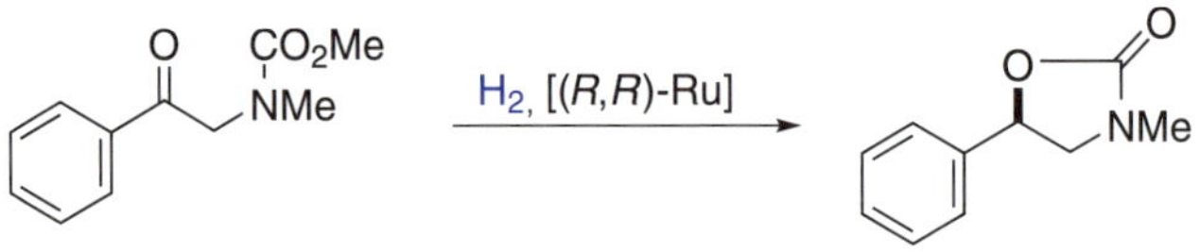

(3)

99% *ee*

from *R*,*R*/*S* to *R*,*R*/*R*. This sense contrasts with that observed with the diamine-free (*R*)-BINAP-Ru catalyst consistently showing the nitrogen-directed *R*/*R* asymmetric induction (**CHAPTER 6**, **Fig. 6.40.**). Certain β- or γ-amino ketones are also used as substrates. α-Alkoxy and α,α-dialkoxy ketones behave in a similar manner.

The applicability of this AH is obvious. **Fig. 7.17.** exemplifies the synthesis of some pharmaceutically important chiral compounds, including denopamine hydrochloride (β_1-receptor agonist) [(*R*)-15], fluoxetine

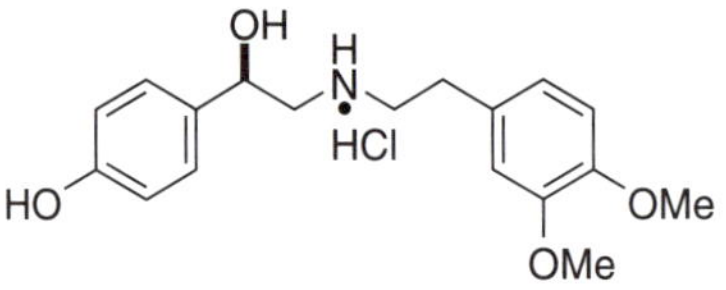

(*R*)-**15**

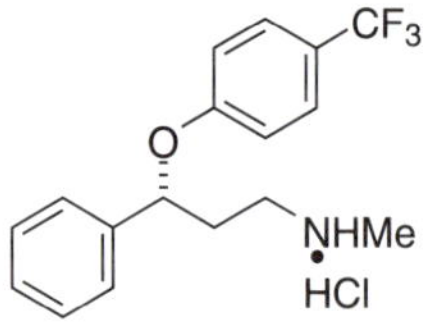

(*R*)-**16**

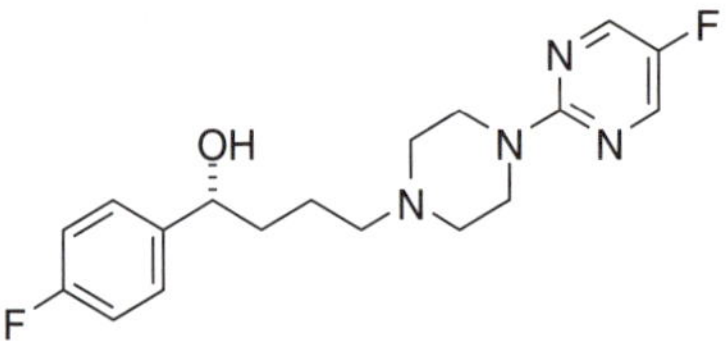

(*R*)-**17**

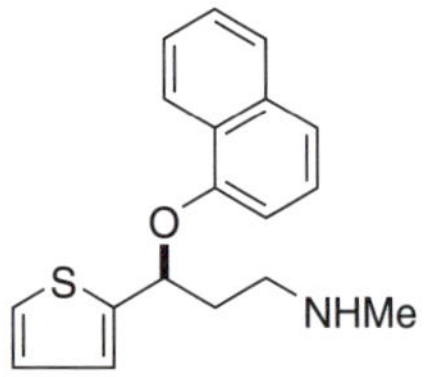

(*S*)-**18**

Fig. 7.17.
Synthetic application of AH of amino ketones.

hydrochloride (antidepressant) [(*R*)-16], BMS 1811100 (antipsychotic) [(*R*)-17], duloxetine (serotonin and norepinephrine inhibitor) [(*R*)-18]. Eli Lilly has sold the racemic hydrochloride as Prozac, an antidepressant, with an annual market value of $2.6 billion (1997), although enantiopure (*R*)-16 thus prepared is more potent.

AH of conjugated enones

Efficient AH of α,β-unsaturated ketones remained an enduring synthetic problem and thus was our cardinal focus. Chiral allylic alcohols are important not only for their own sake but also in connection with the Claisen reaction and various S_N2' substitutions. The ideal stereoselective hydrogenation of alkenyl ketones, however, was difficult because of their conformational flexibility and the presence of two kinds of unsaturated groups, C=C and C=O, the former having higher reactivity in conventional hydrogenation. Furthermore, some simple enones are highly sensitive to bases, hampering the utility of our AH reaction. We thus had to avoid the use of strong bases for this reason.

This pending issue was eventually solved in 1998, when we devised the (*S*)-XylBINAP / (*S*)-DAIPEN-Ru catalyst coupled with the use of K_2CO_3, a simple weak base as a cocatalyst (**Fig. 7.18.**) [293]. This combination significantly widened the synthetic scope. Thus, benzylideneacetone,

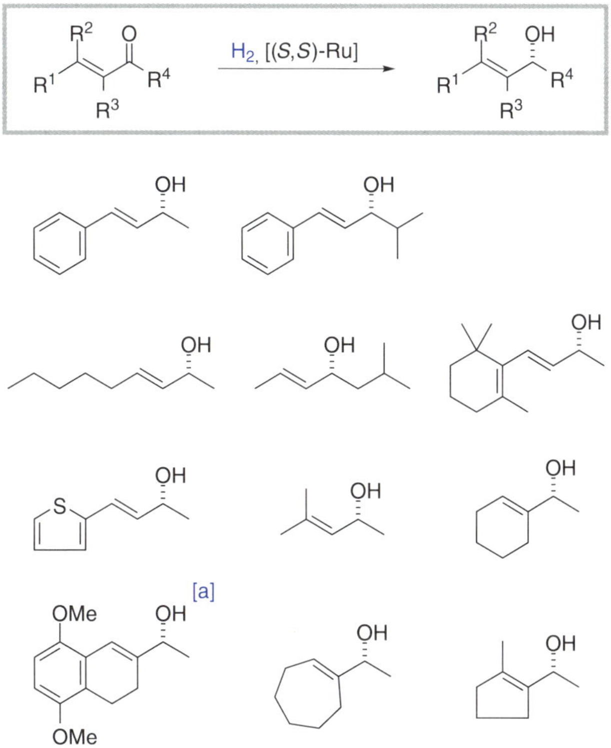

Fig. 7.18. AH of α,β-olefinic ketones. Ru catalyst = *trans*-[RuCl$_2${(*S*)-xylbinap}{(*S*)-daipen}]. Conditions: K_2CO_3 or ^{t}BuOK in 2-propanol. [a] The *R* isomer obtained by the (*R*,*R*)-Ru catalyst is a precursor of anthracycline antibiotics.

a notorious substrate, is hydrogenated in 2-propanol with an S/C ratio of 100,000:1 to give the *R* alcohol in 97% *ee*. The absence of any trace of C=C reduced products is noteworthy, because our diamine-free BINAP-Ru$(OCOR)_2$ catalyst exhibits an eminent activity for AH of the C=C unit of allyl alcohols such as geraniol and nerol (**Fig. 6.33.**). These BINAP-Ru hydrogenation chemistries are thus totally different.

More substituted, less base-sensitive enones can be hydrogenated more rapidly and conveniently by using a strong-base cocatalyst, giving chiral allylic alcohols in 90–100% *ee*. The *S* alcohol formed from (*E*)-6-methyl-2-hepten-4-one is a key building block for the side chain of α-tocopherol (vitamin E). 1-Acetylcycloalkenes were hydrogenated to 1-hydroxyethylcycloalkenes, allowing asymmetric synthesis of a key intermediate for anthracycline antibiotics, adriamycin and daunomycin. β-Ionone, a conjugated dienone, is convertible to β-ionol. Here the combination of the *S* diphosphine and *S* diamine (or *R* and *R*) is crucial for obtaining a high level of enantio-selection, as was observed in AH of aromatic ketones. Differentiation between aromatic and vinyl groups in chalcone was difficult. Simple vinyl ketones appear to undergo both 1,2- and 1,4-hydrogenation competitively.

The entirely base-free AH procedure remained elusive, although the use of K_2CO_3 as a cocatalyst partially solved the problem. In 2002, we eventually devised highly reactive *trans*-[RuH(η^1-BH_4)(binap)(1,2-diamine)], which allows AH without any basic additives [297]. This new method occurring with an S/C ratio of 100,000 in a 3-4 M substrate concentration in 2-propanol is rapid, productive, and enantioselective, and further expanded the scope of AH to base-sensitive substrates such as epoxy ketones and β-dialkylamino ketones.

The high rate of the reaction and carbonyl selectivity of this AH are based on non-classical metal-ligand bifunctional catalysis, which utilizes the "NH effect," as already seen in (η^6-arene)Ru(Ts-dpen)X-catalyzed ATH [298]. We conducted detailed kinetic, structural, and deuterium-labeling studies on AH of acetophenone in 2-propanol containing *trans*-[RuH($^1\eta$-BH_4){(*S*)-tolbinap}{(*S*,*S*)-dpen}] [299]. The result indicated the operation of the catalytic cycle of **Fig. 7.19.a** [300, 301] [C]. The true catalyst is B which is generated via simple ligand dissociation of precatalyst A (X, Y = H, BH_4) or by reaction of A (X, Y = Cl) and a strong base in 2-propanol under conventional conditions. The cationic 16e species B (detected by ESI-TOF-MS and NMR) which is in equilibrium with E is converted to C and then D, which in turn reacts with a ketonic substrate to form E and an alcohol product. The 16e Ru amide E returns mostly to B by protonation, but gives some D due to the reaction with H_2. The reducing 18e Ru dihydride D possesses a hydride and two nitrogen atoms with a *fac* relationship in octahedral geometry. When chiral (*R*)-BINAP and (*R*,*R*)-DPEN ligand are attached to Ru, two hydrogen atoms are asymmetrically activated as shown in D as hydridic RuH and protic NH_{ax}. Most importantly, this coordinatively saturated 18e complex reacts with a simple ketone in its outer coordination sphere, where neither a ketone-Ru nor an alkoxy-Ru complex is involved. The chiral alcohol forms directly from the ketone, and this is the major origin of the high rate and uncommon functional selectivity. This mechanistic scheme is distinct from the conventional catalytic model occurring in the metallic template, as shown in **Fig. 7.19. b** The hydrogenative reactivity originates from the charge-alternating $H^{\delta-}$–$Ru^{\delta+}$–$N^{\delta-}$–$H^{\delta+}$ arrangement. The six-membered, pericyclic transition

Fig. 7.19. Metal-ligand bifunctional mechanism in the Ru-catalyzed hydrogenation of ketones. (a) Catalytic cycle. X, Y = H, BH_4 or Cl, Cl. (b) Reactive RuH_2 species (D, top view) and cyclic TS (F).

(a)

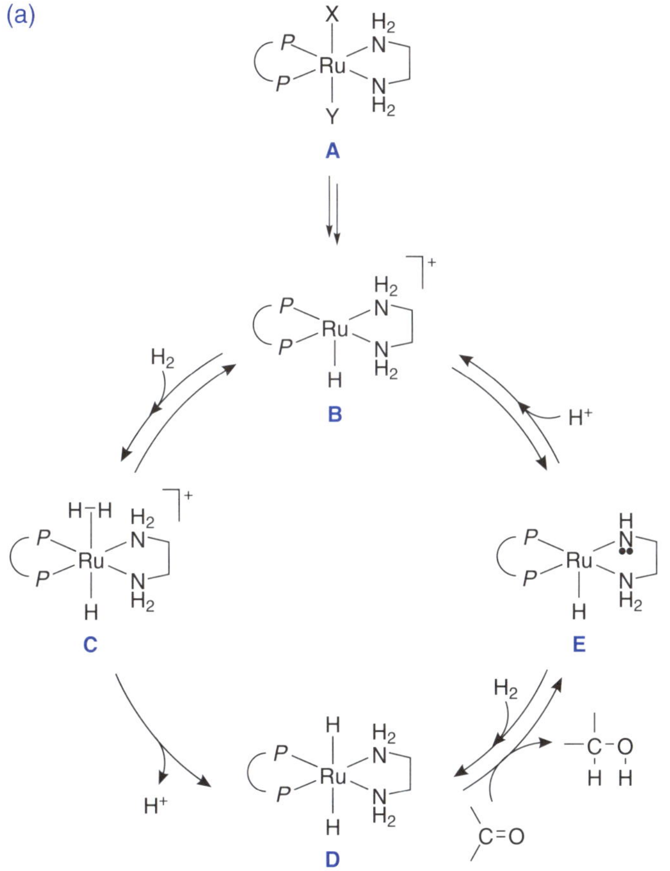

(b)

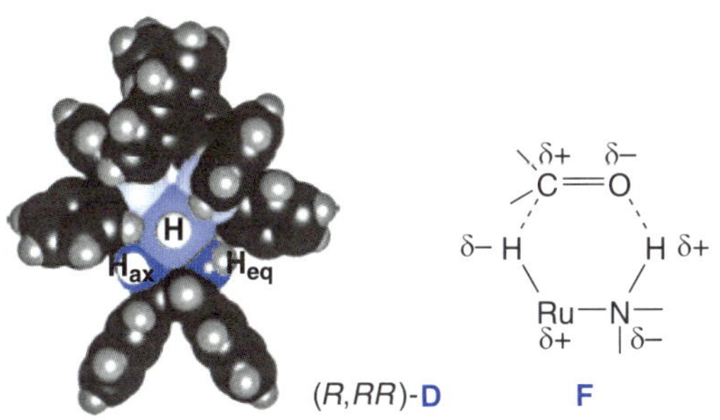

state (TS) is schematically shown by F. This reaction proceeds without any base but is further accelerated by a basic additive. 2-Propanol is the best solvent. This hydrogen donor is considered to assist the recovery of the catalytically active B from the inactive Ru sink that is formed, if any.

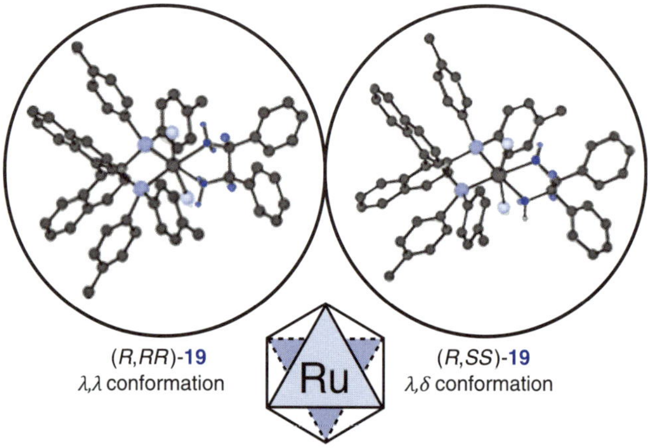

Fig. 7.20. Molecular structures of *trans*-[$RuCl_2${(*R*)-tolbinap}{(*R*,*R*)-dpen}] and *trans*-[$RuCl_2${(*R*)-tolbinap}{(*S*,*S*)-dpen}].
Dark blue: N, medium blue: P, light blue: Cl. (Color figure available in the supporting material. ›Links [6].)

We then intended to reveal the origin of the enantioselectivity. **Fig. 7.20.** exhibits the crystallographic structures of the diastereomeric *trans*-TolBINAP / DPEN-$RuCl_2$ precatalysts, (*R*,*RR*)-19 and (*R*,*SS*)-19 [302]. The analogous *trans*-RuH($^1\eta$-BH_4) complex, enabling AH under base-free conditions, has almost identical geometry [297]. Complex (*R*,*RR*)-19 has a distorted octahedral geometry at the Ru center that is in approximate C_2 symmetry, in which both the TolBINAP and DPEN chelate ligands are coordinated in a λ fashion [302]. In the *R*,*SS* isomer, however, the TolBINAP ligand forms a λ seven-membered ring structure, while the DPEN assumes a five-membered ring structure with a δ conformation. In both isomers, the two phenyl substituents of the N,N-ligated five-membered ring are oriented in the equatorial direction, and the flexible *P*-*p*-tolyl groups adopt similar special arrangements.

The sense and extent of enantioselectivity is highly sensitive to the diphosphine / diamine ligand chirality and all the above described AH requires the *R*/*R*,*R* or *S*/*S*,*S* combination, and both steric and electronic effects have shown to cooperate in determining the stereoselectivity. Thus, in the AH of acetophenones, for example, not only the proton-donating N–H_{ax} of DPEN but also the unreactive N–H_{eq} plays an important role. As illustrated in **Fig. 7.21.**, the ketone aromatic ring avoids nonbonding repulsion with the axially oriented *P*-phenyl group of (*R*)-BINAP but has notable electrostatic NH/π attraction with the N–H_{eq} of (*R*,*R*)-DPEN. Thus, the difference in steric bulk of ketone substituents is not sufficient to well differentiate the enantiofaces of ketones. The diastereotopic H_{ax} and H_{eq} both play a pivotal role in facilitating the AH via the favored *Re* transition state. The unique electronic effects arise from the high NH_2 acidity caused by N to Ru coordination [303].

Fig. 7.21. The (*R*,*RR*)-RuH_2 intermediate and the diastereomeric transition states. The naphthyl rings in BINAP and the equatorially oriented phenyl substituent in DPEN are omitted in the transition states.

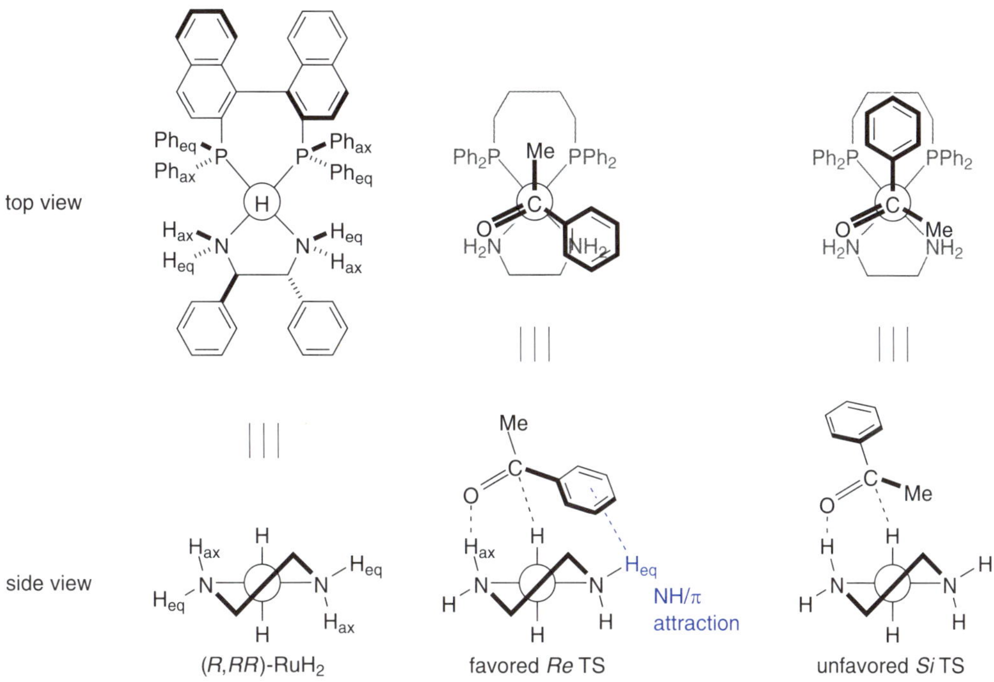

7.3.4 High flexibility of the molecular catalyst system

Catalyst design to cope with the substrate diversity

The mixed ligand complexes, (*R*,*RR*)- or (*S*,*SS*)-$RuCl_2$ 19 and the $RuH(^1\eta\text{-}BH_4)$ analogue enjoy a wide scope allowing AH of a range of prochiral ketones under basic and base-free conditions, respectively. Moreover, the mechanistic study elucidated the key factors for securing high reactivity and stereoselectivity. Electronic attractive forces, if accessible, are to be fully utilized to stabilize the transition states.

However, no universal catalysts exist because of the diversity of prochiral ketone substrates. Every prochiral ketone has its unique three dimensional and electronic characteristics. In order to cope with the reality of the molecular world and further expand the synthetic scope, we had to elaborate the catalyst in a way to kinetically discriminate the enantiofaces of the target ketone substrate. The AH chemistry of acetophenone and its analogues is consistently explained by assuming reactive *trans*-RuHX species (X = H, OR, etc.), analogous to the precursors (**Fig. 7.21.**). However, the geometry of the active species may be dynamically changed, irrespective to the precatalyst structure, to *cis*-RuHX, so that the RuH and NHax interact more comfortably with an enantioface

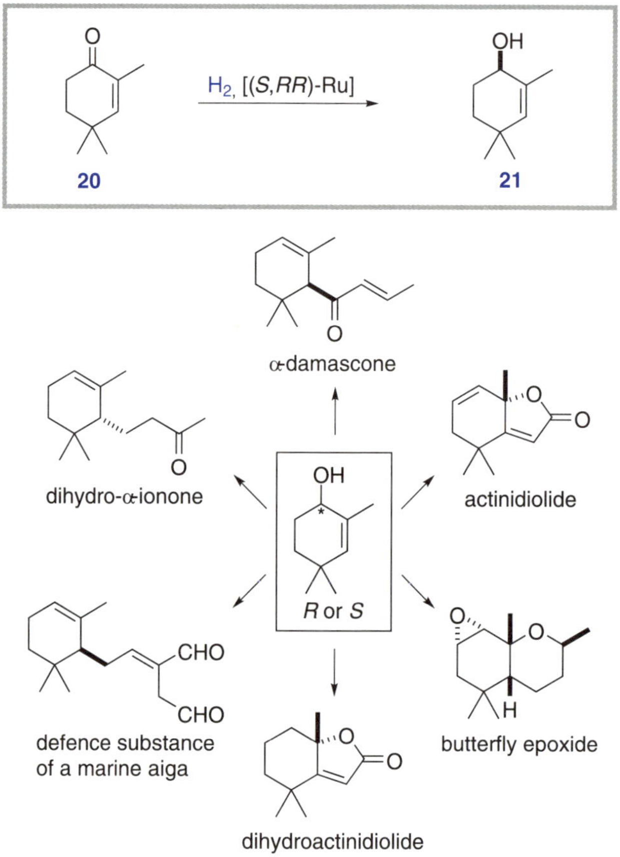

Fig. 7.22. Combination of the Ru-catalyzed AH and Mori's chemistry. Ru catalyst = *trans*-[Ru-Cl_2{(*S*)-tolbinap}{(*R*,*R*)-dpen}]. Reaction conditions: ^{t}BuOK in 2-propanol.

of the ketonic substrate on the 18e molecular surface. We must be free from any preconceptions and remain optimistically open-minded.

Like many standard aryl ketones and enones, (*R*)-puregone, a cyclic ketone with an *s-cis* exocyclic conjugated double bond, is hydrogenated by an (*S*)-BINAP / (*S*,*S*)-DPEN catalyst system resulting in the 1*R*,5*R* alcohol with 98 % selectivity [304].

Notably, however, cyclic enones with an *s-trans* endocyclic olefinic bond are categorized as a different class of substrates. Thus, the use of the *R*,*SS* or *S*,*RR* diphosphine / diamine Ru catalyst, (*R*,*SS*)-19 (see **Fig. 7.20.**) or the enantiomer, is crucial in AH of 2.4,4-trimethyl-2-cyclohexenone (20) [302]. Thus the hydrogenation catalyzed by (*S*,*RR*)-19 with an S/C ratio of 10,000 : 1 enables C=O selective saturation to give *R* alcohol 21 with 94 % *ee* (**Fig. 7.22.**). The resulting *R* and *S* allylic alcohols, coupled with the Claisen reaction, can be converted to a series of carotenoid-derived odorants and other bioactive compounds, which were extensively studied by Kenji Mori (University of Tokyo), a respectable expert in natural products chemistry **[D]**.

[D] K. Mori: Biochemical Methods in Enantioselective Synthesis of Bioactive Natural Products. *Synlett* **1995**, 1097 – 1109.

(*R*)-Carvone (22) possessing three unsaturated bonds and a C(5) stereogenic center poses many selectivity problems (**Fig. 7.23.**). With no

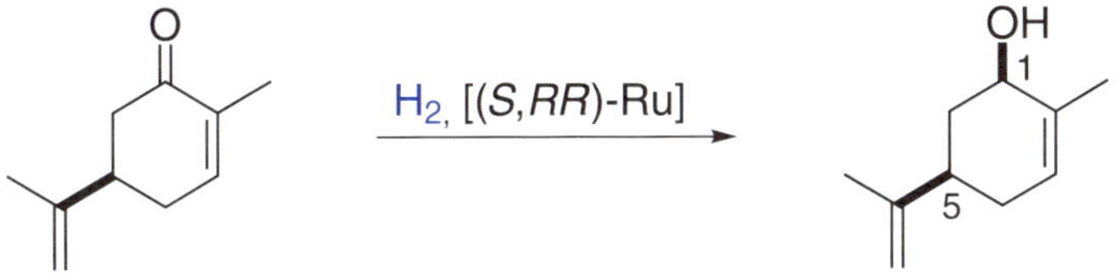

22 (1*R*,5*R*)-**23**

Fig. 7.23. Stereoselective hydrogenation of (*R*)-carvone. Ru catalyst = [$RuCl_2${(*S*)-binap}{(*R*,*R*)-dpen}]. Reaction conditions: KOH in 2-propanol.

control over selectivity, 17 different hydrogenation products are possible. In fact, hydrogenation under 4 atm H_2 with an (*S*)-BINAP / (*R*,*R*)-DPEN combined Ru system resulted in perfect chemo- and diastereoselectivity, giving solely (1*R*,5*R*)-carveol (23) in 100 % yield. The 5*R* configuration of the ketone significantly affects the rate and stereoselectivity. The presence of the methyl group at the C(2) position is crucial. Efficient catalytic access to 2-unsubstituted 2-cyclohexenols was possible only by kinetic resolution by AH [116] or ATH [279] aided by our chiral Ru catalysts.

Asymmetric activation

This AH is accelerated by the cooperation of two chiral ligands, showing an interesting phenomenon. Obviously, racemic catalysts give racemic products. However, an inactive racemic metal complex may be activated in an enantioselective manner by the addition of a suitable non-racemic "vitamer" [305 – 307]. The function is opposite to that of a "poison." The resulting diastereomeric catalyst can promote a highly selective asymmetric reaction, if one diastereomer remains catalytically unreactive as shown in **Fig. 7.24.** Thus, although racemic [$RuCl_2$(tolbinap)$(solv)_2$](polymer) (**CHAPTER 6**) is inert to the hydrogenation of simple ketones, it can hydrogenate enone 20 in the presence of (*S*,*S*)-DPEN (diamine / Ru = 1 : 1) and KOH in 2-propanol to quantitatively give (*S*)-21 in 96 % *ee*, which is very close to the value obtained with enantiomerically pure (*R*,*SS*)-19. The complexation of racemic $RuCl_2$(tolbinap) and (*S*,*S*)-DPEN forms the diastereomeric precatalysts (*R*,*SS*)-19 and (*S*,*SS*)-19. Under these reaction conditions, the *R*,*SS* hydrogenation cycle with a 98 : 2 *S*/*R* enantioselectivity has a turn over, which is 121 times faster than the *S*,*SS* cycle proceeding with a 37 : 63 *S*/*R* selectivity.

The relative significance of the diastereomeric *R*,*SS* and *S*,*SS* cycles, as well as the direction and degree of asymmetric induction, is highly dependent on the structures of the ketonic substrates. When *o* mcthylacetophenone is used in place of the cyclohexanone 20 as a hydrogenation substrate, (*S*,*S*)-DPEN accelerates the *S*,*SS* cycle 13 times more than the diastereomeric *R*,*SS* cycle, giving the *R* alcohol in 90 % *ee* and 100 % yield. Thus, this asymmetric activation is a viable approach for practical asymmetric catalysis when enantiomerically pure ligands are not readily available [E].

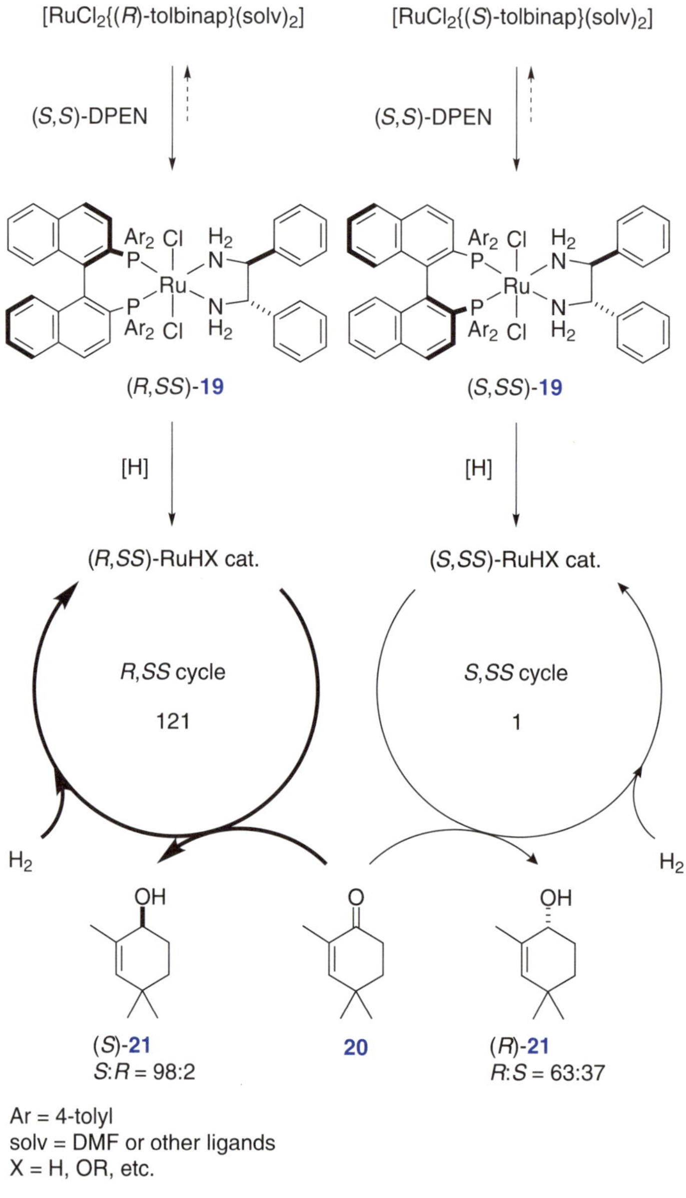

Fig. 7.24. Mechanism of asymmetric activation.

Kinetic discrimination of enantiomeric ketones

The advantage of this AH method is not limited to the flexible molecular design of the mixed-ligand catalysts. Unlike conventional procedures, the reaction conditions are easily tunable. Under base-free conditions with [RhH($^1\eta$-BH_4)((*S*)-xylbinap)((*R*,*R*)-dpen)] as a catalyst, racemic 24 is resolved to afford, at 53 % conversion, unreacted (*S*)-24 in 91 % *ee* together with alcohol (1*R*,2*R*)-25 in 85 % *ee*. This is understandable, but in contrast, under the standard basic conditions, racemic ketone 24 is kinetically resolved with the (*S*)-BINAP / (*R*,*R*)-DPEN Ru catalyst, producing alcohol (1*R*,2*R*)-25 with 96 % stereoselectivity among four

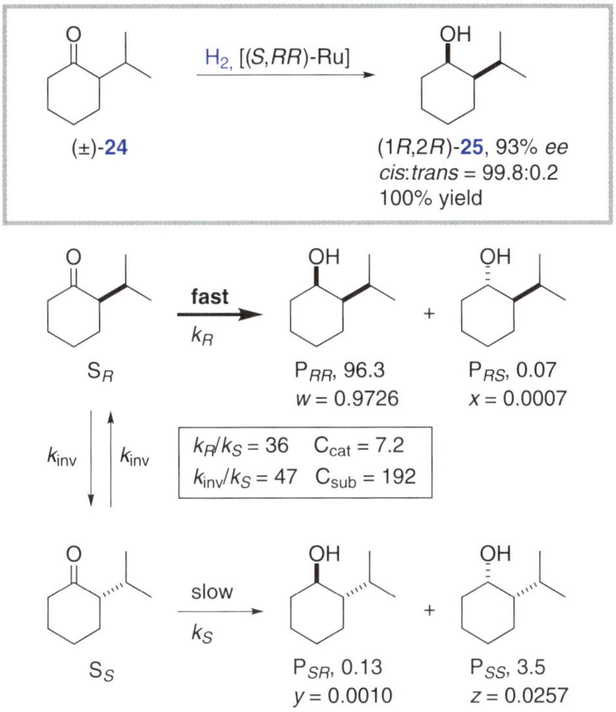

Fig. 7.25. Dynamic kinetic resolution of a racemic cyclic ketone. S = substrate, P = product. w:x:y:z = partition parameter, where the *R* and *S* enantiomers are assumed to exist in equal amounts. Conditions: Ru catalyst = [$RuCl_2${(*S*)-binap}$(dmf)_n$], (*R*,*R*)-DPEN, KOH (1:1:20) in 2-propanol.

possible stereoisomers (Fig. 7.25.) [308]. The key factor is the facile configurational inversion of the α-stereogenic center caused by an alkaline base in a protic solvent. Computer-aided analysis revealed that: (1) the inherent, highest stereoselectivity at time $t = 0$ is 97 % under the reaction conditions; (2) the *R* ketone is hydrogenated 36 times faster than the *S* enantiomer; (3) the slow-reacting *S* substrate undergoes *in situ* stereoinversion 47 times faster than it is hydrogenated; (4) the extent of the substrate-based asymmetric induction (*cis* : *trans*) is 192 in favor of the *cis* isomer; and (5) the catalyst-controlled asymmetric induction (*R*:*S*) is 7.2:1. Thus, the kinetic and stereochemical factors are well coordinated to diastereo- and enantioselective hydrogenation.

This stereoselective method is applicable to α-aryl [309] and -hetero-functionalized cyclohexanones [253] as well. The latter allows the synthesis of building blocks of biologically significant compounds (Fig. 7.26.).

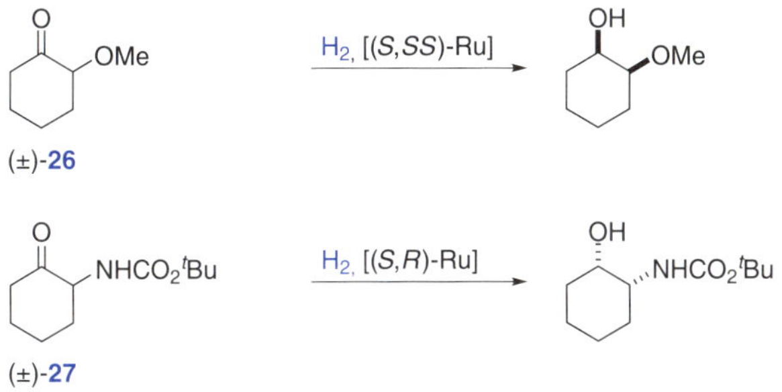

Fig. 7.26. Dynamic kinetic resolution of α-functionalized cyclohexanones. Ru catalyst = [$RuCl_2${(*S*)-xylylbinap}], (*S*,*S*)-DPEN or (*S*)-DAIPEN.

Asymmetric hydrogenation of further difficult substrates

1-Tetralones are neither analogues of acetophenones, cyclohexenones, nor cyclohexanones, and constitute another different class of ketonic substrates. The known asymmetric catalysts did not work for these structurally rigid, cyclic ketones. We solved this problem by replacing the conventional 1,2-diamine catalysts with a chiral 1,4-diamine complexes 28, thereby allowing the AH with an S/C ratio as high as 55,000 : 1 to give 1-tetralols in 99 % *ee* and high yield (**Fig. 7.27.**) **[310]**. The resulting chiral alcohols are useful as intermediates for the synthesis of bioactive compounds. The combination of (*S*)-BINAP and (*R*)-1,4-diamines is essential. Certain alkylated 2-cyclohexenones behave similarly to 1-tetralones, but simple acetophenone does not. These catalysts thus are not substitutes for the standard BINAP / 1,2-diamine Ru catalysts. Because these systems have different characteristics, both are used complementarily, depending on the ketone structures.

[310] T. Ohkuma, T. Hattori, H. Ooka, T. Inoue, and R. Noyori: BINAP/ 1,4-Diamine-Ruthenium(II) Complexes for Efficient Asymmetric Hydrogenation of 1-Tetralones and Analogues. *Org. Lett.* **2004**, *6*, 2681–2683.

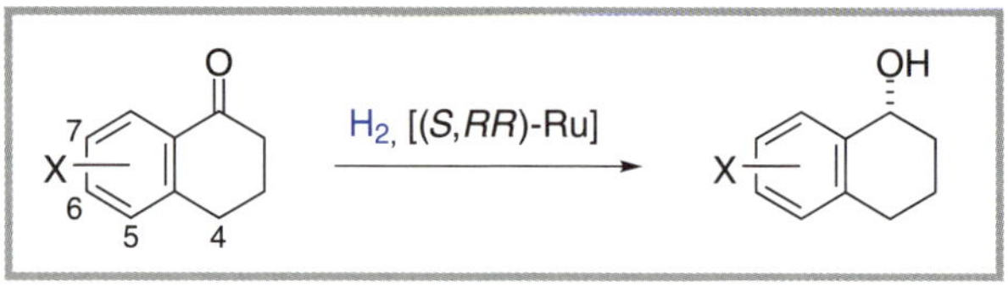

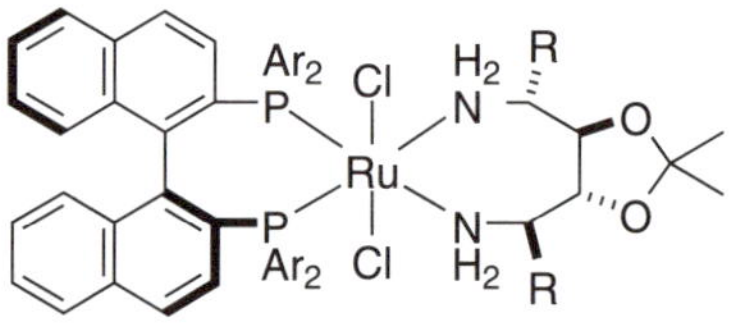

(*S*,*RR*)-**28**
Ar = Ph, 4-tolyl, 3,5-xylyl;
R = H, Me

Examples:

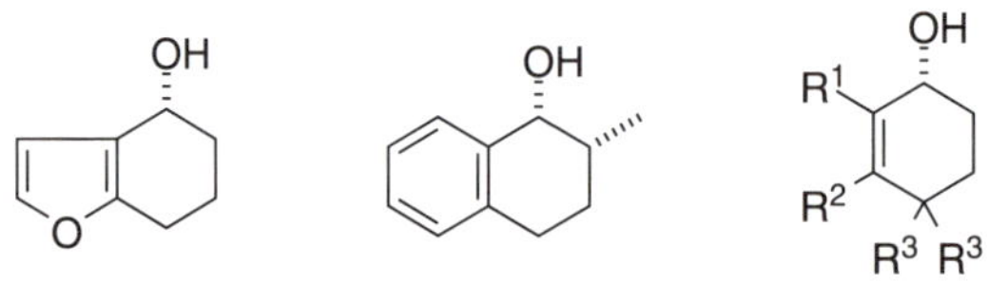

a: $R^1 = R^3 = Me$, $R^2 = H$
b: $R^1 = R^3 = H$, $R^2 = Me$

Fig. 7.27. AH of 1-tetralones and analogues. Conditions: ^{t}BuOK in 2-propanol.

Synthetic challenges are endless. As I reflected, no practical catalyst existed for sterically congested *tert*-alkyl ketones, despite tremendous efforts that were made so far. Our basic idea enjoys extremely high flexibility, but the C_2 chiral diphosphine / diamine catalysts remained unsuitable for obvious steric reasons. Neither simple aliphatic diamines nor bipyridyl worked as a diamine ligand.

Fortunately, we eventually invented highly reactive chiral BINAP-Ru catalysts 29 which possess α-picolylamine (PICA), an asymmetrical NH_2 (functional) / pyridine (unfunctional) hybrid diamine ligand (**Fig. 7.28.**)

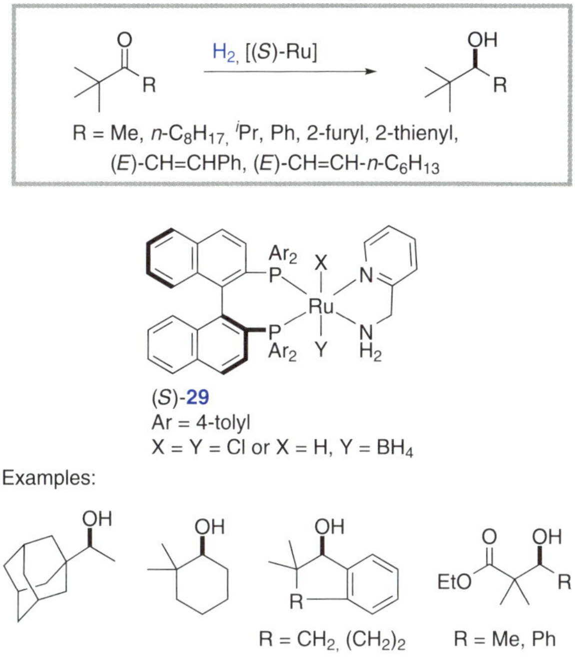

Fig. 7.28. AH of *tert*-alkyl ketones. Base-free conditions with RuH($^1\eta$-BH_4), or in the presence of ^{t}BuOK or phosphazene with $RuCl_2$ catalyst in ethanol.

[311]. This resulted in various *tert*-alkyl ketones with an aryl, heteroaryl, alkyl, or alkenyl substituents, either acyclic or cyclic, being hydrogenated with an S/C of up to 10,000, giving chiral alcohols generally in ca. 98 % *ee* and high yield. Alcoholic solvents affect the extent and even sense of enantioselectivity, suggesting ligation of an alkoxy function to Ru. Overall, ethanol, not 2-propanol, is the solvent of choice. The sense of asymmetric induction is predictable and results simply from the difference in steric bulk of the two substituents flanking the C=O function. Interestingly, the (*S*)-BINAP / PICA catalyst hydrogenates acetophenone giving (*R*)-1-phenylethanol quantitatively but with only 54 % *ee* in ethanol or 14 % *ee* in standard 2-propanol. Conjugated C=C bonds are not saturated, resulting in chiral allylic alcohols with high enantioselectivity.

From this very long lesson, which started in the 1970s, we learned that designing an asymmetric catalyst requires an integrated molecular approach. The desired molecular function emerges by combining various steric and electronic properties and reaction environments. Furthermore, both electronic attraction and steric repulsion kinetically cooperate in obtaining a high level of enantioselectivity and satisfactory reactivity. All of our Ru-based molecular catalysts commonly involve an octahedral BINAP-RuH species; however, the reactivity of these catalysts is greatly influenced by the anionic or neutral auxiliaries present in the three other coordination sites. Deep mechanistic insight is indispensable for developing truly efficient AH reactions. The intellectual challenge is joyful indeed, even if a problem cannot (yet) be fully solved.

Sympathy with Jack Halpern

Professor E. J. Corey (* 1928) at Harvard was, of course, my first American mentor (**CHAPTER 3**). However, readers may be surprised to learn that I was also greatly influenced by Professor Jack Halpern (1925–2018), whatever this organometallic chemist might have thought about my science and academic behavior. He was extremely critical minded (this is very different from myself) and valued clear conclusions in the Occam's razor style.

In the 1960s, I was amazed by his bold comparative concepts of carbene vs. Rh^{I}, carbocation vs. Ru^{II}, and carbon radical vs. Co^{II} (**CHAPTER 2**). Later on, his Rh^{I} chemistry and my Ni^{0} chemistry on strained hydrocarbon molecules resonated with each other, while the intervention of transition metal effects into the Woodward-Hoffmann Rule was being discussed. Furthermore, the mechanism-based limitations of Rh-diphosphine catalyzed asymmetric catalysis were overcome by my Ru-diphosphine catalysis. Halpern probably first recognized me in 1974, when he and I were 49 and 35 years old, respectively, at the US-Japan Organometallic Chemistry Conference in Honolulu. Since then, we have met frequently at international organometallic conferences, and I invited him twice to Nagoya University. I feel he came to understand my chemistry as he looked through my papers as an Associate Editor of *Journal of the American Chemical Society*. He continued to guide me appropriately through his sincere response towards my submissions of manuscripts of various kinds of research.

The University of Chicago was his workplace, and he was my host during my stay as the Morris S. Kharasch Visiting Professor in 1991, which I strongly remember. This famous lectureship was established in 1960 in honor of Kharasch, who was a pioneer in free radical addition and polymerization reactions, a mentor of H. C. Brown, a founder of the *Journal of Organic Chemistry*, and also instrumental in the development of the chemistry department. Starting with Woodward as the first Kharasch Lecturer, eminent researchers who have built the history of organic chemistry have been selected. I was honored to receive this prestigious lectureship and gave a talk on asymmetric organozinc catalysis and nucleic acid synthesis, in addition to my achievements in asymmetric hydrogenation reactions. Among the colleagues there were Philip E. Eaton known as the "founder of cubane," Brice Bosnich, the inventor of CHIRAPHOS and an uncompromising perfectionist, and Takeshi Oka, a structural advocate of H_3^+. I learned that Jack and Helen Halpern had a profound knowledge of culture, and their home was full of artistic crafts largely from China. They loved the performance of the Chicago Symphony Orchestra directed by Daniel Barenboim at the time.

Jack Halpern was more inclined towards inorganic chemistry, structure and reaction studies, and less interested in synthetic organic chemistry. Nevertheless, he was closely associated with William S. Knowles of Monsanto, and fully values the practicalities of catalytic chemistry. He was rigorous in scientific research. He was always in the front row at lectures, and when the speaker finished their talk, he always said, "I have a comment and some questions." He appeared to be true to his own beliefs, but he was a man with a sense of fairness.

At the 2000 IUPAC Conference of Organometallic Chemistry in Shanghai, Halpern happened to listen to my plenary lecture on asymmetric hydrogenation. After I returned to Japan, I received a letter from him with the following words of encouragement: "Your recent works, demonstrating broadly applicable and highly practical applications of stereoselective catalysis, truly raises this field to a new level." Later in 2003, I was awarded an honorary Doctor of Science from the University of Chicago and was also elected as a Foreign Associate of the National Academy of Sciences, which I believe that he must have been involved in the nomination. He passed away in 2018. I offer my heartfelt thanks and condolences.

Fig. 7.29. (left) – 7.30. (right) Author receiving a Doctor of Science Degree at the University of Chicago in 2003. **Fig. 7.29.**: With Professor Jack Halpern (left) and President Don M. Randel (right). **Fig. 7.30.**: Congratulated by author's first chemistry teacher at junior high school in 1951, Kazuo Nakamoto (second from the left, Professor at Illinois Institute of Technology and Marquette University).

7.4 Green chemistry: A responsible science for future generations (1992–)

My ERATO catalysis project on green chemistry was considered unfashionable in Japanese academia at that time, but I thought the status quo was not an option. Traditionally, academics pursued ideal chemistry largely driven by scientific curiosity. Since science is intertwined with society in our age, I started to think that we researchers perhaps should not stay in the traditional ivory tower or modern research laboratory. Obviously, science and technology exist to enrich our individual lives, secure the safety, peace, and prosperity of nations, and eventually sustain human civilization. Chemists and chemical engineers must be mindful of how their work connects with society and should not leave things to chance. Anything that goes against the law of nature or deprives us of our humanity is inappropriate and wrong. Looking back on the 20th century, chemical synthesis technology contributed enormously to achieving an affluent civilization. Still, the current standard of technology is to be much improved so as to minimize the energy needed to transform substances and also to reduce the burden on the environment. Green chemistry, I consider, is not a mere catch phrase but an indispensable principle that will sustain our civilized society [F]. At the same time, this is not a matter of clear-cut scientific or technical expertise, but is rather a serious, complex social issue. In order to materialize the goals in the real world, we need a rational, long-term program supported by public understanding, governmental policies, and worldwide cooperation. At both the individual and institutional level, all members of our society must recognize the necessity of green chemistry. Chemists must take responsibility for the current and future generations to create a sustainable society in this century [312].

Many existing organic syntheses, which are achieved in a step-by-step fashion, are beneficial but produce undesirable side products along with the target products. On this occasion, it seemed the right time for some serious soul-searching in Japan to respond to the needs of the society [313]. In this regard, every reaction in a multi-step synthesis should proceed in a high atom-economy, a term raised by Barry M. Trost **[G]**, and the overall synthesis is to be accomplished with a low E-factor, a term defined by Roger Sheldon [H]. As such, chemists of our days are asked to develop "perfect chemical reactions" that proceed with 100 % yield and 100 % selectivity without any side products. In practice, the 3Rs (reduction, recycling, and reuse) of resources are particularly important. Taking these factors into account, molecular catalysis, together with traditional heterogeneous catalysis, significantly contributes to realizing this goal.

[G] B. M. Trost: Atom Economy—A Challenge for Organic Synthesis: Homogeneous Catalysis Leads the Way. *Angew. Chem., Int. Ed.* **1995**, *34*, 259–281.

7.4.1 Replacing metal hydrides by hydrogen (1995 –)

To begin with, I questioned how significant our hydrogenation is in modern organic synthesis. Despite a century of efforts in hydrogenation chemistry, the scope and selectivity profiles remained empirical and unclear. The applicability of our methods goes far beyond asymmetric transformation. Prior to our findings, selective reduction of ketones heavily relied on the use of boron hydrides or organoboranes pioneered by Herbert C. Brown, a 1979 Nobel Laureate in chemistry. Even now, $NaBH_4$, $LiAlH_4$, or some silicon hydrides are used for C=O reduction in many conservative laboratories. We aimed at replacing these conventional methods by clean hydrogenation technology, particularly for large-scale synthesis [253].

Chemoselective hydrogenation

Hydrogenation must proceed selectively at a specific reaction site in multifunctional molecules. Our catalytic methods utilizing the metal-ligand bifunctional mechanism (**CHAPTER 7.3**) tolerate many sensitive functionalities including F, Cl, Br, I, CF_3, OH, $OSiR_3$, OMe, $CO_2{}^tBu$, $CONR_2$, NO_2, NH_2, NR_2, SR, SO_2R, and most heteroaromatic groups. Prior to our challenge, there were no general catalysts that enable preferential reduction of a C=O linkage over an olefinic or acetylenic linkage. Most available catalysts, either homogeneous or heterogeneous, are selective for C=C bonds with some special exceptions. Our goal has been to solve this long-standing problem [253].

$[RuCl_2(PPh)_3]$ is an excellent catalyst for the hydrogenation of olefins, as first reported by Wilkinson in 1965 [I]. However, this catalyst is very inactive for the reduction of aldehydes; 1-octene, a terminal olefin, is hydrogenated in 2-propanol / toluene 250 times faster than heptanal. We found that the selectivity is dramatically reversed, when one equivalent of ethylenediamine and two equivalents of KOH are added to the Ru catalyst [314]. Now, hydrogenation of the aldehyde takes place 1500 times faster than the olefin, where the C=O / C=C selectivity profile is changed by a factor of 375,000! Likewise, benzaldehyde is hydrogenated 450 times faster than styrene. As exemplified in **Fig. 7.31.**, the selective hydrogenation is applicable to a variety of carbonyl compounds containing a C=C or C≡C bond. Moreover, most α,β-unsaturated ketones and aldehydes are convertible to the corresponding allylic alcohols. **Fig. 7.32.** shows structures of the Ru precatalyst 30 and possible reducing species 31.

The hydrogenation does not proceed via an enol form, as proven by the reduction of benzophenones with an S/C (substrate / catalyst) ratio of 20,000 : 1 giving diphenylmethanol in pure form [294]. Unlike other conventional hydrogenation methods, this reaction does not produce any diphenylmethane products because of the presence of a base in this system.

(a) Aliphatic series

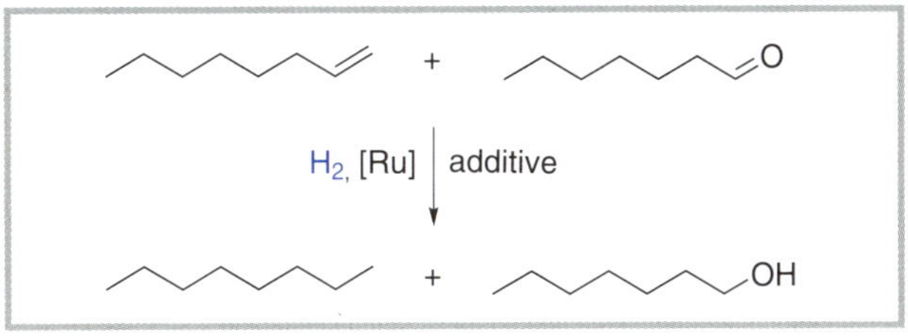

additive	t [min]	C=C:C=O
——	150	250:1
$NH_2(CH_2)_2NH_2$ + KOH (Ru:diamine:KOH = 1:1:2)	10	1:1500

(b) Aromatic series

X = O, CH_2 — $k_{C=O}/k_{C=C}$ = 450

X = O, CH_2 — $k_{C=O}/k_{C=C}$ = 1500

(c) Intramolecular competition

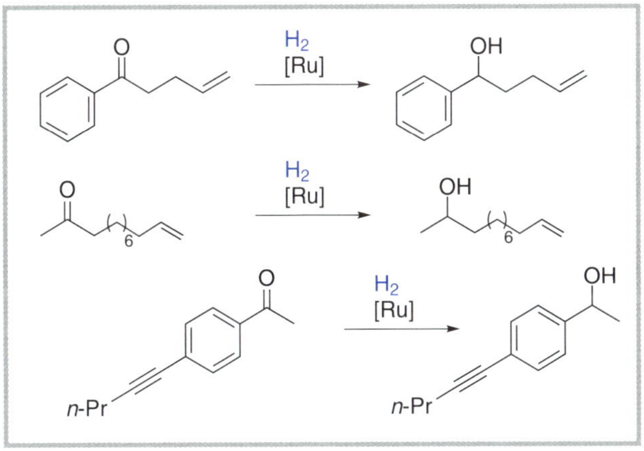

Fig. 7.31. The ethylene-diamine effect on carbonyl-selective hydrogenation. a) [Ru] = $[RuCl_2(PPh_3)_3]$; b) $[RuCl_2(PPh_3)_3]$, $NH_2(CH_2)_2NH_2$, KOH. 4 atm H_2.

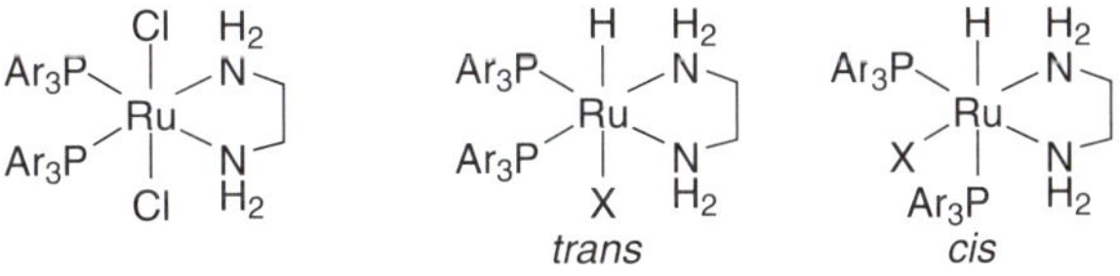

Ru precatalyst **30** Ru catalyst **31**

Fig. 7.32. A catalyst precursor and possible catalytic species. X = H, OR.

Diastereoselective hydrogenation

The extraordinary reactivity and productivity of this catalyst system surprised us [302]. For example, in the presence of Ru precatalyst 30 (Ar = 4-$CH_3C_6H_4$) and ^{t}BuOK in 2-propanol (10 atm and 60 °C), cyclohexanone is hydrogenated to cyclohexanol at S/C = 100,000 and with an initial TOF (turnover frequency) of 563,000 h^{-1} or 156 s^{-1}. Such high reaction rate was unprecedented in reduction chemistry.

In a similar manner, this standard system hydrogenates conformationally fixed 4-*tert*-butylcyclohexanone to give the *cis* alcohol with a very high diastereoselectivity (**Fig. 7.33.**). The hydrogen atom on the Ru species 31 (Ar = C_6H_5) acts as a bulky hydride, approaching the C=O carbon from the less crowded equatorial direction. For the same reason, the hydrogenation of 2-alkylated cyclopentanones and cyclohexanones shows high *cis* selectivity to a similar extent. The level of diastereoselectivity better compares with the best boron-based hydride reagents. The hydrogenation of open-chain, conformationally flexible aryl 1-phenylethyl ketones gives the Cram-selective, *syn*-alcohols similarly to the stoichiometric L-Selectride reagents (**Fig. 7.33.**).

Fig. 7.33. Diastereoselective hydrogenation of cyclic and acyclic ketones. [Ru] = 30, Ar = C_6H_5, KOH. 4–10 atm.

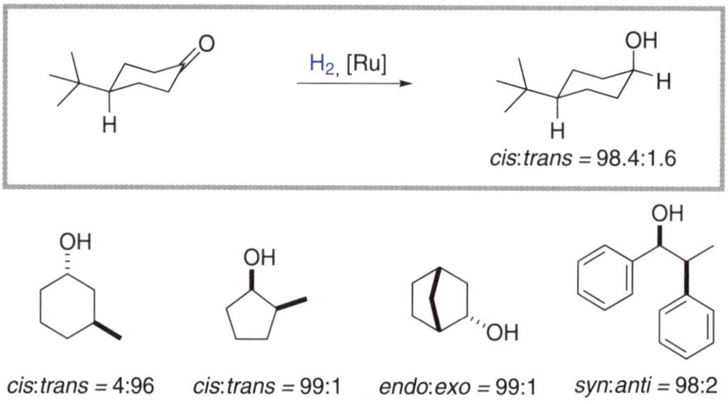

Unlike the conventional stoichiometric reduction using reagents with a given structure, this hydrogenation method enjoys a more flexible molecular design of the phosphine / diamine mixed-ligand Ru catalysts. The reaction conditions are highly flexible as well. These properties allow us for systematic variation of reactivity and selectivity.

Obviously, stereoisomeric ketones react at different rates. Now, when a configurationally labile α-substituted ketone is used, dynamic kinetic discrimination of the stereoisomers is achievable [308]. This is based on the facile configurational inversion caused by an alkaline base in a protic solvent. Thus, hydrogenation of 2,6-dimethylcyclohexanone with 30 under the standard conditions gives the *cis*,*cis*-alcohol with a 98.7 % selectivity out of three stereoisomers. The RuH species 31 undergoes the equatorial approach to the major, reactive *cis* dimethyl ketone rather than the coexisting *trans* diastereomer as explained in **Fig. 7.34.**

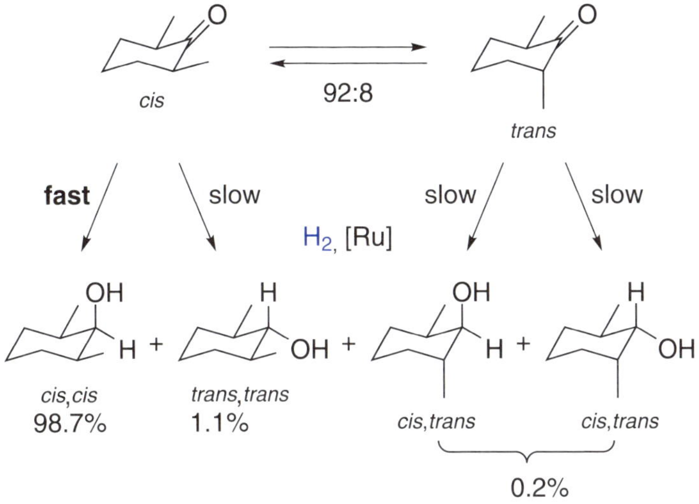

Fig. 7.34. Dynamic kinetic discrimination of diastereomeric ketones. [Ru] = 30 (Ar = C_6H_5), KOH, 4 atm, S/C = 500.

Likewise, certain epimers can be kinetically discriminated by the precatalyst 30 (**Fig. 7.35.**). (–)-Menthone has a stable 1*R* stereogenic center and an unstable 4*S* center, in equilibrium with isomenthone, the 4*R* epimer. Since the substrate is chiral and non-racemic, the stereo-distinction becomes more evident in the hydrogenation conducted with an enantiomeric (*R*)-BINAP / (*S*,*S*)-DPEN Ru catalyst, giving (+)-neo-menthol exclusively.

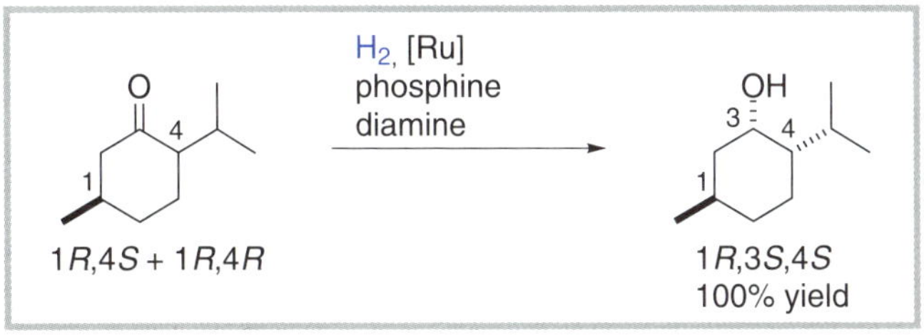

phosphine	diamine	1*R*,3*S*,4*S* : 1*R*,3*R*,4*R* : 1*R*,3*R*,4*S*
PPh_3	$NH_2(CH_2)_2NH_2$	93.7:6.1:0.2
(*R*)-BINAP	(*S*,*S*)-DPEN	100:0:0

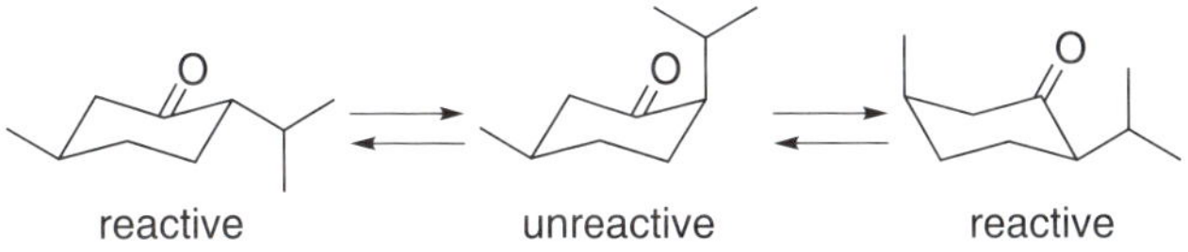

Fig. 7.35. Dynamic kinetic discrimination of epimeric ketones. KOH, 8 atm, S/C = 500.

Enantioselective reduction of simple ketones is an essential process in organic synthesis. This had long been achieved with a stoichiometric reagent including our BINAL-H (CHAPTER 5, Fig. 5.18.) [101, 102], DIP chloride [J], Alpine-borane [K] or the Corey-Bakshi-Shibata asymmetric hydroboration using diborane or catecholborane [L]. These reactions are convenient but produce large amounts of aluminum- or boron-containing wastes. We were able to successfully replace most of these methods by clean asymmetric hydrogenation (AH) as described in CHAPTER 7.3.

I would like to add that, unlike the stoichiometric $LiAlH_4$ reagent, this standard catalyst does not reduce esters to primary alcohols. However, the NH effect can be utilized for this purpose. Later, a $RuHCl(CO)[HN(CH_2CH_2PPh_2)_2]$ / MeONa combined system was found to efficiently hydrogenate methyl (*R*)-lactate to (*R*)-1,2-propanediol without any racemization [M]. This method is applicable for the production of antibacterial levofloxacin.

7.4.2 Supercritical carbon dioxide as green catalysis medium (1992–1996)

Without green media, industrial chemistry is not sustainable. We should avoid the use of harmful solvents such as chlorinated hydrocarbons. A common question is: why do we perform synthetic reactions and catalysis in a liquid solution phase? The stereotypical answer is: reactants dissolve in solvents to achieve homogeneity and facilitate mass transport and molecular collisions. This is just as we thought.

In the liquid phase, molecules are packed densely to form a solvent sphere, as schematically shown as a simple diagram in Fig. 7.36. In order to realize an intermolecular reaction between the substrate and catalyst molecules, the solvent molecules must be removed, or the

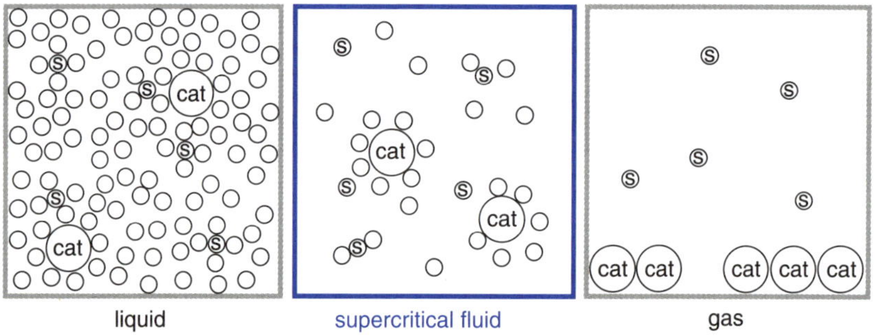

	liquid	supercritical fluid	gas
density, g cm^{-3}	1	0.1–0.5	10^{-3}
viscosity, Pa s	10^{-3}	10^{-4}–10^{-5}	10^{-5}
diffusivity, cm^2 s^{-1}	10^{-5}	10^{-3}	10^{-3}

(cat): catalyst (S): substrate O: medium molecule

Fig. 7.36. Phases of catalytic reactions and typical liquid, supercritical (sc) fluid and gas properties.

shell structure must be reorganized. We know certain chemical reactions occur very rapidly in the gas phase due to the high molecular mobility and the lack of interaction between the solvent and reactants. Unfortunately, however, many organic compounds are not volatile enough to occur in gas phase. Then we noticed the utility of supercritical (sc) fluids for synthetic reactions. These fluids have the beneficial effects of both liquid and gas phases because their physical properties, such as density, viscosity, and diffusivity, are an intermediate between the two. An early example includes a patent for polymerization of scC_2H_4 or scC_3H_6 with a dissolved $AlClEt_2$ / $TiCl_4$ catalyst, where the advantage was the ability to selectively precipitate the product fractions by lowering the pressure. Industrial production of linear low-density polyethylene by homogeneous metallocene-catalyzed polymerization of scC_2H_4 was commercialized.

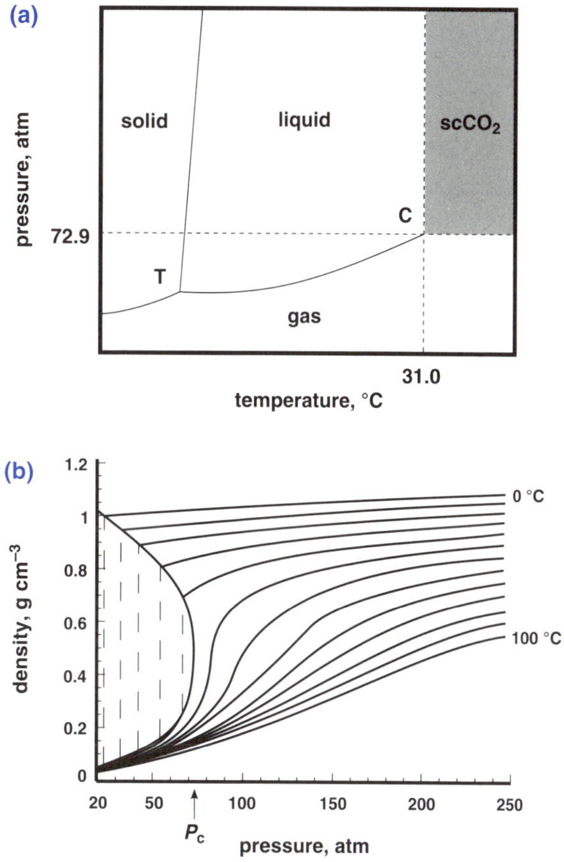

Fig. 7.37. (a) Phase diagram of pure CO_2. T: triple point, C: critical point. (b) Density of CO_2 depending on pressure and temperature, Pc: critical pressure.

We became interested in carbon dioxide (CO_2) from both a scientific and a technical point of view. CO_2 has a critical temperature of 31 °C and a critical pressure of 73 atm (Fig. 7.37.) [315, 316]. When CO_2 is heated above 31 °C, it cannot liquefy, even if a high pressure is applied, but instead forms a supercritical state, $scCO_2$. The physical properties vary dramatically near the critical point under the influence of pressure and temperature. In addition, since CO_2 interacts only weakly with reactants under such environments, some unique molecular clustering may take place locally. Furthermore, CO_2 has several technical benefits: it is nontoxic, nonflammable, and very cheap; it can be separated easily from the reaction system without leaving any harmful residues.

The chemistry of homogenous catalysis may be improved by the use of supercritical media. In 1994, our ERATO team acted as a pioneer by using $scCO_2$ as a medium for homogeneous catalysis [317]. Consequently, we could achieve high reaction rates, improved selectivity, and elimination of mass transfer.

[317] P. G. Jessop, T. Ikariya, and R. Noyori: Homogeneous Catalytic Hydrogenation of Supercritical Carbon Dioxide. *Nature* **1994**, *368*, 231–233.

Figs. 7.38.–7.39. show hydrogenation of $scCO_2$ in the presence of a Ru complex, where CO_2 acts as both the reactant and reaction medium. Since the reaction of H_2 with CO_2 giving formic acid is endothermic, the presence of triethylamine is needed to stabilize the product. The initial rate of formic acid production, a turnover frequency (TOF) of up to 1,400 moles per mole of catalyst per hour, is much higher than any existing catalytic system in an ordinary reaction medium. In the presence of methanol, formic acid is further converted to methyl formate by dehydration [318]. When triethylamine is replaced by dimethylamine, DMF is formed with a high turnover number (TON) up to 420,000 [319, 320]. The excellent efficiency is ascribed to the very high solubility of H_2 in $scCO_2$

$$CO_2 + H_2 \xrightarrow[\text{NEt}_3,\ 50\ °\text{C}]{[\text{Ru}]} HCO_2H \quad \text{(7,200 TON)}$$

$$CO_2 + H_2 + MeOH \xrightarrow[\text{NEt}_3,\ 80\ °\text{C}]{[\text{Ru}]} HCO_2Me \ \text{(3,500 TON)} + H_2O$$

$$CO_2 + H_2 + NHMe_2 \xrightarrow[100\ °\text{C}]{[\text{Ru}]} HCONMe_2 \ \text{(420,000 TON)} + H_2O$$

Figs. 7.38. Hydrogenation of CO_2 in supercritical phase. Synthesis of formic acid and its derivatives.

and the high reactivity of the Ru catalyst under such conditions. Thus, $scCO_2$ has clear advantage as a catalytic reaction medium. Although pure CO_2 can dissolve only nonpolar substances, the solubility is greatly enhanced by addition of certain protic or fluorinated "entrainers."

The substrate is not limited to CO_2 [321, 322]. The BINAP-Ru catalyzed asymmetric hydrogenation of α,β-unsaturated carboxylic acids was achieved in $scCO_2$ [323]. In addition, we examined the selectivity for the hydrogenation or hydroformylation of olefins by a Mn(I) hydride species in $scCO_2$, leading to the suggestion of a nonradical pathway for aldehyde formation [324]. In $scCO_2$ containing a Pd-phosphite complex, *o*-iodobenzyl alcohol undergoes efficient carbonylation in high yield [325].

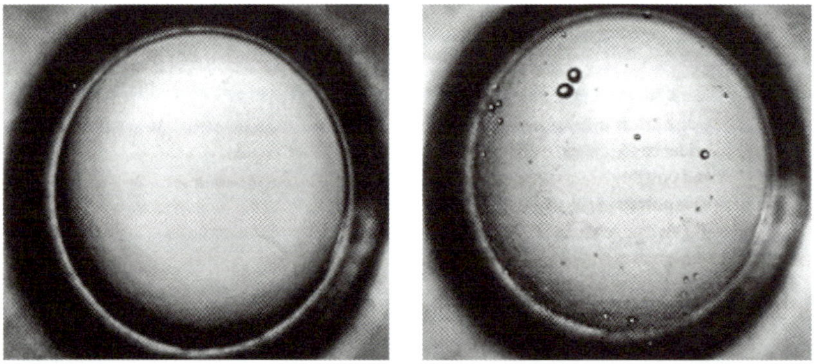

Figs. 7.39. The interior of the reaction vessel. Initial conditions (left): [Ru] = $[RuCl_2(PMe_3)_4]$, H_2 (80 atm) in $scCO_2$ (total 210 atm). The drops in the right are liquid complexes of formic acid and NEt_3.

This hydrogenation method is rapid, high-yielding, selective, and productive, and its potential applications in industry are numerous. We provided an overview on this important subject in *Chemical Reviews* and elsewhere [321, 326, 327]. Philip G. Jessop, who acted as a main researcher in this ERATO project, became a leader in green chemical technology at Queen's University in Canada, expanded the technical utility of $scCO_2$ as a medium[N] and further developed "switchable materials."

7.4.3 Green oxidation with hydrogen peroxide (1996 – 2003)

Oxidation is one of the most fundamental reactions in laboratories and chemical manufacturing. However, it remained among the most problematic processes. Many textbook methods are unacceptable because of the low atom economy. Furthermore, conventional heavy metal oxidants form toxic wastes and known organic stoichiometric oxidants are usually very expensive. Nitric acid, the most conventional industrial oxidant, is inexpensive but unavoidably forms various nitrogen oxides, including N_2O which has a strong greenhouse effect and also causes catalytic stratospheric ozone destruction as well as acid rain and smog. We conceived that O_2 (or air)[O] and H_2O_2 are complementarily useful oxidants in practical synthesis. Molecular oxygen (O_2) is an ideal oxidant, only if (1) the process is controlled safely without over-oxidation or combustion and (2) both O atoms are used for oxidation achieving 100 % atom-economy. Aerobic oxidation often uses only one O atom out of two and requires certain reducing agents to capture the extra oxygen, resulting in 50 % atom-economy.

Hydrogen peroxide (H_2O_2), produced effectively from H_2 and O_2 in a quantity of some 3 million metric tons per year, is another attractive oxidant for liquid phase reactions. It costs only 50 cents to one US dollar per kg, as 100 % H_2O_2. Since H_2O_2 is viewed as an adduct of an O atom and an H_2O molecule, water is the only theoretical waste product in its use as an oxidant with a 47 % atom-economy. Since the structures of chemical substances are diverse and reaction volumes are variable depending on whether the purpose is research or production, there are no universal synthetic methods. One of the major benefits of H_2O_2 oxidation is the high tunability of the reaction parameters. The reaction should be achieved with an H_2O_2 concentration of < 60 %, because the use, storage, and transportation of higher concentrations of H_2O_2 are not desirable for safety reasons. The oxidation process may further switch to *in situ* or on-site technology using H_2 and O_2 in the future. To my knowledge, H_2O_2 oxidation was initiated by G. B. Payne (alcohol oxidation, 1959) and S. E. Jacobsen (olefin epoxidation, 1979), and developed largely by C. Venturello and Y. Ishii among others. But much room remained for improvement.

Alcohol oxidation

In the mid-1990s, we developed a series of clean oxidation procedures using aqueous H_2O_2, a physiologically harmless tungsten catalyst, and a phase-transfer catalyst (PTC), which proceeds safely *without organic solvents and halides* [328]. This procedure, if achievable, stands at advantage over the existing methods for many practical reasons.

We chose a combined system of Na_2WO_4 and methyl-trioctylammonium hydrogen sulfate ($Q^+HSO_4^-$) as an acidic phase-transfer catalyst (PTC). This novel system catalyzes efficiently dehydrogenation of secondary alcohols with 30 % H_2O_2 to give the corresponding ketones in high yields at a 100 g scale as generalized in **Fig. 7.40.** [329 – 331]. The organic-aqueous biphasic oxidation is run at or below 90 °C, for safety reasons, without using halide salts or hazardous organic solvents such as conventional 1,2-dichloroethane. Only a slight excess of H_2O_2 is necessary because its unproductive decomposition is negligible. The pure product is obtainable simply by separation from the aqueous phase, washing with saturated $Na_2S_2O_3$, and distillation. Thus, the reaction of 2-octanol showed a turnover number (TON) of 77,000, which is two orders of magnitude higher than any previously reported H_2O_2 oxidation. The oxidation of 1-phenylethanol gives an even higher TON of 179,000. If necessary, toluene could be used as solvent, for example with crystalline substrates. This simple oxidation procedure is suitable for medium- to large-scale synthesis.

The presence of the acidic, lipophilic PTC is essential, since Na_2WO_4 is only soluble in aqueous phase. **Fig. 7.40.** presents our mechanistic hypothesis. In aqueous phase, the catalyst precursor is rapidly oxidized by H_2O_2 according to $Na_2WO_4 + 2H_2O_2 \rightarrow Na_2[WO(O_2)(OH)_2] + H_2O$. The resulting A is equilibrated with mono-protonated B and di-protonated C. The dianion A is feebly active, while B and C are sufficiently reactive. The alcohol oxidation is considered to occur by utilizing the organic-aqueous biphasic system. The alkoxy ligand in F is dehydrogenated by the adjacent hydroperoxyl ligand via a six-membered transition state (TS), where the hydridic α-hydrogen migrates to the positively polarized oxygen atom, which is consistent with the relative reactivity of *para*-substituted benzyl alcohols (< 10 % conversion) showing a ρ value of −0.31. The resulting monoperoxo tungsten ion G is reoxidized by H_2O_2 at an organic-aqueous interface or after returning to the aqueous phase.

Fig. 7.41. illustrates the behavior of primary alcohols under the present conditions. The oxidation of primary alkanols with 2.5 molar amounts of H_2O_2 afforded carboxylic acids by way of aldehyde intermediates (Eq. 1). 1-Octanol is only three-times less reactive than 2-octanol, while β-branched primary alcohols are much less reactive. Proper selection of the reaction conditions is needed to convert substituted benzylic alcohols to the aldehydes or carboxylic acids. The efficiency is highly

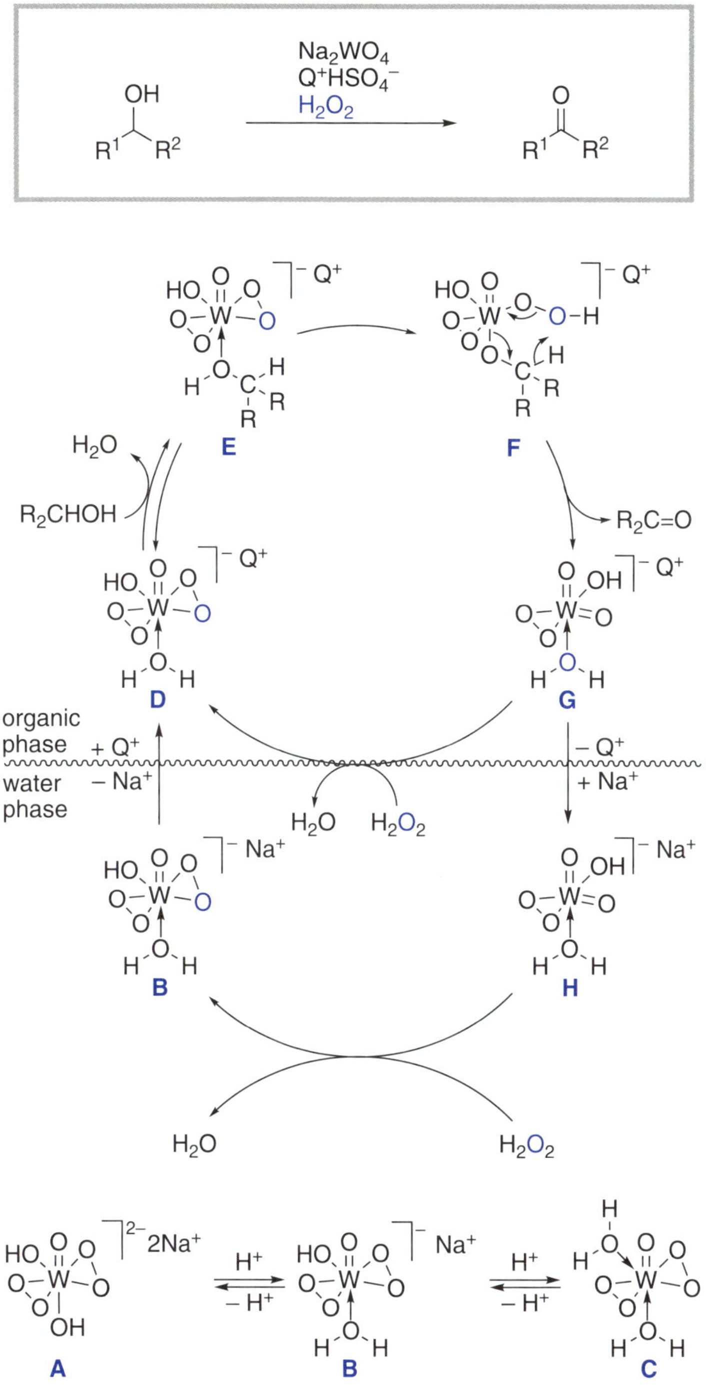

Fig. 7.40. Catalytic oxidation of secondary alcohols under organic-aqueous biphasic conditions. $Q^+ = CH_3(n\text{-}C_8H_{17})_3N^+$.

Fig. 7.41. Oxidation of primary alcohols and aldehydes with aqueous H_2O_2. $Q^+ = CH_3(n\text{-}C_8H_{17})_3N^+$, X = electron-withdrawing substituent, Y = electron-donating substituent.

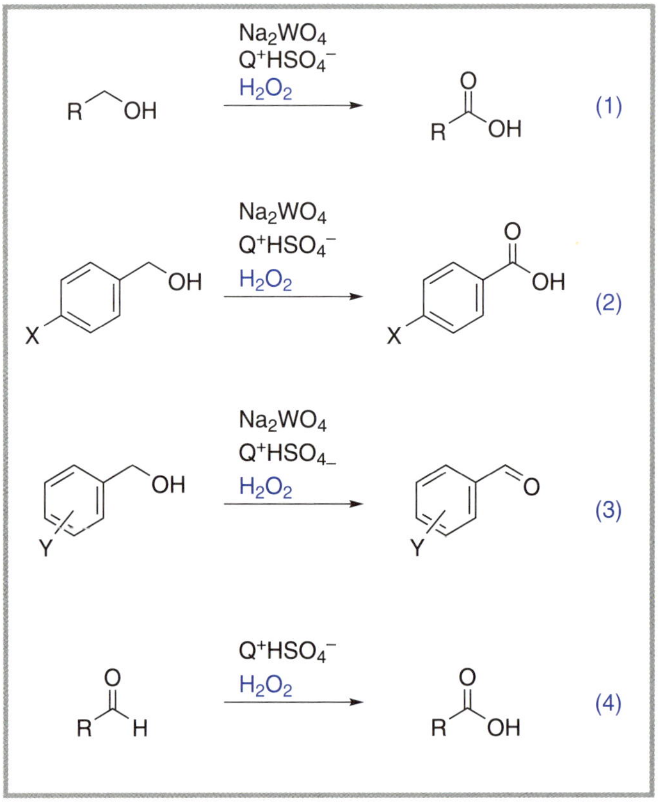

dependent on the electronic properties of the substituents, indicating the operation of dual mechanisms in the aldehyde oxidation, via dehydrogenation of the hydrate and a Baeyer-Villiger type fragmentation of the peroxyhydrate. Thus, the parent benzyl alcohol and the substrates bearing an electron-withdrawing substituent can be converted to the benzoic acids, simply by changing the amount of H_2O_2 (Eq. 2). Benzyl alcohols with an electron-donating group do form the benzaldehydes in good yields (Eq. 3) but the products undergo further complicated oxidative conversion via aryl formate intermediates.

This finding urged us to rethink of a seemingly simple aldehyde to carboxylic acid oxidation. Experienced practitioners, I am afraid, are often fettered by nonscientific common sense. Conceiving without proof that aqueous H_2O_2 has only a weak ability to oxidize aldehydes, chemists have tried to develop efficient metal catalysts facilitating the reaction. Our reinvestigation revealed that various unbranched carbaldehydes and aromatic aldehydes, but not *p*-methoxybenzaldehyde, are effectively oxidized *without any metal catalysts* or organic solvents under aqueous-organic biphasic conditions, but with assistance of a lipophilic acidic PTC and rapid stirring (**Fig. 7.41.**, Eq. 4) **[332]**. Thus, heating a mixture of octanal, 30 % H_2O_2, and $Q^+HSO_4^-$ (200 : 300 : 1 molar ratio)

[332] K. Sato, M. Hyodo, J. Takagi, M. Aoki, and R. Noyori: Hydrogen Peroxide Oxidation of Aldehydes to Carboxylic Acids: An Organic Solvent-, Halide- and Metal-Free Procedure. *Tetrahedron Lett.* **2000**, *41*, 1439–1442.

at 90 °C while stirring at 1,000 rpm (revolutions per minute), followed by workup and distillation gave octanoic acid in 85 % yield. The oxidation takes place via a tetrahedral perhydrate intermediate, where the acidic Q^+ salt facilitates both the addition of H_2O_2 to the C=O function and the elimination of water accompanied by a 1,2-hydride shift. Because of the absence of metallic catalysts, primary and secondary alcoholic and olefinic functions survive the oxidation conditions.

Oxidation of sulfides

In a similar manner, sulfides are oxidized to sulfoxides or sulfones with an equimolar amount of 30 % H_2O_2 at 35 °C. In the presence of a catalyst system consisting of Na_2WO_4, phenylphosphonic acid, and $Q^+HSO_4^-$, various aromatic and aliphatic sulfides are convertible to sulfones in a high yield without organic solvents or, if necessary, with some hydrocarbons or ethyl acetate [333]. Thus, diphenyl sulfide is oxidized with a TON of 122,000, two orders magnitude higher than the previous best. **Fig. 7.42.** exhibits examples of some "difficult" sulfide substrates, which are oxidized in 93 – 98 % yield. Coexisting primary or secondary alcohols or olefinic moieties are unaffected under such conditions.

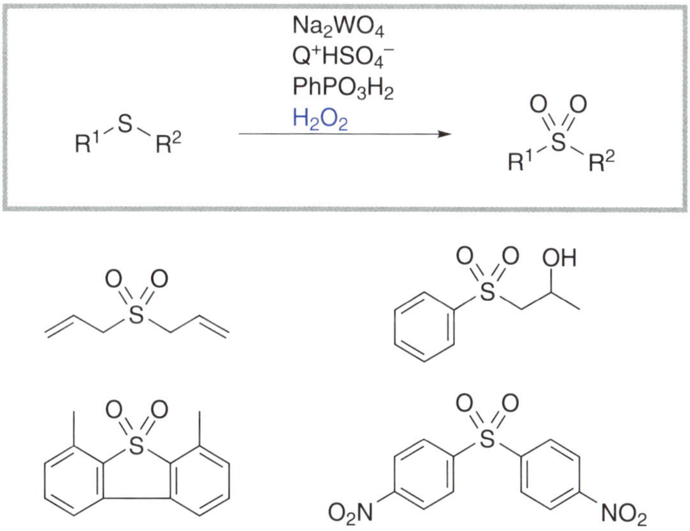

Fig. 7.42. Oxidation of sulfides to sulfones. $Q^+ = CH_3(n\text{-}C_8H_{17})_3N^+$.

Olefin epoxidation

A clean epoxidation method is indispensable in modern chemical synthesis, where the alcohol / olefin chemoselectivity is always a major concern. In our oxidation, as illustrated in **Fig. 7.40.**, the bisperoxo W species D dehydrogenates alcohols via E and then F but D would epoxidize olefins as well. In fact, the selectivity in the oxidation of olefinic alcohols is highly affected by the relative location of the two functionalities and the olefin substitution pattern [329]. As shown in Fig. 7.43., under the standard oxidation conditions, secondary alcohols with a terminal olefinic bond gives the unsaturated ketone preferentially, whereas a highly substituted allylic alcohol affords the epoxy alcohol with a near perfect selectivity leaving the hydroxyl intact.

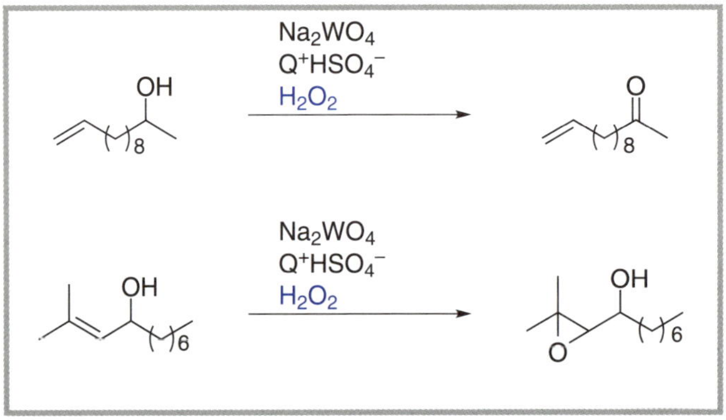

Fig. 7.43. Chemoselective oxidation of olefinic alcohols. $Q^+ = CH_3(n\text{-}C_8H_{17})_3N^+$.

No practical procedure for epoxidation of terminal olefins existed in the 1990s. The problem was very clear. We had to develop a reliable method to epoxidize the least reactive olefins such as 1-dodecene with 30 % H_2O_2 without organic solvents and any halides. Searching for suitable conditions solving this fundamental subject, we could discover that a ternary system consisting of Na_2WO_4, (aminomethyl)phosphonic acid, and $Q^+HSO_4^-$ efficiently catalyzes epoxidation of simple olefins. As illustrated in Fig. 7.44., the epoxidation proceeds consistently in a high yield without organic solvent. Toluene could be used as a cosolvent, when necessary. Notably, both the *phosphonic acid* and lipophilic acidic PTC are essential for the general use [334, 335]. This is in contrast to the standard catalytic conditions that effectively epoxidize certain classes of olefins but are not applicable to terminal olefins.

This cost-effective method is usable for many otherwise unreactive substrates including terminal olefins. Internal olefins with a high electron density are more reactive. This catalyst system displays a TON as high as 150 – 400 per W atom with terminal olefins and 1,600 with cyclooctene. (*Z*)- and (*E*)-1,2-disubstituted olefins produce epoxides with retention of the configuration and higher reactivity for the *Z* isomers, ranging from a *Z*/*E* preference of 7.3 : 1 for 3-octene to > 27 : 1 for β-methylstyrene. In the absence of an (aminomethyl)phosphonic additive, the structural sensitivity remains rather moderate.

This method allows the epoxidation of a range of functionalized olefins containing an ester, ether, or α,β-enone linkage. A unique reactivity and selectivity are seen in the oxidation of allylic alcohols. Notably,

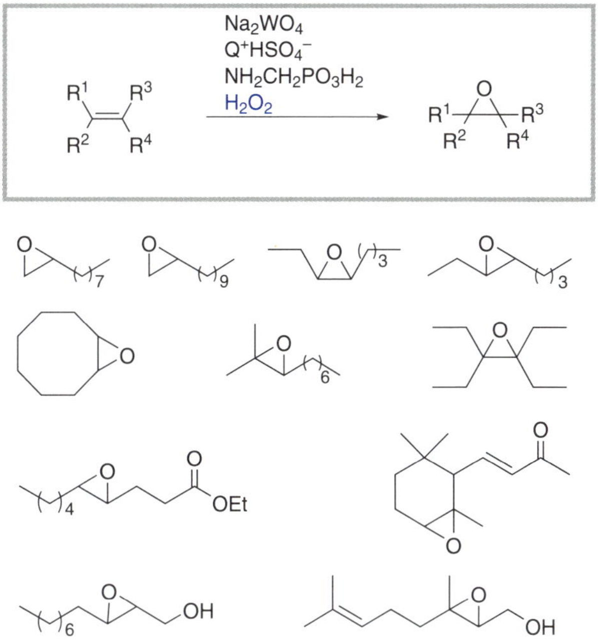

Fig. 7.44. Epoxidation of simple and functionalized olefins. $Q^+ = CH_3(n\text{-}C_8H_{17})_3N^+$.

geraniol was oxidized rapidly even at 0 °C to give only C(2)–C(3) epoxide, leaving the C(6)–C(7) double bond and the hydroxyl untouched. Oxidation of 2-cyclohexen-1-ol occurs selectively from the face of the hydroxy function (*syn*:*anti* = 9:1), while the pivalate of cyclohexen-1-ol exhibited an opposite diastereoselectivity (*syn*:*anti* = 1:12). The relative reactivity of 2-cyclohexen-1-ol, cyclohexene, and pivalate of 2-cyclohexen-1-ol is 170:1:0.05.

The presence of (aminomethyl)phosponic acid accelerates the epoxidation of 1-octene by a factor of 31 compared to its absence. Phenylphosphonic acid and even phosphoric acid, produced from (aminomethyl)phosphonic acid during the reaction, can be used as an accelerator of epoxidation. Although the origin of the marked additive effects remained unclear, we proposed the mechanistic cycle of **Fig. 7.45.** The coordination of a phosphonate or phosphate anion to the W center would increase the electrophilic nature of the peroxo moiety in A, which undergoes the oxygen transfer via transition state E, which is consistent with the conspicuous *Z* olefin preference. With very reactive allylic alcohols, the hydroxy and olefinic bonds are linked by a methylene unit, facilitating epoxidation via transition state F.

We found that, as illustrated in **Fig. 7.46.**, cycloolefins with a *Z*-configuration are particularly reactive under the standard biphasic conditions, resulting in the oxidative cleavage products [336]. In most cases, the (aminomethyl)phosphonic acid additive is unnecessary.

The utility was best illustrated by its application in the synthesis of adipic acid. Currently, its industrial production mostly uses nitric

Fig. 7.45.
Mechanism of epoxidation. $Q^+ = CH_3(n\text{-}C_8H_{17})_3N^+$, X = HO or RP(O)(OH)O. (a) Catalytic cycle. (b) Transition states.

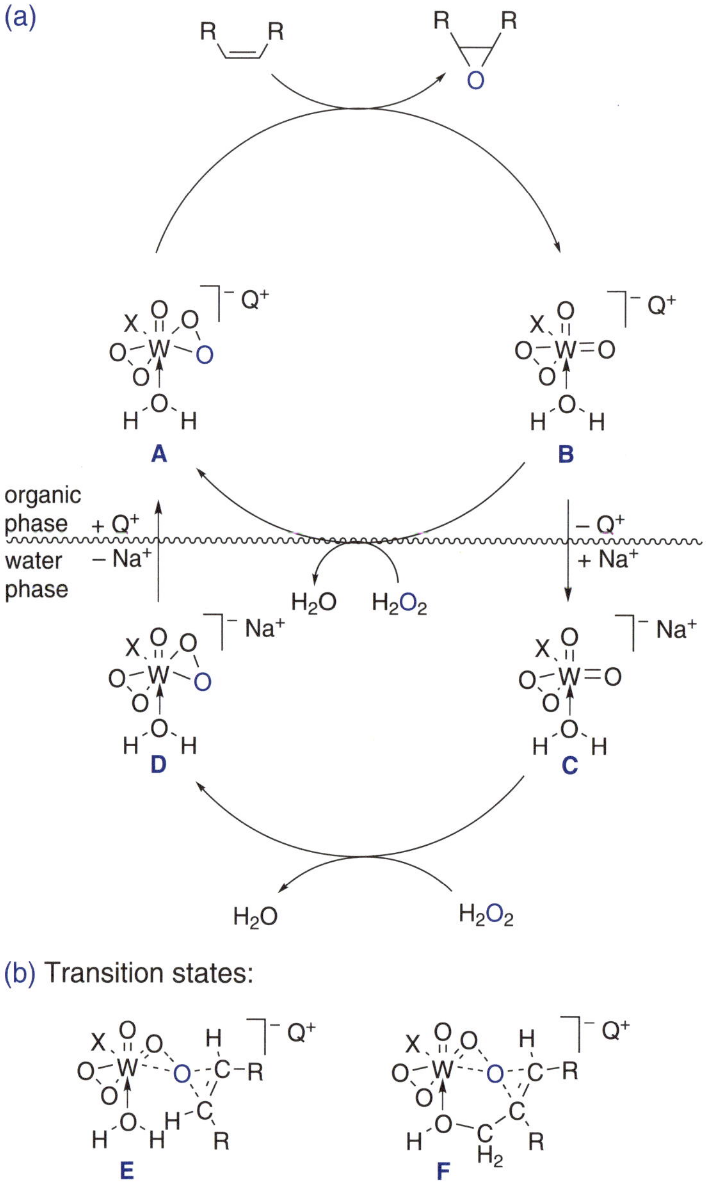

acid oxidation of cyclohexanol or a cyclohexanol / cyclohexanone mixture (Alcohol / Ketone (AK) oil), which is accessible by a low-conversion aerobic oxidation of benzene. The worldwide production of up to circa three million metric tons per year is mostly used to manufacture nylon-6,6 [P]. However, this process produces an inevitable, stoichiometric waste, nitrous oxide (N_2O) that causes global warming and ozone depletion [Q]. Thus, 400,000 metric tons of N_2O are emitted each year (out of 730,000-ton total anthropogenic emission),

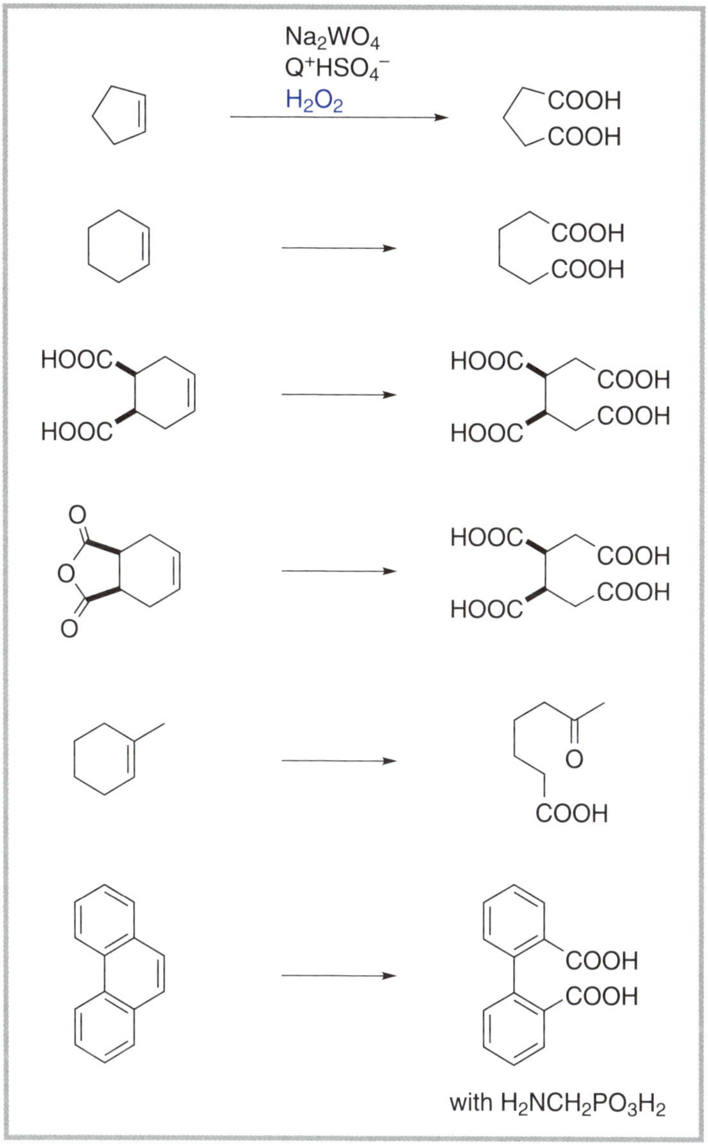

Fig. 7.46. Oxidative cleavage of cycloolefins. $Q^+ = CH_3(n\text{-}C_8H_{17})_3N^+$.

which is to be effectively recovered and decomposed at a high temperature using a huge amount of energy.

In this regard, cyclohexene is a very attractive C_6 starting material, and Asahi Chemical Industry (now Asahi Kasei Corporation) in Japan established an excellent technology for its production by selective hydrogenation of benzene. Thus, when a mixture of cyclohexene, circa 4 molar equivalents of 30 % H_2O_2, and a small amount of Na_2WO_4 and $Q^+HSO_4^-$ was stirred at 75 – 90 °C, adipic acid is directly obtained as shiny, colorless,

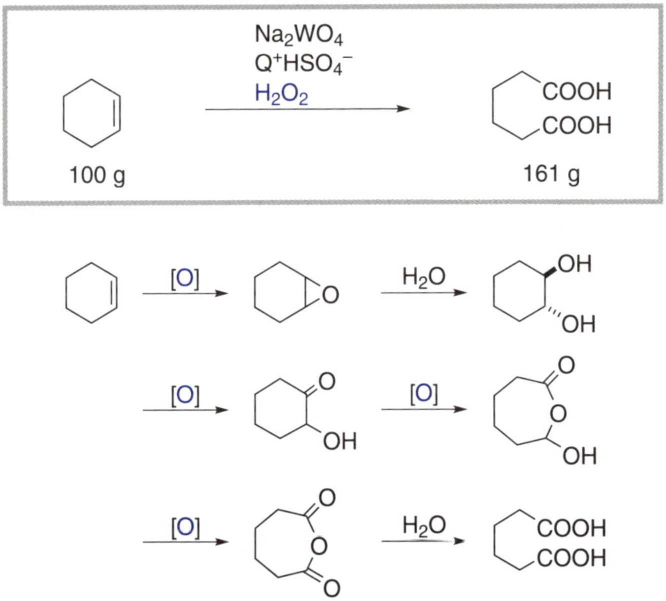

Figs. 7.47. Green route to adipic acid.

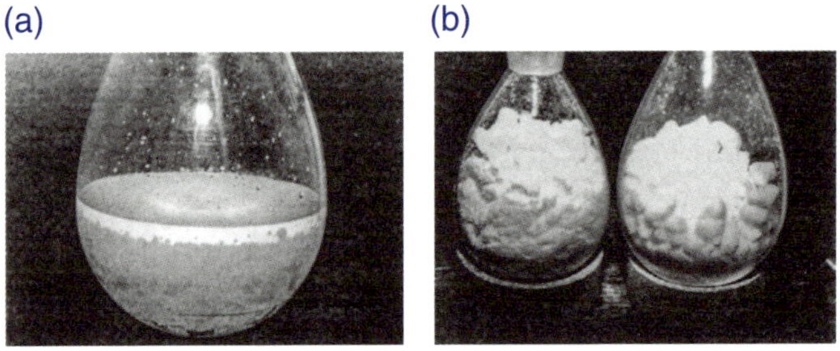

Figs. 7.48. (a) Reaction mixture and (b) filtered products, first run (left) and second run (right).

analytically pure crystals in high yield, as illustrated in **Figs. 7.47.–7.48.** The only byproduct is water. This one-pot transformation is achieved via a six-step scheme involving three kinds of oxidations and two hydrolytic reactions, most of which are facilitated by acidic conditions.

This H_2O_2 oxidation is clean, safe, and reproducible with conditions that are less corrosive than those of nitric acid oxidation. No operational problems are foreseen for scaling up the reaction, except for the energy-intensive recovery of purified water. The practicality of this green process depends largely on the cost of H_2O_2 relative to nitric acid and the expenses for waste disposal / recovery / recycling, as well as the stringency of restriction regarding N_2O emission.

In any event, the general advantage of the halogen- and organic solvent-free oxidation is obvious. It is used in practical production of epoxides for synthesizing various materials with high values added, such as ultra-long-life resistance for manufacturing semiconductors. Kazuhiko Sato, a major contributor at Nagoya and the former Director of Research Center of Catalytic Chemistry at the National Institute of Advanced

Industrial Science and Technology (AIST) has been trying to improve this already attractive method by inventing a continuous flow apparatus.

Our general 30 % H_2O_2 oxidation not only enjoys a wide synthetic scope but also promotes green chemistry. This clean technology was highly educational to convince students of the significance of the environmentally benign chemistry. The adipic acid synthesis protocol was demonstrated at universities classes in the US, Australia, and Japan.

7.5 Asymmetric catalysis in the real world

Fig. 7.49. Kotaro Honda (1870 – 1954).

I spent most of my research career in academia, but I strongly sympathize with Kotaro Honda (1870 – 1954), called a father of metallurgy in Japan, who said, "Industry is an academic *dojo* (training hall for Japan's traditional military arts)." We must bear in mind that any technologies (not scientific knowledge) are to meet with technical, economic, and environmental requirements. Without doubt, chemical products tend to be increasingly dependent on catalytic synthesis, and better catalytic efficiency leads directly to a number of scientific and economic benefits. Hydrogenation represents about 14 % of all chemical reactions in the pharmaceutical industry. Therefore, asymmetric catalysis has enormous potential in the chemical manufacture of pharmaceuticals, animal health products, agrochemicals, fungicides, pheromones, flavors, and fragrances. The BINAP-transition metal catalytic systems have offered unprecedented scientific generality. However, we were unable judge ourselves whether our inventions met the established industrial standard of the time.

Takasago International Corporation provided a *dojo* to approve the technological potential of our science. This long-term industrial collaborator of mine accomplished numerous large-scale AH and ATH (asymmetric transfer hydrogenation) processes using various architecturally manipulated catalysts for the synthesis of biologically important chiral compounds (**Fig. 7.50.**). Takasago Corporation uses our green methods to manufacture more than 3,500 tons of optically active compounds, including menthol (**CHAPTER 6**), per annum and produces more than 350 tons for pharmaceutical use. Some examples are listed in **Fig. 7.51.** The antibacterial levofloxacin and carbapenem antibiotics were commercialized at Daiichi Pharmaceutical (now Daiichi-Sankyo) and Merck Company, respectively.

I am no longer able to follow the whereabouts of these general catalytic processes. The AH technology was applied worldwide in contract manufacturing, sales of catalysts and ligands, contract research, and patent licensing. The patent rights owned by Japan Science and Technology Agency (JST), a public agency, which supported the ERATO Noyori project in 1991 – 1996, were transferred to Takasago, Chirotech

Technology in the UK, Dow Pharma in the US, and Dr. Reddy's Laboratories in India, among others. Obviously, the endeavor of our academic laboratories at Nagoya was very limited, although many talented young colleagues and students contributed to these accomplishments. The cooperation with industry truly helped to maximize the utility of our original scientific findings, and AH technology is now indispensable in the chemical industry [R]. For example, the Novartis Company in Switzerland produced a famous herbicide Metolachlor in 2 million tonnage quantities by Ir / XYLIPHOS-catalyzed AH of an imine substrate [S].

Fig. 7.50. (opposite page) A large-scale BINAP-Ru catalyzed hydrogenation at Takasago International Co. Volume: 10 m^3, Pressure resistance: 50 atm.

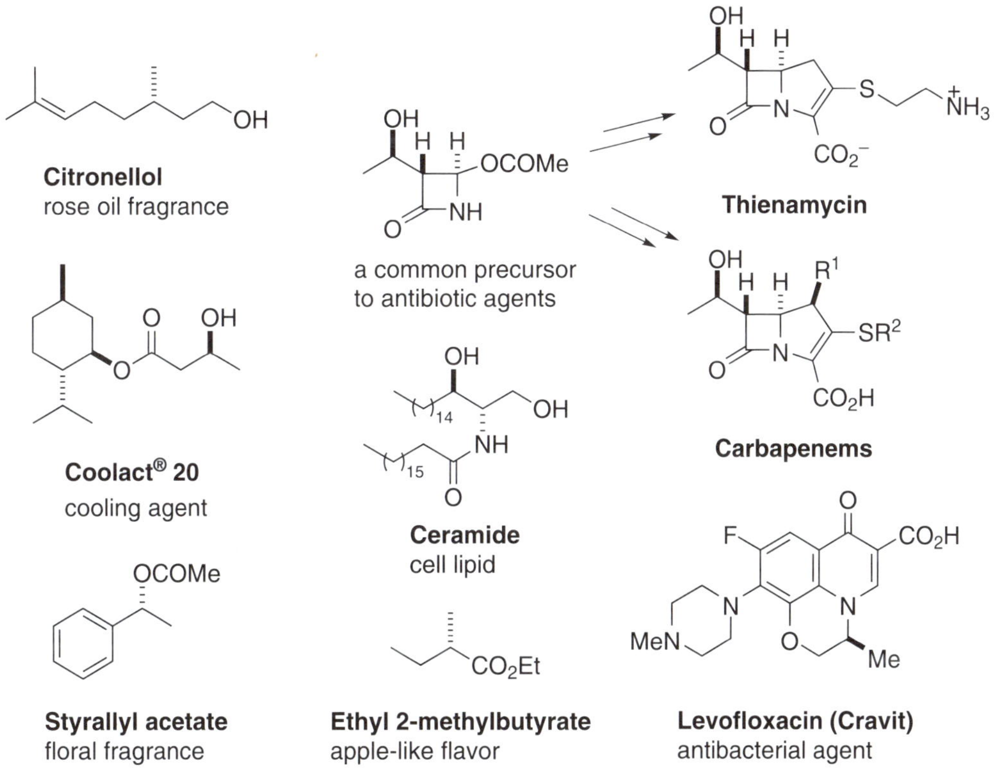

Fig. 7.51. Industrial applications of the asymmetric hydrogenation.

Looking back on the past five decades, I have seen AH grow from a single, bright point of light (Rh) to a long, thick line (Rh to Ru, Ir, and others), and anticipate the continued extension of this remarkable science / technology as the thick line becomes a plane or even a three-dimensional space [255 – 257, 337]. The possibilities of this core chemistry / technology are truly unlimited.

In 1983, prior to establishing these AH technologies, Japan's academia-industry collaboration made (–)-menthol production possible (**CHAPTER 6**). Now what then was the major social impact of our science, aside from the obvious economic benefits? In the early 1990s, reflecting the difficulty of practical asymmetric synthesis, 88 % of synthetic chiral drugs were still racemic mixtures of both enantiomers [T]. This situation was obviously irrational from a scientific point of view. Any problems arising from inappropriate chirality recognition in the human body should be avoided at all costs. Thus, in 1992, the US Food and Drug Administration (FDA) set out guidelines for "racemic switches."[U] The new regulations strongly urged pharmaceutical companies to manufacture and commercialize purely single-handed compounds. However, no such large-scale asymmetric synthesis as (–)-menthol production existed anywhere else. I imagine that this successful chiral technology was a contributing factor to this important science-based decision, which later resulted in major improvement in medicine.

We see that the revolutionary advances in asymmetric catalysis technology have entirely changed the synthetic approach to producing performance chemicals that require a high degree of structural precision. Thanks to the progress in asymmetric synthesis based on the endeavors of many researchers of our days, the proportion of single-enantiomer drugs increased dramatically, and the worldwide sales of such drugs reached 123 billion USD in 2000 [V], and even 225 billion USD in 2005, representing 37 % of the total pharmaceutical market of 602 billion USD [W]. The distribution of new pharmaceuticals approved by the US FDA in 2008 was: 63 % single enantiomers, 32 % achiral compounds, and only 5 % racemates. I believe this trend certainly continued to accelerate.

Chemists' ability to devise straightforward and practical chemical synthesis is indispensable to shape the modern civilized society. In retrospect, simple small organometallic molecular catalysts could achieve reaction rates and stereoselectivities rivaling enzymes. This became a core technology in modern chemical synthesis complementary to ever-developing biotechnology. A diverse array of chiral compounds is accessible in a predictable manner by these purely chemical means. Science and technology continue to make a progress. I recall no one could foresee the current stage of organic synthesis when we started to study this subject in the 1960s.

References for chapter 7

[A] H. F. Fisher, E. E. Conn, B. Vennesland, and F. H. Westheimer: The Enzymatic Transfer of Hydrogen. I. The Reaction Catalyzed by Alcohol Dehydrogenase. *J. Biol. Chem.* **1953**, *202*, 687 – 697.

[B] (a) B. Rosenberg, L. Van Camp, E. B. Grimley, and A. J. Thomson: The Inhibition of Growth or Cell Division in Escherichia Coli by Different Ionic Species of Platinum(IV) Complexes. *J. Biol. Chem.* **1967**, *242*, 1347 – 1352; (b) J. Reedijk: The Relevance of Hydrogen Bonding in the Mechanism of Action of Platinum Antitumor Compounds. *Inorg. Chem. Acta* **1992**, *198 – 200*, 873 – 881; (c) *Platinum and Other Metal coordination Compounds in Cancer Chemotherapy 2* (eds.: H. M. Pinto and J. H. Schornagel). Plenum, New York, **1996**; (d) *Cisplatin: Chemistry and Biochemistry of a Leading Anticancer Drug* (ed: B. Lippert). Wiley-VCH, Weinheim, **1999**.

[C] K. Abdur-Rashid, S. E. Clapham, A. Hadzovic, J. N. Harvey, A. J. Lough, R. H. Morris: Mechanism of the Hydrogenation of Ketones Catalyzed by *trans*-Dihydrido(diamine)ruthenium(II) Complexes. *J. Am. Chem. Soc.* **2002**, *124*, 15104 – 15118.

[D] K. Mori: Biochemical Methods in Enantioselective Synthesis of Bioactive Natural Products. *Synlett* **1995**, 1097 – 1109.

[E] K. Mikami, M. Terada, T. Korenaga, Y. Matsumoto, M. Ueki, and R. Angelaud: Asymmetric Activation. *Angew. Chem., Int. Ed.* **2000**, *112*, 3532 – 3556.

[F] P. T. Anastas and J. C. Warner: *Green Chemistry. Theory and Practice*. Oxford University Press, Oxford, **1998**.

[G] B. M. Trost: Atom Economy – A Challenge for Organic Synthesis: Homogeneous Catalysis Leads the Way. *Angew. Chem., Int. Ed.* **1995**, *34*, 259 – 281.

[H] R. A. Sheldon: Organic Synthesis; Past, Present and Future. *Chem. Ind.* **1992**, *23*, 903 – 906.

[I] D. Evans, J. A. Osborn, F. H. Jardine, and G. Wilkinson: Homogeneous Hydrogenation and Hydroformylation using Ruthenium Complexes. *Nature* **1965**, *208*, 1203 – 1204.

[J] P. V. Ramchandran and H. C. Brown: *Reductions in Organic Synthesis: Recent Advances and Practical Applications* (ed: A. F. Abdel-Magid). (ACS symposium series 641, chapter 5.) Am. Chem. Soc., Washington, D. C., **1996**.

[K] M. M. Midland: Asymmetric Reductions with Organoborane Reagents. *Chem. Rev.* **1989**, *89*, 1553 – 1561.

[L] E. J. Corey and C. J. Helal: Reduction of Carbonyl Compounds with Chiral Oxazaborolidine Catalysts: A New Paradigm for Enantioselective Catalysis and a Powerful New Synthetic Method. *Angew. Chem., Int. Ed.* **1998**, *37*, 1986 – 2012.

[M] W. Kuriyama, T. Matsumoto, O. Ogata, Y. Ino, K. Aoki, S. Tanaka, K. Ishida, T. Kobayashi, N. Sayo and T. Saito: Catalytic Hydrogenation of Esters. Development of an Efficient Catalyst and Processes for Synthesising (*R*)-(*R*)-1,2-Propanediol and 2-(*l*-Menthoxy)ethanol. *Org. Process Res. Dev.* **2012**, *16*, 166 – 171.

[N] *Chemical Synthesis using Supercritical Fluids* (eds.: P. G. Jessop and W. Leitner). Wiley-VCH, Weinheim, **1999**.

[O] C. Parmeggiani and F. Cardona: Transition Metal Based Catalysts in the Aerobic Oxidation of Alcohols. *Green Chem.* **2012**, *14*, 547 – 564.

[P] J. Rios, J. Lebeau, T. Yang, S. Li, and M. D. Lynch: A Critical Review on the Progress and Challenges to a More Sustainable, Cost Competitive Synthesis of Adipic Acid. *Green Chem.* **2021**, *23*, 3172 – 3190.

[Q] *World emission of nitrous oxide (N_2O) 2020 Report published*. Press release. National Institute of Environmental Studies, October 5, **2020**. https://www.nies.go.jp/whatsnew/20201005/20201005.html (last accessed Feb. 20, 2025).

[R] *Asymmetric Catalysis on Industrial Scale. Challenges, Approaches and Solutions* (eds.: H.-U. Blaser and E. Schmidt). Wiley-VCH, Weinheim, **2004**.

[S] H.-U. Blaser: The Chiral Switch of (*S*)-(*S*)-Metolachlor: A Personal Account of an Industrial Odyssey in Asymmetric Catalysis. *Adv. Synth. Catal.* **2002**, *344*, 17 – 31.

[T] (a) Instrumentation '90. *Chem. Eng. News* **1990**, *68*, 26 – 66; (b) S. Borman: Chirality Emerges as Key Issue in Pharmaceutical Research. *Chem. Eng. News* **1990**, *68*, 9 – 14.

[U] S. C. Stinson: Chiral Drugs: In Wake of New FDA Guidelines, Most Drug Firms are Developing Single Enantiomers, Spawning a "chirotechnology" Industry. *Chem. Eng. News* **1992**. *70*, 46 – 79.

[V] S. C. Stinson: Chiral Chemistry: Driven by the Needs of the Drug Industry and Fueled by the Ingenuity of Chemists, Sales of Single-enantiomer Chiral Compounds Keep Accelerating. *Chem. Eng. News* **2001**, *79*, 45 – 57.

[W] G.-Q. Lin, J.-G. Zhang, and J.-F. Cheng: *Chiral Drugs: Chemistry and Biological Action* (eds.: G.-Q. Lin, Q.-D. You, and J.-F. Cheng). First Edition, chapter 1. John Wiley & Sons, **2011**.

THE NOBEL PRIZE IN CHEMISTRY (2001)

8

 DOI.ORG/10.47261/1564-8

In 2001, the Royal Swedish Academy of Sciences recognized my accomplishment at Nagoya. I shared the Nobel Prize in Chemistry with William S. Knowles of Monsanto Company and K. Barry Sharpless of the Scripps Research Institute for our pioneering achievement on asymmetric catalysis. Arising from the fundamental significance of asymmetric synthesis, after 20 years, the 2021 Prize was awarded to Benjamin List and David MacMillan who both pioneered the development of metal-free, asymmetric "organocatalysis".

8.1 A phone call from the Swedish Academy of Sciences

At around 6:30 p.m. on October 10, 2001, I received a phone call from Erling Norrby, who was the Secretary General of the Royal Swedish Academy of Sciences, and was informed that I had been selected to receive the Nobel Prize in Chemistry together with Knowles and Sharpless. I had just come back to Nagoya after a two-week trip to Europe and the United States and was trying to organize myself in my office. The first thing I did was call my wife, Hiroko.

At around 6:45 p.m., our names appeared on the Nobel Foundation's website. Masato Kitamura and my collaborators rushed to my office. Unable to understand my own situation, I had no choice but to follow the instructions of the university's administrative staff. Although I was unprepared for this situation, a press conference was held at 7:00 p.m. The conference room was filled with excitement. Among the many reporters in attendance was my elder son Eiji, who was a reporter for the Yomiuri Shimbun and the representative of the press to ask questions. I started by expressing my gratitude to Nagoya University for allowing me to freely conduct research for more than 30 years, and to the students and collaborators who have cooperated with me. During the Q & A session, I talked about creativity in research, the high level of chemistry research in Japan, the importance of international exchange, and the application of basic scientific research into industry. While the conference took place, various television news reports began to appear, and special editions of newspapers were distributed, but I was unaware of any of these developments apart from the phone call from the Swedish Academy. The Prime Minister at the time, Junichiro Koizumi, congratulated me on the phone saying, "Today is October tenth, and the tenth Japanese person (10 – 10 – 10) will be awarded the Nobel Prize. It will give encouragement to researchers." A phone conference was set up with the ministers and other government officials, as well as the past Nobel Laureates Leo Esaki and Hideki Shirakawa. Before I knew it, a banquet had been prepared for the celebration. I returned home after midnight, but many reporters were waiting for me, and my house was filled with an unusual hustle and bustle.

Fig. 8.1. Author raising a late-night toast with Hiroko after receiving the call from the Nobel Foundation.

After the reporters left, Hiroko, Eiji, and I shared champagne and wine and stayed up talking until the early hours about everything that had happened to each of us since the news broke (**Fig. 8.1.**).

Norrby of the Royal Academy of Sciences sent me an official award notification and an explanation of my achievements by fax dated October 10. The fax stated that William S. Knowles and I were recognized for "their work on chirally catalysed hydrogenation reactions" and K. Barry Sharpless was awarded for "his work on chirally catalysed oxidation reactions." The letter also stated that the prize would be presented by the King of Sweden in Stockholm on December 10, that the 45-minute prize lecture is to be given on December 8, that the audience for the lecture would include not only scientists but also students and the general public, and that the content of the lecture would be published in the "Les Prix Nobel" yearbook, so a manuscript of the lecture must be submitted.

Soon after, the Nobel Foundation sent me the details of the related events in Stockholm. The year 2001 was actually the 100th anniversary of the establishment of the Nobel Prize. In addition to the usual events, there would be special commemorative events, and an extremely busy and demanding official schedule had been set. Participants were divided into the Laureates, their spouses and families, and invitees. The Laureates were asked to quickly select their invitees, and I asked my wife Hiroko, my sons Eiji and Koji (a painter), my mother Suzuko (who lived in Kobe and was 87 years old), and my sister Yoshiko Murahashi to assist our mother.

To represent my many collaborators, I invited Masato Kitamura, Yoshihiro Hayakawa, Masaaki Suzuki, Takao Ikariya, and Takeshi Ohkuma, who served as Associate Professors in my lab, as well as my secretary, Yuriko Nakamura to accompany me. The number of participants in the Nobel Concert, Prize Award Ceremony (1200 seats), Nobel Banquet, Nobel Foundation's Reception, and other events was limited, and I believe I caused some inconvenience to many of my guests. Each event had its own dress code, including formal attire, business suits, and casual wear.

During this period, I was under inexperienced pressure from the general public, both publicly and privately, including media interviews by television, newspapers, and other agencies, along with roundtable discussions with publishers. Although my initial response would be "no" to each request, there were many that I had no choice but to accept. At the time, I was also extremely busy with my positions at the Ministry of Education, Culture, Sports, Science and Technology (MEXT) and the Chemical Society of Japan (CSJ). Prime Minister Koizumi invited me to lunch at his official residence, where I spoke about the current state of the scientific community and asked for his support.

What was most troublesome was the interview with a Swedish television program. They said that they wanted to come to Japan to film not only my office and laboratory at the university, but also my family's daily life and other personal aspects. They came to visit my home and filmed not only my study and parlor, but also the *tatami* (Japanese traditional straw mat) room where we sleep and get up. They said it was to arouse public interest in science by showing the private lives of Nobel Laureates, but honestly speaking, this was not so much in my favor.

In any case, my greatest duty as a Nobel Laureate was to give a lecture on December 8 in front of the past Laureates, scientists, students, and other members of the public, and I could not ask anyone to help me prepare for this event. I managed to find time in my overcrowded schedule to work on the manuscript, but I was unable to concentrate as I wished. I was also asked to write a short autobiographical essay, but I lacked experience in this matter as well, and my writing did not move on at all. At the same time, I had to prepare a manuscript of an article for publication in the annual report of the Nobel Foundation.

8.2 The Nobel Week in Stockholm

The day of departure for the Nobel Prize week arrived in no time. I had traveled abroad countless times, but this time, I was extremely nervous. On November 30, a send-off party was held at the Swedish Embassy in Tokyo prior to my departure, and on the morning of December 5, I left for Stockholm with my family. Our gracious hosts provided us with rooms at the famous Grand Hotel, a competent local guide, a limousine with a private chauffeur, and we were told any wishes we had could be arranged. But the schedule was very busy, especially because of the 100th anniversary celebrations, with more than 160 past Laureates in attendance. The schedule for the Nobel week was explained to us, and although we were told to make any requests we wished, we were given invitations and admission tickets to all events, so we had no choice but to follow the instructions of our hosts. I was also asked to attend unfamiliar press conferences and televised debates. The winter days in the north were extremely short, and many events were held under electric lamps.

Notably, this commemorative Nobel event took place only three months after the 9/11 terrorist attacks in the US. Safety inspections for the gathering of dignitaries were extremely strict, and everything was carried out on high alert. However, those involved were determined that the Nobel Prize would last for another century, even if the country was destroyed.

The day after our arrival, I met with many renowned scientists at the Bolinderska Palatset. On December 7, the 2001 Laureates met to prepare for the television discussion "Nobel Minds," there was an orientation for the Laureates in Chemistry, as well as a press conference and a reception.

Along with the award ceremony, the most nerve-wracking event was the commemorative lecture in the main auditorium of Stockholm University on December 8. In the presence of many senior Nobel Laureates in Chemistry, I gave a 40-minute lecture titled "Asymmetric Catalysis: Science and Opportunities." E. J. Corey, my mentor at Harvard, congratulated me on my lecture, and we shook hands, smiling at each other. Flowers were delivered to my hotel room by his courtesy. After an intensive evening, I was able to relax at the centennial Nobel Concert. Its final piece was "The Impossible Dream (The Quest)" from *Man of La Mancha*, which seemed to me as if they were celebrating the reckless life of the Japanese Don Quixote.

On the evening of December 10, the death day of Alfred Nobel, the award ceremony was held at the Royal Concert Hall in Stockholm. It was a solemn and splendid ceremony backed by the 100-year history of the prize. At each point in the strictly prescribed program, the music by the Royal Stockholm Philharmonic Orchestra, conducted by Alan Gilbert, and the vocal by Anne Sofie von Otter, created a brilliant atmosphere. The stage was set up with Nobel's statue, and the seats were occupied by members of the royal family, the Nobel Foundation, the selection committee, and many senior Laureates. The King of Sweden, Carl XVI Gustaf appeared on stage with a fanfare of orchestral music and a standing ovation. Subsequently, the Laureates entered the hall and took their seats in the front row. We were bathed in dazzling lights from above. The seats on the first three floors were filled with people in tails and evening gowns, including my family.

Per Ahlberg, an organic chemist, was in charge of explaining the achievements of the Nobel Prize in Chemistry on behalf of the selection committee with Bengt Nordén as the chairman. Ahlberg was considerate to first congratulate me in Japanese and then stated, "The man who led the further development resulting in today's more selective and general hydrogenation catalysts was Ryoji Noyori. A single one of his catalyst molecules can yield millions of product molecules." Concluding, Ahlberg said, "I have focused especially on the industrial role of the Laureates' discoveries, but they also constitute extremely important tools in academic research, thereby contributing to more rapid advances in research – not only in chemistry but also in materials science, biology, and medicine … This year's Nobel Prize in Chemistry not only makes it easy to answer what science is good for, but is also a prize entirely in the spirit of Alfred Nobel … The beneficial consequences of your achievements to mankind are already plentiful and will certainly be amplified by the work of your numerous followers in the field of chiral catalysis." I felt much obliged by these warm compliments and the perspective on the future. After explaining the Laureates' accomplishments, the Laureates stepped up to the N-shaped mark in the center of the stage to receive their prize. The King of Sweden congratulated each of us individually on our research achievements, shook our hands, and handed us the Nobel Prize diploma and a gold medal (**Fig. 8.2.**). I felt that my life of research had been rewarded. My heart was flooded with emotion as I was on the stage of the award ceremony to receive the prize with the pride of my country and with the Order of Culture on my chest.

I wonder what my family were thinking as they attended the award ceremony. After the ceremony, the stage was opened to all the attendees, and my family and I took a memorial photo in front of the podium with Nobel's statue (**Fig. 8.3.**). My 87-year-old mother, Suzuko, looked satisfied. For some reason, she had a habit of not sharing her thoughts directly with me on such occasions, nevertheless, I was pleased to pay her back for her wit, patience, and devotion by sharing this rare occasion with her.

The renowned artist Nils G. Stenqvist created the design of the award certificate (**Fig. 8.4.**). Each certificate is different, and I wonder how he was able to complete all of them in just two months. My certificate is illustrated with a beautiful right-handed shell mirrored by an ugly left-handed shell, symbolizing the importance of the difference between right- and left-handed molecules. In the background are many circles, each with a dot in the center. This is the atomic symbol for hydrogen, invented in 1803 by John Dalton, the founder of modern atomic theory, and represents the achievement of "asymmetric hydrogenation reactions" as a whole. The gold medal is plated in 18 k and 24 k gold, weighs 200 grams, and features the portrait of Alfred Nobel on the front and the goddess Isis holding the horn of fertility on the back.

Fig. 8.2. (opposite page) Author conferred the Nobel Prize in Chemistry from King Carl XVI Gustaf. December 10, 2001.

Fig. 8.3. On the stage of the award ceremony. From the left: Hiroko, Koji, Eiji, and the author.

At the grand dinner at Stockholm City Hall after the award ceremony, I had a chance to talk with the people sitting next to me, Nadira Naipaul, the wife of Vidiadhar Naipaul, Laureate of the Literature Prize, and Mary Collins, the wife of Tim Hunt, Laureate of the Physiology or Medicine Prize. I also had the honor of sitting with Queen Silvia at the royal banquet in the Palace. On December 12, the United Nations Secretary-General Kofi Annan, who had just received the Peace Prize in Oslo, also arrived at the Nobel Foundation, bringing all the Nobel Laureates together.

The rest of the week was spent visiting several academic institutions. On December 13, I moved to Uppsala to present a lecture hosted by Bengt Långström at the Chemical Society of Uppsala and to tour the University Hospital with PET (Positron Emission Tomography) equipment and then went back to Stockholm for a Lucia Dinner hosted by the Student Union of Stockholm University. Finally, on December 14, I flew to Gothenburg to give a lecture at Gothenburg University / Chalmers University of Technology, and my wife and I were invited to the home of Per Ahlberg.

I managed to complete the entire Nobel week and indeed it was a series of events condensed with the glorious history of science, and I was relieved to have finally fulfilled my major role. However, this was not the end of the road. I felt that this was still another starting point. On December 15, I flew back to Tokyo, thinking about the unpredictable life that awaited me.

Fig. 8.4. Diploma of the 2001 Nobel Prize in Chemistry. Designed by Nils G. Stenqvist.

8.3 Nobel Symposia on Asymmetric Synthesis

Readers of this book might wonder when the Nobel Foundation started to recognize the importance of our research subject, "asymmetric synthesis." In the mid-1970s, the production of single-handed molecules had long been monopolized by microbial processes, but the chemical community began to acquire confidence about the practicality of asymmetric synthesis and the area eventually became the center of synthetic organic chemistry. However, it was not until 1984 that the Foundation first made it the subject of a Nobel Symposium. This important meeting organized by Göran Bergson and Salo Gronowitz was held at the beautiful Björkborn Manor, Alfred Nobel's former residence and brought together a diverse group of researchers in this field. Invited speakers were Stephen Hanessian, Teruaki Mukaiyama, Albert I. Meyers, Torbjörn Norin, Clayton H. Heathcock, Brice Bosnich, Reinhard W. Hoffmann, Tamio Hayashi, K. Barry Sharpless, Mathias Nilsson, Meir Lahav, Henri B. Kagan, Svante

Fig. 8.5. The Nobel Symposium on Asymmetric Synthesis in 1984.

Brandänge, Bertram Fraser-Reid, Hans Wynberg, Martin F. Semmelhack, Dieter Enders, Guy Solladié, Gary H. Posner, and Herbert C. Brown and his wife were at the center of this gathering (**Fig. 8.5.**). We discussed not only catalytic reactions but also trends in asymmetric synthesis in general, and I talked about the chiral BINAL-H asymmetric reagent (**CHAPTER 5**) and the newly invented BINAP-Rh catalyst (**CHAPTER 6**). As a consequence, the practicality of this subject was finally recognized.

However, this Nobel Foundation symposium did not end there and was held again 11 years later in 1995 in Tammsvik, west of Stockholm. This time Torbjörn Norin was the chairman and the event focused on Catalytic Asymmetric Synthesis, a subject of greater practical value. Tamio Hayashi, Henri Kagan, Sei Otsuka, and others talked about metal catalysts, Julius Rebek, Jr. discussed about molecular recognition, Richard Lerner, Peter Schulz, Ehud Keinan, Anthony Kirby, and Donald Hilvert talked about the development of catalytic antibodies, and Chi-Huey Wong, J. Bryan Jones, Steven Benner, and Hiromichi Ohta emphasized the importance of enzyme reactions. K. Barry Sharpless gave an opening lecture titled

Fig. 8.6. The 2001 Nobel Laureates in Chemistry. K. Barry Sharpless (left) and author. William S. Knowles is standing behind.

"Noncovalent Binding Phenomena in Asymmetric Catalysis" and I gave the closing talk on "Asymmetric Hydrogenation" [290]. The symposium was a great success, and this chemistry became the subject of the Nobel Prize six years from then, though the behind-the-scenes details of how it came to be are not known to me (Fig. 8.6.).

[290] R. Noyori: Asymmetric Hydrogenation. *Acta Chem. Scand.* **1996**, *50*, 380–390.

8.4 A tribute to K. Barry Sharpless

K. Barry Sharpless has been my friend for over 50 years. I first met him at Harvard in 1969 (CHAPTER 3). My respected colleague received his second Nobel Prize in Chemistry in 2022. Only the biochemist Frederick Sanger received the Nobel Prize in Chemistry twice (in 1958 and 1980). I would like to express my sincere congratulations to this marvelous accomplishment.

Youth in science is the ability to create new knowledge based on one's intrinsic motivation and is not to be measured by age. Although the first fruits of Nobel Prize achievements are often seen around the age of 40, Sharpless's achievements demonstrate that creativity is not necessarily the exclusive preserve of young people. Laureates later go on to take different paths, but awards are never the end of the road for a researcher, but an inspiration to set new propositions and work toward solutions. Sharpless was not limited by his successful experiences in "asymmetric catalytic reactions," but boldly challenged the possibilities of new concepts on "click chemistry" and revealed their validity. I recall that during the 2001 Nobel Prize Award Lecture, while Knowles and I talked about our "accomplishments," Sharpless shared his passion for a new project that was already underway to develop acetylcholinesterase inhibitors (AchEIs), in addition to his achievements in asymmetric oxidation reactions. I see Sharpless as a person of intuition, full of curiosity, whose research is driven not by logic but by passion, by action rather than by precise planning. He has realized his big dreams one after another, including the epoxidation of olefins, dihydroxylation, and aminohydroxylation reactions. Based on an idea that he came up with around the age of 60, I would like to pay tribute again to his fresh curiosity and courage in taking on a new goal.

I am reminded that in science, everything and everybody is linked together and that no chemistry can ever be based solely on cutting-edge knowledge. The 2022 Nobel Prize is based on the fundamental science of the 1,3-dipolar cycloaddition reaction, which Rolf Huisgen invented in 1961 and established over the course of his entire life. Since meeting Huisgen in 1983 at the Hoechst conference held at the Reisensburg Castle in Germany, I have always admired this perfectionist for his intellect and his neat, cultured behavior. Once again, I would like to pay tribute to the soul of this true scholar.

Fig. 8.7. Rolf Huisgen (1920–2020)

8.5 How I was honored

Upon receiving this honor, I recalled the decades-long journey to Stockholm that I had traveled together with my mentors, colleagues, and family, and how much my country and I had changed in that time. I began my primitive but particular brand of chemistry while my country still suffered from the aftermath of WW II.

My generation survived poverty in childhood and struggled to overcome economic difficulties through our early university years. The living standard first recovered to a reasonable level only in the mid-1970s. Our harsh situation induced us to think differently from researchers in the US and Europe. I do not want any of our next generations to have to undergo the same kind of difficulties as we did, but I am convinced that the poverty that we experienced during our youth helped to make us creative and mentally strong. We knew how to persevere. We were patient, diligent, and hardworking. Secondly, we fully enjoyed our academic freedom without any kind of external pressure or restriction. Curiosity is the true key for promoting academia, not issue-oriented institutional research.

In addition, I believe that Japanese culture permeates every aspect of our research in science and technology. I fully agree with the statement of Louis Pasteur, "Science has no borders, but scientists have their father land." It is obvious that my background is distinctly different from that of European and American scientists. Science cannot stand alone; it must be properly integrated with indigenous national cultural heritage and value ethnic and regional characteristics in approaches to research [A].

8.6 Molecular beauty of BINAP

In the 1984 Nobel Symposium on Asymmetric Organic Synthesis (**Fig. 8.5.**), I delivered a talk titled "Binaphthyls: The Beauty and Chiral Uses" [94]. I emphasized the structural characteristics of the binaphthyl motif relevant to discriminating three-dimensional space and facilitating the reactions. I tried to convince the audience of the relationship between the craft-like beauty of thin bamboo strips (rather than wire or rubber bands) and chemical catalytic activity using pictures of bamboo crafts in Kyoto, such as bamboo groves and *chochin* (Japanese lanterns). I wonder if this non-scientific, artistic postulate was understood by top-notch Western scientists. I suppose not, but I am grateful to my colleagues around the world who tried to understand my senses with generous patience. In any case, this approach appeared to work well, and I presented the scientific / technological fruit in the 1995 Nobel Symposium [290].

The molecular architecture of chiral BINAP-transition metal catalysts is simple and beautiful. The BINAP ligand is a fully aromatic diphosphine, composed simply of four benzene rings, two naphthalene units, and two phosphorous atoms. Both benzene and naphthalene are fundamental organic compounds that are highly symmetrical, rigid, and static in nature. However, the newly emerged BINAP is axially dissymmetric, "pliant" and dynamic, and also functional. Being fully composed of sp^2 carbons, this molecule is architecturally well ordered yet flexible (neither rigid nor flaccid) (Fig. 8.8.).

Fig. 8.8. Molecular flexibility in stereoselective catalysis.

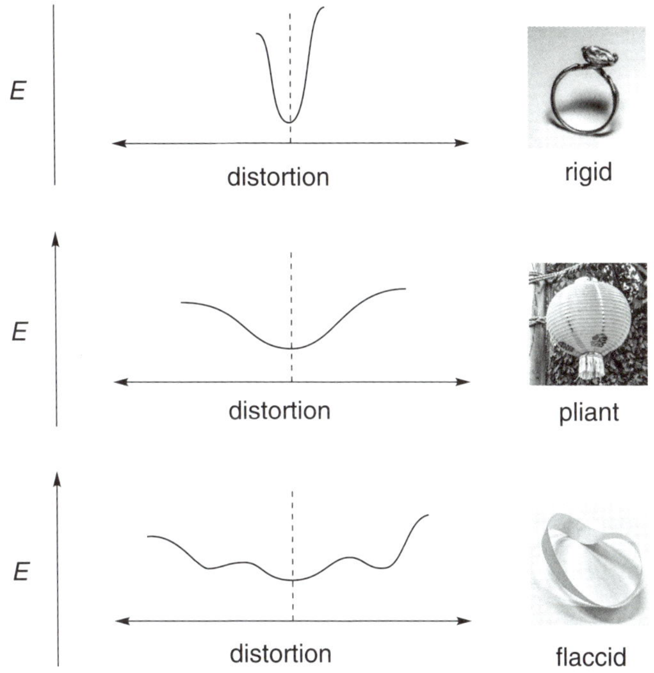

As is seen in Fig. 8.9., BINAP can accommodate various transition metals M with a different P–M bond length in an induced-fit manner through rotating about the binaphthyl Csp^2(1)–Csp^2(1') pivot (not a normal Csp^3–Csp^3 bond) without any serious increase in torsional strain, yet maintaining its skeletal unambiguousness. X-ray crystallographic studies showed that, in going from free BINAP[B] to the metal complexes [221 – 232, 252, 297, 302], the dihedral angle of the two BINAP naphthalene planes (β) is reduced from 93° to even 74 – 66°. The resulting seven-membered metal chelate, in turn, can accept a range of atoms and functionalities by modifying its conformation and rotating the phenyl Csp^2–P bonds. Here, the four phenyl groups are not a simply abbreviated "Ph" but, as a chemical hexagon, actively play a steric or even an electronic role. The square planar BINAP-Rh(norbornadiene) and octahedral

Fig. 8.9.
Pliant structure of BINAP-transition metal complexes.

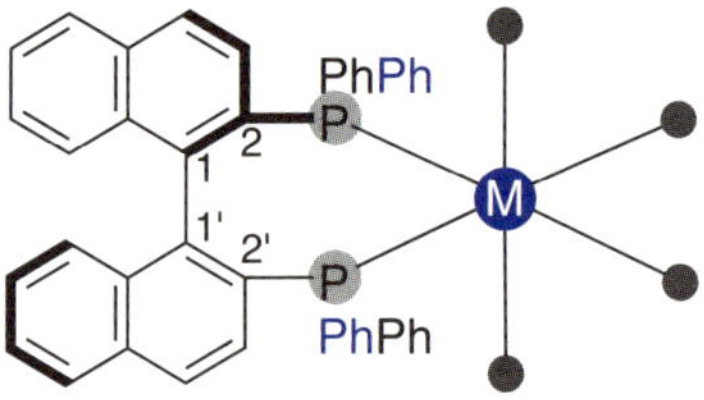

(*R*)-BINAP–M complex (M = Rh, Ru)

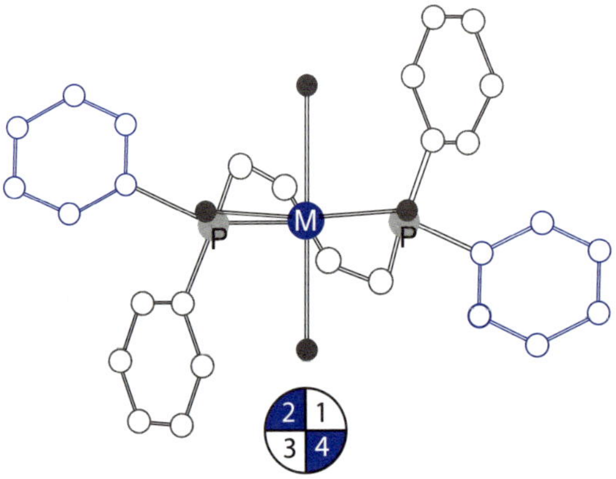

side view (naphthyls are omitted)

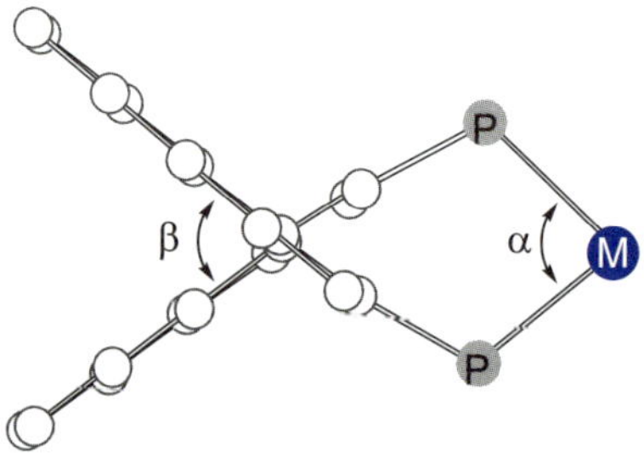

top view (*P*-phenyls and other ligands are omitted)

M	geometry	M–P, Å	α, deg	β, deg
–	–	–	–	93.2
Rh(I)	square planar	2.31	91.8	74.4
Ru(II)	octahedral	2.24	90.6	65.6

BINAP-Ru(OCO^{t}Bu)$_2$ complexes (**CHAPTER 6**) indicate that the BINAP-metal complexes form a highly skewed seven-membered chelate ring with a similar, near-rectangular P–M–P bite angle (α), but a different β dihedral angle which reflects the difference in the M–P bond length.

Overall, the coordination sphere is sufficiently spatial (more room compared with the similar five-membered chelates) and distinctly chiral from a chemists' perspective. Thus, BINAP-based metal catalysts clearly define the asymmetric sense of inter-ligand reaction in the molecular cavity, or sometimes the reaction occurring on its surface. Notably, catalytic reactions often proceed stepwise via several intermediates with a different structure. The skeletal pliancy or flexibility in solution tends to offer the transition-state stabilization, accelerating the catalytic cycle, and differs from structural rigidness, which does not cope with this situation, or flaccidness, which generally leads to inferior stereoselectivity.

8.7 Cultural heritage nurtures scientific creativity

I have been contemplating the origin of unique creativity in scientific research, and I now strongly believe that ethnic culture is important in nurturing scientific originality.

Culture is a strategy of our species for survival. Science and art are undertakings common to all humanity, as seen in the prehistoric drawings of hunting scenes and dynamic renditions of deer and bison in caves and cliff facings around the world. They are both important elements of culture; in fact, they are essential to our very existence. In other words, science and art can be viewed as humanity's dialogue with nature (For example: see [C]). Both the scientist and artist seek through their work to elevate and express themselves, and both enrich our lives spiritually and materially. Science and art originate from the same source, and the only difference is in emphasis [D]. As such, it puzzles me to why, as Charles Percy Snow wondered in his famous Rede Lecture, "The Two Cultures," at Cambridge in 1959 **[E]**, that scientists and artists do not share their intellect, sensibility, and technology to a greater extent. Science and art are said to be frequently at odds with each other, two cultures in conflict that never communicate. If this is true, I am afraid it represents a serious threat to research culture, because excellence in scientific research certainly requires intellect, aesthetic sensitivity, and superior technology.

[E] C. P. Snow: *The Two Cultures and the Scientific Revolution*. Cambridge University Press, Cambridge, **1959**.

The artist is not a scientist and vice versa. Nevertheless, artistic inspiration is strongly linked to the cultural background and scientific and technological events, which are grounded in the same rational foundation. In 1907, Pablo Picasso painted his famous *Les Demoiselles d'Avignon*, replacing traditional perspective with Cubism. According to Arthur Miller, the unmatched originality of this "simultaneous vision" was inspired by a series of modern technologies invented at the end of the 19th century [F]. The advent of the telephone, radio, automobile, airplane, and other inventions convinced Picasso that a whole subject could be portrayed as an entirety only by moving himself to view it from all possible angles. The discovery of X-rays, radioactivity, and electrons were also thought to have played their part in emboldening him to explore this new concept. In both art and science, "seeing may be believing," but the truth is not always visible to the naked eye, as I myself have frequently experienced in catalysis science. Science is objective, but scientific research is a human activity, and a scientific endeavor is a work of art. Science is descriptive out of necessity. However, I have long felt it difficult to properly represent its innermost essence, particularly in a foreign tongue. Regretfully, I have avoided developing stronger and more lucid forms of expression. I feel that the expression of scientific essence is found in both its abstractions and in its concrete representations. The abstractions give science its universal character while its concrete representations have a powerful immediate impact.

References for chapter 8

[A] E. Norrby: *Nobel Prizes and Life Sciences.* World Publishing Co., Singapore, **2010**.

[B] X. Wu, C. Y. Huang, D.-G. Chen, D. Liu, C. Wu, K.-J. Chou, B. Zhang, Y. Wang, Y. Liu, E. Y. Li, W. Zhu, and P.-T. Chou: Exploiting Racemism Enhanced Organic Room-temperature Phosphorescence to Demonstrate Wallach's Rule in the Lighting Chiral Chromophores. *Nat. Commun.* **2020**, *11*, 2145.

[C] (a) I. Stewart: *What Shape is a Snowflake?* The Ivy Press, **2001**; (b) B. Atalay: *Math and the Mona Lisa: The Art and Science of Leonardo da Vinci*. Smithsonian Institution, **2004**; (c) J. D. Barrow: *Cosmic Imagery – Key Images in The History of Science*. The Random House Group Ltd., **2008**; (d) *Medicine and Art: Imaging a Future for Live and Love – Leonardo da Vinci, Okyo, Damien Hirst*. Mori Art Museum, Tokyo, **2009**.

[D] S. Zeki and Balthus: *Balthus ou La Quete de l'essentel*. Les Belles Letters, Paris, **1995**.

[E] C. P. Snow: *The Two Cultures and the Scientific Revolution*. Cambridge University Press, Cambridge, **1959**.

[F] A. I. Miller: *Einstein, Picasso – Space, Time, and the Beauty That Causes Havoc*. Basic Books, **2001**.

NOYORI, THE CHEMIST AND JAPANESE CITIZEN

9

DOI.ORG/10.47261/1564-9

9.1 Born and raised in Japan

I would like to repeat the quote by Louis Pasteur: "Science has no borders, but scientists have their fatherland." Life in chemistry and chemistry in life are inseparable. Without understanding this unique situation, it would be impossible to understand the way of life of researchers. Few people in the wider international research community are aware of the circumstances and events that I have encountered in Japan. I would like to express my gratitude to the editors for their generosity in allowing me to disclose this part.

I am a chemist who was born and raised in Japan. In recent years, English, the lingua franca of the modern world, has become the language of science education in emerging Asian countries, but since the Meiji era (1868 – 1912), Japanese people have learned and studied science in Japanese, which is a language significantly different from English, German, and French. In fact, our pioneers invented words for science, physics, chemistry, hydrogen, carbon, oxidation, reduction, catalyst, and others, using ideographic *kanji* characters to deepen our understanding of nature. While it is undeniably disadvantageous from the perspective of maintaining conformity with international standards and promoting internationalization to not use English, different languages may also give rise to original ideas. Personally, I am not very good at writing or speaking in English, and I am especially nervous when I stand in front of the public. Fortunately, I am an experimental chemist, and I have acquired scientific facts that are not readily available in native English and let the science speak for itself. I would like to recall and share some of the moments in my life in which local cultural approaches and international cooperation in science contributed to the development of my country's universities, scientific community, and hopefully society at large.

1
一
one

10
十
ten

10^2
百
hundred

10^3
千
thousand

10^4
万
ten thousand

10^8
億
hundred million

10^{12}
兆
trillion

10^{16}
京
ten quadrillion

The numbers are each represented by a single *kanji* character.

Being an "only one" creator

Chemistry is the science of matter, and there are an infinite number of interesting research subjects. In my laboratory at Nagoya University, I have always tried to address fundamental topics. I have long said that research should be fresh, simple, and clear, and I have strived to make clear-cut decisions and seek the most straightforward answers. At the same time university researchers need original creativity and should aim to be either "number one" or "the only one."

If you are working on the same problem as others, then your research must be the best to be meaningful. Rarer yet is to do something no one else is doing – to be the only one. Often research that cannot be compared to others is ignored, but it is in fact the most innovative ideas that achieve this status and is something to be proud of.

My lab always followed my own "simple and clear" concept; I would set up a concrete but rough plan and left the subsequent revisions and improvements to the ingenuity of the young researchers. My interests were wide-ranging and rather fluid, and they went far beyond my work on AH reactions that brought me the Nobel Prize. I had a diverse group of collaborators, and together we studied synthetic organic chemistry, organometallic chemistry, molecular catalysis, asymmetric synthesis, new reaction media, green chemistry, physical organic chemistry, and *in vivo* molecular science. Although my main objective has been the synthesis of useful organic substances, under the influence of the surrounding natural product organic chemists, the scope of my work has expanded to include terpenes, alkaloids, antibiotics, prostaglandins, carbohydrates, nucleic acid-related substances, and their artificial derivatives

Research in the natural sciences consists simply of a problem and its solution. A clear answer to an essential problem is highly valued, but this is not easy to achieve. This is because the giants and authorities in any given field have already identified a significant problem, have a research system in place, and are thus ahead of the curve, so it is unlikely that the young people would be able to catch up and overtake them. What is even more difficult, however, is to set up a good problem. It must be one's original problem, not one borrowed from someone else. Vidiadhar Surajprasad Naipaul, who received the Nobel Prize in Literature in 2001, said, "a writer's greatest task is to find a subject to write about. That's half of the job." Like literature, the subject of academic research can be about anything. Therefore, it is very difficult to set up a problem, but it should be selected by one's own experiences, guided by intuition. Finding a problem using only logic is ordinary; the ability to attract good fortune, or serendipity, is necessary. Although it depends solely on personal qualities, there are limits to any person's ability to think. It is the multiplication of different ways of thinking and the accumulation of skills that creates new power. I believe that a good mix of people with different abilities and roles is the best organization for academic research.

Unlike in other countries, Japanese professors often want to keep their students in their own labs for long. In principle, I have avoided this practice and recruited young people from other labs and universities. After working with them for a while, I advise them to move on from my lab and become independent. When they are together with me for a long time, their way of thinking unconsciously becomes similar to mine, and it is time for them to leave the nest. Young talent will not flourish if it remains in a comfortable environment. Each of my young collaborators have been recognized by academic societies with awards of encouragement, and they have broken new ground. Even as my younger students moved on from Nagoya University, they remained dedicated to chemistry. We worked as hard as anyone else (except for E. J. Corey in the 1960s as far as I know) with a concentration that was second to none, and we continued to work autonomously and happily. I look back on it as an extremely healthy experience from a mental perspective.

During my tenure at Nagoya University, I supervised about 160 students and collaborated with 60 postdoctoral researchers and research students. Japanese universities were slow to internationalize, and the initial members of my lab were all Japanese. When my lab began, I conducted research with assistant professors, instructors, graduate and undergraduate students, and a few researchers sent by companies under the organizational system of the time. Around 1980, I began to receive requests from young scholars from other countries who wanted to come to work in my lab as postdoctoral researchers, but there was no system to accept them. It was only in 1985 that my first postdoctoral researcher arrived in my lab — Hans-Joachim Weintz, who came from the Max-Planck-Institut für Kohlenforschung (coal research). Interacting in English was difficult for the students in my lab, and I remember making sure that our German researcher had no problems living his daily life in Japan. There were 44 overseas researchers in both the university and ERATO project labs. I was a pioneer in the rapid internationalization of Japanese universities in the years that followed.

Working with a group of talented collaborators

As I detailed in **CHAPTER 3**, I created the Organic Reaction Research Group at Nagoya University in 1968 with two young researchers. One of them, Masao Katô, who specialized in photochemistry, also had business sense and after studying in the US he played an active role in industry. The other researcher, Hidemasa Takaya, had an academic mind, and we collaborated for a long time on transition metal catalytic chemistry. After he studied under James P. Collman at Stanford University, he worked at the Institute for Molecular Science; his talent blossomed, and he became Professor at his alma mater, Kyoto University. He is a man of sincerity and my ally with whom I shared my younger days.

Soon a group of young researchers with backgrounds in natural product organic chemistry began to gather in my lab. In 1971, Yoshihiro Hayakawa, who came from the Hirata Laboratory at Nagoya University, developed a method for synthesizing five- and seven-membered carbocyclic compounds using Fe-carbonyl compounds. He studied in the Corey lab at Harvard University, and after returning to Japan, he developed new synthetic methods for nucleic acid-related compounds using a variety of metal compounds. In 1977, I invited Masaaki Suzuki, who was also a student of Hirata, to come back to Japan from the Woodward lab at Harvard University, where he was working at that time. He had participated in the synthesis of the antibiotic erythromycin as a

Fig. 9.1. Noyori group alumni in May 1981. In the first row, from the left: Masashi Yamakawa, Hidemasa Takaya, Yoshihiro Hayakawa, Masao Katô, Keiko Marunouchi, the author, Masaaki Suzuki, Mugio Nishizawa, and Hideo Asai.

Fig. 9.2. Noyori group in 1991. Front row, from the left: Masato Kitamura, Kazuhiko Sato, Yuriko Nakamura (long-term secretary), Masaaki Hirose, Mami Sano, and the author.

postdoctoral researcher, and when I went to the US to directly negotiate with Woodward, I promised that I would take good care of him. After returning to Japan, he developed an efficient prostaglandin synthesis method employing organometallic chemistry. He became Professor at Gifu University, and through a collaboration with Bengt Långström in Uppsala, he worked at the RIKEN Molecular Imaging Science Center. In the same year, Mugio Nishizawa of Paul Grieco group at Pittsburgh joined my group and completed the most efficient BINAL-H reduction reaction of the time. He was a cheerful and open-minded man who had many friends and was admired as the older brother of the younger generation in the field of natural product organic chemistry. Just as he left, Shunichi Hashimoto, also an alumnus of the Corey lab, joined my group and found a silicon-based catalytic method for the transformation of carbohydrate compounds. He was a perfectionist by nature. In 1983, Masato Kitamura joined my group. Although his background had made him an expert in natural product synthesis, he changed his way of thinking and achieved results in asymmetric hydrogenation (AH) reactions and asymmetric organozinc chemistry. He was later promoted to Professor at the Research Center for Materials Science at Nagoya University, where he also established and managed the Noyori Materials Science Laboratory.

I was an early advocate of the importance of environmentally friendly chemical processes (green chemistry) in Japan and led the ERATO project on this subject from 1991. Since 1996, as the Committee Chairman, I also led the Japan Society for the Promotion of Science's (JSPS) "Research for the Future Program on Advanced Processes." In fact, prior to this, in 1990, I had invited Kazuhiko Sato, who had obtained his degree with Hideki Sakurai at Tohoku University, to develop oxidation reactions using a hydrogen peroxide solution (**CHAPTER 7**). Later, he led a large unit at the National Institute of Advanced Industrial Science and Technology (AIST) to develop green chemistry. In 1993, Takeyuki Suzuki, who was also specialized in asymmetric chemical reactions with Masakatsu Shibasaki, joined my group.

Takeshi Okuma participated in the ERATO Noyori project and also contributed to making great progress in AH research. He was a very stubborn fellow, but a dedicated researcher who worked from five in the morning and was a reliable collaborator who never gave up even under difficult circumstances. Although I was an experimental chemist, I was also interested in theoretical chemistry to learn more about the catalytic reactions we had invented. I would like to express my gratitude to Masashi Yamakawa of Kinjo Gakuin University, an alumnus and an expert in computational chemistry, for his assistance. During the hustle and bustle upon receiving the Nobel Prize in 2001, my research was supported by Masanori Kataoka from Kochi University and Akira Takeda from Tohoku University.

Fig. 9.3. Noyori's associates gathered to congratulate the author's receiving of the 2000 Order of Culture. Front row, from the left: Yoshihiro Hayakawa, Masato Kitamura, the author, Hiroko Noyori, Masaaki Suzuki, and Takeshi Okuma. Back row, from the left: Mitsuyo Hayakawa, Midori Kitamura, Yoko Suzuki, Yuriko Nakamura, Akira Takeda, and Masanori Kataoka.

In 2003, I left the research field to take up the presidency at RIKEN. However, I also received the title of University Professor at Nagoya University, where I studied the latest developments in chemistry with Kenichiro Itami (now at RIKEN), Susumu Saito, Hiroshi Naka, Jieun Jung, Naoki Noto, Shogo Mori, and others. I am grateful for the dedication of everyone involved, especially my secretary, Yuriko Nakamura, who has run my lab since 1990. She has supported many students in times of struggle in both their studies and personal lives.

Fig. 9.4. Charter members of the Noyori ERATO project. Front row from the left: Shin-ichi Inoue and Tadahito Nobori. Second row from the left: Takao Ikariya, Tatsuya Miyatake, and Masami Oya. Back row from the left: the author, Shohei Hashiguchi, Kenzo Sumi, Yuriko Nakamura and Hiroko Noyori.

Fig. 9.5. From the left: Takao Ikariya, Robert Grubbs, and the author in 1996, Tokyo.

Research seminars on Saturdays

Our lab consisting of up to 20 members has always aimed to find something out of nothing, an impetus for further development, and hopefully, to make ground-breaking achievements. In fact, while careful planning is important in research, one should not overlook the good fortune that comes our way by chance. Unlike the military's or a corporation's objective-based research, I greatly value uncertainty in creative academic research. Despite the recent surge in data-based AI-driven research, I have always believed that unexpected discoveries bring about scientific advancement. For this reason, the significance of students is immeasurable. They have a fresh way of thinking and young energy that professional researchers, indoctrinated with a variety of information and knowledge, do not possess. What is most gratifying is that they sometimes challenge absurd ideas and make unimaginable "mistakes." While a professor must have a research mindset, my day-to-day role was more like that of an organization's conductor, a field director, or a theater director. The main performers are curious young people, and although the Noyori group has always had multiple research topics, only one out of three or four of the students were working as planned, and others were still in search for details on how to implement them.

From the 1970s, the research environment at Japanese universities began to improve rapidly, but government's investment was still extremely low compared to Europe and North America. Although I was able to successfully obtain competitive funding, all professors were given a certain amount of basic funding with no defined research objectives. This was the driving force of curiosity-driven research at national universities. At the same time, scholarships for students were scarce, and it was a pity that they had to spend much of their time working outside the university to earn their living.

Research was indeed joyful and was not hard for me at all. During our time, Japan had a bad reputation, and every organization had to work every day except Sundays and on national holidays. In my group, we had a research seminar on Saturday mornings at ten, when all group members would gather. Several students would present their progress based on the results of their experiments. This was followed by a thorough discussion. The more unexpected the result, the more interesting it was. Basically, I wanted to make the distinction between what we understood and what we did not understand as clear as possible on the spot. We were scheduled to finish at noon, but we kept going until we knew. It often took us until 2:00 p.m. without lunch, but that was unavoidable—science does not know about lunch time. Sometimes imaginative ideas come to mind. I taught them that the minimum duty for a student is to make an effort to exceed the mentor and that this will advance science.

Every autumn, we took a two-day overnight group excursion. The area around Nagoya is rich in natural beauty of the sea and mountains. It was very useful for the smooth operation of the lab to observe the students' hobbies, personalities, and relationships to each other over food and drinks. I am sure that the students were relieved to know that my true character is kind and gentle.

I am somewhat uncomfortable with the modern research trend of having a computer think and a robot work to come up with the answers, simply to search for results that can be presented. This may be effective, but it does not bring the real pleasure of research. In academia, which is driven by intrinsic motivation, I would like to see distinctive ideas rather than efficiency. I feel that it's fun to think about the true nature. Although I have a certain degree of respect for the preconceived notions written in textbooks, I have a habit of trying to convince myself with my own logical system and different perspectives. I am pleased when I find consistency, but what is more important for me are the contradictions I encounter. The nature of science is objectivity, but originality means to be creative on your own, and researchers should be proud to be in the minority rather than to be in the majority. I believe that individual subjectivity is essential, and especially, the power of intuition based on experience is decisive.

Meeting Hirata's expectations

I have never forgotten the expectation set for me by Yoshimasa Hirata on my arrival at Nagoya in 1968: to explore a new frontier of organic chemistry and to make the rather small department visible to the world (CHAPTER 3). Although I was determined to fulfill his wishes, institutional recognition is hardly achieved by an individual, and I was fortunate to work together with enthusiastic colleagues who shared the same goal.

In 2006, the American Chemical Society's Division of the History of Chemistry, launched a unique celebration program called the Citation for Chemical Breakthrough Award, which acknowledges groundbreaking publications or inventions since the late 18th century. Indeed, short-sighted citation indexes or journal impact factors are not taken into account. As such, Antoine Lavoisier's 1787 paper on the method of chemical nomenclature was honored in 2015, 228 years after its publication. In 2021, our paper on the Ru-BINAP catalyzed AH of β-keto carboxylic esters in a 1987 issue of the *Journal of the American Chemical Society* [248] was added to this great list. Most importantly, as emphasized by the ACS Division, this credit does not go to myself or my coworkers but is conferred to the institutions that hosted the researchers, namely Nagoya University, the Institute for Molecular Science, and the Takasago International Corporation. I would like to dedicate this departmental honor to the spirit of the late Professor Hirata, who had set his expectations for the department over half a century ago.

The tradition of natural products organic chemistry

While my research group together with Hisashi Yamamoto's group focused on creating a new trend of chemistry in Nagoya centered on OMCOS, natural products organic chemistry, the origin of Nagoya's unique chemistry, also continued to develop soundly. Toshio Goto, who succeeded Yoshimasa Hirata as a pioneer in this field, emphasized the importance of dynamic aspects of natural products as well as conventional structural studies. Unfortunately, he died suddenly at the climax of his research. He had found a solution to why morning glories are blue, a question which had been around for nearly a century. On August 29, 1990, right after he saw the X-ray diffraction image that led directly to the structural elucidation, he suffered a myocardial infarction. In fact, the day before, I hosted Samuel and Sarah Danishefsky at our university, and in the evening we had a vegetarian dinner with them and enjoyed talking together with Goto, Hisashi Yamamoto, Kiyoyuki Yamada and others. Fifteen hours later, this tragedy occurred, and Japan lost one of its greatest treasures in this exciting field. In a short letter from Danishefsky dated September 6, he wrote, "the death of Toshio was a great blow. He was a great gentleman and a great scholar." The study of flower pigments was successfully taken over by Kumi Yoshida.

Figs. 9.6. Toshio Goto of Nagoya University.

The Goto Memorial Lectureship was established the following year upon the suggestion of his mentor Yoshimasa Hirata, and was organized by Goto's successor, Minoru Isobe. Prominent organic scientists such as Derek Barton, Albert Eschenmoser, E. J. Corey, Gilbert Stork, Jean-Marie Lehn, and Yoshito Kishi, who knew Goto well, visited the university in his memory, and their outstanding lectures impressed young researchers.

The highly respected Hirata became Professor Emeritus in 1979 and continued his research as a Professor at Meijo University. He passed away in 2000. However, his spirit of experimentalism under his distinctive leadership is passed on to the next generation through the Hirata Memorial Foundation, which was established in 2002 with Koji Nakanishi as Chairman. Daisuke Uemura led the foundation and invited leading researchers to give wonderful lectures in memory of Hirata.

Tradition and innovation. Our generation has created a new stream of organic chemistry in Nagoya by developing unique

Figs. 9.7. Samuel and Sarah Danishefsky were entertained by Nagoya organic chemists on the evening of August 28, 1990.

Fig. 9.8. First Goto Memorial Lectureship in Nagoya in May 1992. From the left: Yoshimasa Hirata, the author, Derek Barton, Minoru Isobe, and Takayuki Shioiri.

Fig. 9.9. From the left: Gilbert Stork, Minoru Isobe, the author, and Amos B. Smith in July 1992.

Fig. 9.10. Goto Memorial Lecture in 1994. Front row from the left: The author, wife of Toshio Goto, E. J. Corey (lecturer), and Shô Itô. Back row from the left: Mrs. and Minoru Isobe, Hisashi Yamamoto, and Kiyoyuki Yamada.

OMCOS, inspired by encounters with many people, while paying respect to organic chemistry focusing on natural product compounds. I wonder what kind of great development will be brought about in the future. I am confident that our successors, Kenichiro Itami (RIKEN since 2024), Shigehiro Yamaguchi, Takashi Ooi, Susumu Saito, and others, who are much brighter than us, will open up even greater horizons.

Fig. 9.11. Author and Jean-Marie Lehn (right), who visited Nagoya University to give a Goto Memorial Lecture in May 1997.

Figs. 9.12.–9.31. Nagoya natural products chemistry school (top three rows) and gifted researchers pursuing sustainability chemistry (bottom row).

Fig. 9.12. Yoshimasa Hirata.

Fig. 9.13. Koji Nakanishi.

Fig. 9.14. Toshio Goto.

Fig. 9.15. Osamu Shimomura.

Fig. 9.16. Shosuke Yamamura.

Fig. 9.17. Kiyoyuki Yamada.

Fig. 9.18. Yoshito Kishi.

Fig. 9.19. Minoru Isobe.

Fig. 9.20. Daisuke Uemura.

Fig. 9.21. Toru Fukuyama.

Fig. 9.22. Kumi Yoshida.

Fig. 9.23. Hideo Kigoshi.

Fig. 9.24. Toshio Nishikawa.

Fig. 9.25. Hirokazu Arimoto.

Fig. 9.26. Masaki Kita.

Fig. 9.27. Kenichiro Itami.

Fig. 9.28. Susumu Saito.

Fig. 9.29. Shigehiro Yamaguchi.

Fig. 9.30. Takashi Ooi.

Fig. 9.31. Kazuaki Ishihara.

Creating a good education and research environment for Nagoya University

Since I started my own laboratory at Nagoya University, I have received consistent support from my senior colleagues, especially during my frequent trips abroad, for which I am incredibly grateful. As a tenured professor, I have tried to give back to the university and my colleagues by improving the facilities and bringing opportunities for future scientific generations.

When I arrived in 1968, calculators and automatic balances became available, but small measuring instruments that are now considered standard were not widely accessible, and it was not uncommon to request measurements from corporate laboratories or other universities. Yoshimasa Hirata's long-held belief that major research equipment should be widely and effectively shared within the university, rather than owned by individuals, was eventually implemented; in 1979, the Chemical Instrument Center was established as an intra-university joint-use facility, and I served as Director of the center for 12 years. I was especially proud of the 220 MHz superconducting nuclear magnetic resonance (NMR) spectrometer, the most advanced one at the time and which cost 200 million yen. The Ministry of Education (ME) was cautious about introducing the equipment to Nagoya University, but I was able to persuade them of the great potential to advance chemical research, and its success has since be proven.

I have been continuously successful in obtaining competitive government research funding for my own research, as well as smaller unrestricted donations from a number of companies. However, the research facilities of Japanese national universities were as miserable as ever, including the cramped and unsafe conditions of the laboratories. I was embarrassed to show my Western friends my laboratory, who I imagine wondered how we could conduct high-quality experiments in such an environment. Finally, in the mid-1990s, the Japanese Ministry of Education launched a Center of Excellence (COE) Program to foster university researchers with original ideas. In the first year, only six out of 194 applications were accepted for this program, including my proposal for a Molecular Asymmetry Research Unit. From 1995 to 2001, I received a total funding of around 2.4 billion Japanese yen in part to develop a center for chemical research at Nagoya University. Thanks to the efforts of Masato Kitamura and others, the Research Center for Materials Science (RCMS) was established in 1998. Later, the Noyori Memorial Research Hall for Materials Science was also established. The center has brought about the internationalization of chemistry at Nagoya University, especially thanks to exchanges with the University of Münster in Germany arranged by the close friends Kazuyuki Tatsumi and Gerhard Erker.

Fig. 9.32. The Nagoya COE–RCMS Conference in 2000. Front row, from the left: Andreas Pfaltz, Walter Kaminsky, Osamu Yamauchi, the author, Fred Wudl, Sunggak Kim. Standing, from the left: Carsten Bolm, Kazuyuki Tatsumi, Kazuhiko Sato, Yoshio Okamoto, Kohei Tamao, Kazuhiko Seki, Yoshihiro Hayakawa, Paul Knochel, Masato Kitamura, Masahiro Irie, Takeshi Nakai, Toyoko Imae, Eiichi Nakamura, Kenji Itoh, and Matthias Beller.

During the course of this COE project, in my absence, I was assigned to serve as Dean of the School of Science for three years. In the fall of 1996, Wolfgang Herrmann, the President of the Technical University of Munich at the time, bestowed on me an honorary degree, and I was in Garching. This very unfortunate news came to me via fax, while I was staying in the office of my host Horst Kessler. This well-known scientist with experience as Dean of the Faculty of Chemistry, Biology, and Earth Sciences was deeply sympathetic to my puzzled expression. In Germany as well, this seemed a position that researchers prefer to avoid. Deans in Japan at that time had no discretionary authority, and I had no definite achievements to mention. I am pleased to say that I managed to simplify faculty meetings and minimize other meetings that were consuming our valuable time. This measure was highly welcomed by my faculty colleagues.

In 2002, the Nagoya University Institute for Advanced Research was established, as an academy within the university that would disseminate the best research achievements of the university. I was its first Director, and since 2007 I have served as Honorary Director together with Yuan-Tseh Lee of Taiwan. In these roles, I have worked toward the development of Nagoya University. Looking back, I can only say that these were small contributions, but I hope that these improvements will foster the creative research and self-determination of young scientists.

9.2 Expressing my gratitude to my allies

In order to widely disseminate Nagoya University's research achievements to the world, it was not only necessary for us to publish, but also to create a network of colleagues worldwide. Since the 1970s, I have often visited international conferences and universities held in North America and in Europe, and in later years I have cooperated in various ways with research societies in Asia. In addition to these overseas activities, we have invited many prominent organic chemists to Nagoya to give lectures and had individual discussions with them about the chemical world, society, and culture with them over Japanese meals. These guest lectures were particularly important in developing our network because the scale of organic chemistry research at Nagoya University was small

Fig. 9.33. The author with Xi-yan Lu of SIOC in China (left) who liked our de novo *C*-nucleoside synthesis (structure shown in the frame) at Nagoya in 1981.

compared to other national universities, which was determined by the Japanese Ministry of Education. The university was thus unable to handle in entirety the wide range of developing fields. By inviting guests and collaborating with other departments and with countries, my colleagues and I were able to strengthen Nagoya University's standing. Kiyoyuki Yamada, Hirata's successor, and Hisashi Yamamoto in the Faculty of Engineering, as well as Toshio Goto and Minoru Isobe, both in the Faculty of Agriculture helped me achieve this common goal. There were no barriers between these faculties, and they actively cooperated with me not only in research but we also met socially. Occasionally, we were also assisted by researchers from the nearby Nagoya Institute of Technology, Nagoya City University School of Pharmaceutical Sciences, Meijo University, and the Institute for Molecular Science. Students were also involved in this exchange, which further benefited the research.

In those difficult times more than half a century ago, it was not easy to demonstrate the presence of a region in the Far East to the world. Without a doubt, Japan's chemistry community had developed with the guidance and advice of researchers from many advanced countries. Our own efforts and ingenuity, as well as the cooperation of our friends in other countries, have been invaluable and made Nagoya's chemistry what it is today. In some way, I wanted to show our appreciation for their support. From the brilliant idea by Hisashi Yamamoto and with the support of the Banyu Life Science Foundation

Fig. 9.34. The author and Roald Hoffmann in Nagoya in 1997.

(now known as the MSD Life Science Foundation), the Nagoya Medal of Organic Chemistry was first awarded in 1995 to a researcher who had contributed in some way to the advancement of organic chemistry in Nagoya, and it now includes a number of distinguished names. In 1999, we began awarding the Silver Medal to early career chemists in Japan, who have since gone on to lead the field of organic chemistry in Japan. I would like to acknowledge the efforts of Minoru Isobe, Masato Kitamura, Kenichiro Itami, and Takashi Ooi, who have been involved in the management of this award in recent years.

In 2006, we presented a special award to Princess Chulabhorn of Thailand, who was active in natural products organic chemistry and medicinal chemistry. In fact, in 1987, I attended the Princess Chulabhorn Science Congress to celebrate the 60th anniversary of the birth of King Bhumibol Adulyadej. In 2005, I had the rare opportunity to be invited to give a special lecture to celebrate the fourth Birth Cycle of the Princess herself, as well as attend dinner at the Royal Palace.

9.3 Leading The Society of Organic Synthesis, Japan

I am a member of many national and international societies, but one of my home bases in Japan is the Society of Organic Synthesis, Japan. This society was founded in 1942 during World War II at the suggestion of Riko Majima as a forum for exchange between academia and the chemical industry in the field of organic synthetic chemistry.

In 1992, under the leadership of Teruaki Mukaiyama, an extremely enthusiastic scholar, and joined by pheromone chemist Kenji Mori and others, a grand 50th anniversary celebration of the society was held. With great pride of my country, I took on the role of communicating to the rest of the world of the history of the society's activity. With the cooperation of leading figures in industry and academia, I published *Organic Synthesis in Japan – Past, Present, and Future*, which was well received. I served as the president of the society in 1997, and as chairman of the organizing committee for the 60th anniversary of the society's foundation.

In 2002, to commemorate my 2001 Nobel Prize in Chemistry, my long-time acquaintances, Yoshinari Niimura, Haruki Tsuruta, Tetsuo Hiraoka, and Takeshi Nakai, kindly established the Ryoji Noyori Prize. The prize sponsored by Takasago International Corporation is awarded to individuals, regardless of nationality, who have made significant contributions to the field of asymmetric synthetic chemistry in the broadest sense. Since Henri Kagan received the first prize, 23 senior researchers have been selected for the prize. I look forward to the award ceremony every February, listening to their lectures, and reflecting on my past joyful encounters with them.

Fig. 9.35. The 50th Anniversary of the Society of Synthetic Organic Chemistry, Japan, 1992. Front row, from the left: E. J. Corey, Teruaki Mukaiyama, Yukichi Kishida (President of the Society). Back row, from the left: Hisashi Yamamoto, Isao Kuwajima, and the author.

Fig. 9.36. Henri Kagan receives the first Ryoji Noyori Prize. From the left: the author, Kagan, and Yoshinari Niimura (President of Takasago Corp.).

Fig. 9.37.
2002
H. B. Kagan.

Fig. 9.38.
2003
G. Stork.

Fig. 9.39.
2004
D. Seebach.

Fig. 9.40.
2005
T. Katsuki.

Fig. 9.41.
2006
D. A. Evans.

Fig. 9.42.
2007
T. Hayashi.

Fig. 9.43.
2008
A. Pfaltz.

Fig. 9.44.
2009
Y. Okamoto.

Fig. 9.45.
2010
E. N. Jacobsen.

Fig. 9.46.
2011
H. Yamamoto.

Fig. 9.47.
2012
M. Shibasaki.

Fig. 9.48.
2013
B. M. Trost.

Fig. 9.49.
2014
D. Enders.

Fig. 9.50.
2015
L. E. Overman.

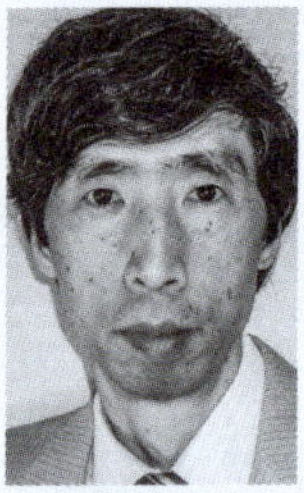

Fig. 9.51.
2016
K. Maruoka.

Fig. 9.52.
2017
D. MacMillan.

Fig. 9.53.
2018
Y. Kishi.

Fig. 9.54.
2019
S. E. Denmark.

Figs. 9.37.–9.59.
Recipients of the Ryoji Noyori Prize in 2002–2024.

Fig. 9.55.
2020
T. Imamoto.

Fig. 9.56.
2021
E. M. Carreira.

Fig. 9.57.
2022
G. C. Fu.

Fig. 9.58.
2023
K. Soai.

Fig. 9.59.
2024
T. Bach.

9.4 The role of awards in science

When we were young, the gap between Japan and the West was still wide (**CHAPTER 2–4**), and international commendations were like dreams to us. It was in the late 1980s, around the time I turned 50 that I started to receive international commendations. I have been deeply moved and humbled but sometimes I felt undeserving, compared to the past respected recipients. The purpose of each award was different, and each time I received one, it provided me with encouragement. While the quantity and quality of research achievements do not necessarily correspond to value, I am nevertheless always grateful that someone understood me, whether in Japan or abroad, recommended me for an award, and that the selection committee agreed after evaluating my work.

Unlike winning gold medals in athletic competitions, the honors bestowed on scientists are always awarded based on the evaluation of "others" for their dedication to science. This thought was shared with Svante Lindqvist, a former President of the Swedish Royal Academy of Sciences who recently visited us in Tokyo. In general, any award cannot be objective, because each of them includes a specific intent of the commendation. No matter how good the researchers themselves think their results are, if they do not match the awarding party's purpose and subjectivity, they will not receive the prize. On the other hand, scientific research is diverse, and evaluation is multifaceted. It is not a matter of swinging between joy and sorrow, as awards can come from unexpected sources. The Nobel Prize Selection Committee hears the opinions of responsible individuals from various nominators around the world and does not request nominations as an organization. Furthermore, individual nominators are not to consult with others or solicit support from others. Therefore, to be a researcher with integrity, one must have a certain modesty and readiness to accept a high scholarly distinction. In tune with the thought of Jacques Monod (1965 Nobel Laureate in Physiology or Medicine), the October 1988 issue of *The Scientist* magazine warned, "Prestigious awards like the Nobel strongly discourage cooperation and foster all sort of shady ethics." I completely agree to this statement. Most awards are the result of chance, and I feel that the overly competitive state of the research community in recent years is detrimental to a healthy academic culture.

This inconvenience applies for nation states as well. Shortly after my Nobel Prize was announced in October 2001, Japan's Council for Science and Technology Policy set a target of producing 30 Nobel Laureates in the next 50 years in relation to the national promotion of science and technology. At the press interview, I commented that "numerical targets for the Nobel Prize are injudicious," which caused some controversy. I simply thought that setting such a goal was extremely undignified for a country. I still believe that my opinion is correct.

9.5 The Emperor as a scientist

Japanese academia has a close and proud relationship with the Imperial Family. I have lived in three generations of Japanese emperors, the Showa (1926 – 1989), Heisei (1989 – 2019), and Reiwa (2019 –) eras, and scientists have traditionally received support by them. In 1995, I received the Japan Academy Prize, the highest academic award in the country. It was my greatest honor as a Japanese citizen to stand on the stage of the award ceremony in front of Emperor Akihito and Empress Michiko (current Emperor Emeritus and Empress Emerita). In the past, when deciding to move from Kyoto University to Nagoya University, my late father's following words were always in my mind: "an academic is meaningless unless they are first in class. A half-hearted attempt will not work. You must be prepared to do it" (**CHAPTER 3**). Twenty-eight years later, I felt that I had finally lived up to his expectations. My 80-year-old mother made the long trip to Tokyo from Kobe to attend the ceremony despite her mobility difficulties, along with my wife, Hiroko, and my elder son, Eiji. Prior to the ceremony, I was given an opportunity to explain to Their Majesties the achievements related to asymmetric synthetic reactions. Although I was able to use a large board, it was extremely difficult to do so within the allocated three minutes. In order to show the difference in the effects of right- and left-hand substances on living organisms, I presented (–)-menthol (fragrance) and its (+)-enantiomer, which were industrially produced by our method, and had Their Majesties smell each.

Fig. 9.60. The author is congratulated by Professor Tetsuo Nozoe (center) and the author's mother, Suzuko (left), at the ceremony of the 1995 Japan Academy Prize in Tokyo.

Fig. 9.61. The Order of Culture was conferred to the author in 2000.

I was nominated as a Member of the Japan Academy in 2002, and in June 2010, I had the honor of addressing the Japan Academy's Centennial Celebration as a past prize recipient. I recalled that the first award ceremony was held in 1911, the birthyear of my father, who led me to the path of science. He passed away at the age of 71, but I was deeply moved by the thought that if he had lived, he would have been almost 100 years old. I was informed that my father's maternal grandfather, Chiaki Watanabe, who was then Minister of the Imperial Household, had delivered a congratulatory address at the first award ceremony, and I again felt a special connection.

Good news always comes out of the blue, but the most surprising of all was the award of the Order of Culture in 2000. When I returned to Tokyo Narita airport from a chemistry conference in Rennes, France, I was stopped by an airport attendant and informed that the Ministry of Education had decided to award me the honor. I had already been selected as a Person of Cultural Merit in 1998, and my seniors had told me that if I took good care of my health and lived long enough, there was a possibility that I would eventually receive the honor, but this happened much quicker than was expected. It is the highest cultural distinction given by the state, and on November 3 (Culture Day in Japan), I shared the honor with prominent people, including a researcher of contemporary China, a leather craftsperson, a calligrapher, a theater artist, as well as Hideki Shirakawa, who had just been selected to receive the Nobel Prize in Chemistry in the same year. Each of us was awarded a medal and a certificate by His Majesty the Emperor in the Pine Room of the Imperial Palace's Main Hall. This great honor is "conferred upon those who have made remarkable achievements in the development and improvement of science, technology, art, and other aspects of culture," and while I am truly honored, I felt again a heavy responsibility upon the acceptance of this award.

The scientific community was very proud that Emperor Akihito was a scientist. His Majesty was an internationally renowned biologist, who was engaged in the taxonomy of gobies for 40 years. In fact, many academic societies, researchers, and educational institutions have received encouragement from the Imperial Family.

In 2002, I was appointed as the President of the Chemical Society of Japan (CSJ). In the following year, with the help of many people, I served as the Chairman of the 125th Anniversary Project Committee of the CSJ. We were able to hold a grand commemorative ceremony with the presence of Their Majesties the Emperor and Empress of Japan and

Prince Masahito Hitachinomiya, who was a Distinguished Member of the CSJ. Due to its unique geographical and cultural conditions, Japan has a short history of modern science, but in spite of this, it has rapidly gained strength since the Meiji era. His Majesty spoke highly of Japan's chemistry predecessors and the efforts of industry in the harsh environment and thanked foreign countries for fostering Japanese chemists. He quoted Kenichi Fukui's words when he received the Nobel Prize: "We pray that every field of science will bring happiness to mankind and never bring disaster." He also expressed his concern for environmental preservation. His persuasive words, "We must admit that the progress of science and technology has brought mankind many benefits, but the results have not always brought only happiness," deeply impressed not only the members of the CSJ, but also the experts gathered from all over the world. In addition to Saburo Nagakura, President of the Japan Academy, leading figures from various fields in Japan and presidents of foreign chemical societies, the Nobel Laureates Roald Hoffmann, Yuan-Tseh Lee, Jean-Marie Lehn, and Harold Kroto were invited to attend this memorable ceremony. A commemorative lecture by Koichi Tanaka, who had just received the Nobel Prize in Chemistry, added color to the ceremony.

In 2005, I had the special honor to present The New Year's Ceremonial Lecture at the Imperial Court together with other two renowned scholars in non-scientific fields. I talked to the Royal families on the "Significance of Asymmetric Synthesis" without using slides but with the aid of specially prepared molecular models.

In 2006, Their Majesties the Emperor and Empress visited the Wako campus of RIKEN, where I was working. I explained the history of RIKEN, which was established in 1917, and showed them around the latest research facilities. His Majesty's questions at the site were always frank and sharp. When Kosuke Morita explained about the discovery of the new element 113 (later named nihonium), which is the pride of RIKEN, His Majesty commented that since this element did not exist in the first place, it should be explained as newly synthesized rather than discovered. He was exactly right.

In 2007, Their Majesties Emperor Akihito and Empress Michiko made a 10-day official visit to Sweden, Estonia, Latvia, Lithuania, and the United Kingdom primarily to attend events commemorating the 300th anniversary of the birth of Carl von Linné, the father of modern taxonomy. I had the honor of accompanying Their Majesties as the Head of the Official Suite, representing the scientists of Japan. I was completely uninformed and inexperienced in such formal interactions, and I received private lessons in advance to prepare for the trip. In the course of my duties, I accompanied His Majesty to official luncheons and dinners and heard many stories about plants and animals from him. I was

Fig. 9.62. Author attending the celebration dinner at Uppsala Castle, Sweden, to commemorate the 300th anniversary of Linné's birth. From the left: Crown Princess Victoria of Sweden, Emperor Akihito, Queen Silvia of Sweden, and the author, May 23, 2007.

thrilled and nervous at the same time to be in the company of Queen Silvia at Uppsala Castle in Sweden and Queen Elizabeth at Buckingham Palace in the UK. Each country took every possible precaution to accommodate their guests. Their Majesties Emperor Akihito and Empress Michiko, despite the extremely tight schedule, energetically and carefully engaged in a wide range of exchanges with the royal families, government officials, scientists, cultural figures, and members of the general public that they met on the trip.

The Japanese Imperial Family and the Swedish Royal Family have a particularly close relationship, and they spent three very enjoyable days together. His Majesty Emperor Akihito was also treated as a scientist, receiving an honorary fellowship from Uppsala University. Erling Norrby, former Secretary General of the Royal Academy of Sciences, who informed me of the Nobel Prize award in 2001, also spoke highly of the importance of the presence

Fig. 9.63. At the event celebrating the 300th anniversary of Linné's birth in Uppsala, May 23, 2007. From the left: Queen Silvia of Sweden, Emperor Akihito, King Carl Gustaf of Sweden, Empress Michiko, and Crown Princess Victoria. In the back, the author in his chief entourage.

of the "emperor scientist." Science was a place for His Majesty to escape from all the complexities of society and was a place for the pure pursuit of truth. During their visit to Europe, His Majesty always looked back objectively on historical facts, respected the culture and pride cultivated by each nation, and paid respect to the difficulties and efforts of each nation to achieve independence. He then shared his thoughts on international cooperation for the future. His words were always full of hope.

Akihito: Early Cultivators of Science in Japan. *Science* **1991**, *258*, 578–580.

His Majesty has consistently emphasized the importance of transnational cooperation, as is seen in his article published in *Science* in 1991, which describes his own research on fish and reflection on the development of science in Japan at the end of the 16th century during its infancy. The highlight on this trip was the commemorative event of the Linnaean Society in London. The lecture was professional, elegant, and rich in historical narrative, and was met with thunderous applause at its conclusion. The scientific journal *Nature* also highly praised the lecture and published a lengthy abstract. In response to His Majesty's speech, Richard Fawlty, President of the Geological Society, who was a speaker at the event said, "Problems such as global warming and climate change cannot be solved by scientists alone; people all over the world must cooperate. At a time when cooperation between the general public and science is extremely important, it is a great encouragement to the world that a science-based nation like Japan has an emperor who is a scientist." I was so proud and deeply moved when I heard this.

His Majesty The Emperor of Japan: Linnaeus and taxonomy in Japan. *Nature* **2007**, *448*, 139–140.

9.6 Roles in academic administration and my Presidency at RIKEN

I see most of my university friends in the US are devoted to education and research and are not involved in academic administration, for which special expertise is needed. The same is true in Europe. In Japan, however, the system does not allow for a separation of these roles and instead places all the burden on a single faculty member. I am one of the victims of this inconvenient university system. As an academic, my daily activities had to be multiple-track, and I believe I would have been more creative and productive if I had been able to concentrate more on research. However, my surrounding environment did not allow it.

After spending a comfortable 10 years or so focusing on teaching and research as a young professor at Nagoya University, it became inevitable that I would assume administrative positions within the university, such as Dean of Faculty of the Graduate School of Science. Outside of the campus, I was also involved in a wide variety of activities, including serving as a Committee and Council Member of the Ministry of Education (ME), a Board Member and Councilor for many private foundations, a Member of Advisory Boards of various international scientific journals among others. In addition, I acted as a technical advisor for corporate research institutes. I lamented the fact that this was my fate, as these tasks consumed both energy and time.

My involvement in academic administration began in 1992 with my appointment as a Science Advisor to the ME. I am not a fan of compromise and prefer to argue for what is right. I do not know who recommended me for this appointment, but this was the beginning of a long association with the ME.

The leader of the researcher side of the ME administration at that time was Saburo Nagakura (IUPAC President in 1981 – 1983), an authority in physical chemistry. I was fortunate to have had the opportunity to get

Figs. 9.64.–9.65. Author with Saburo Nagakura in 1978 (left) and 1998 (right).

acquainted with authorities in various academic fields, presidents of major universities, and ME's high officials. At the Science Council, which I participated in for the first time as a Science Advisor, Kenichi Fukui was the chairman, but he was quiet and rarely took part in discussions. Each member was well versed in academic administration, and there was an atmosphere of mutual respect in the community spirit, and tension in dealing with the bureaucracy. Eventually, in 1996, at the end of my four-year term, I was elected as a member of the Scientific Council. It was a very important job, and I felt that I must not cut corners. However, I realized that it would be difficult to improve the existing structure of the academic world. Rather, it was much easier to create a new one. Thus, under the guidance of Saburo Nagakura, I became more involved in national education / science policy.

My Presidency at RIKEN (2003 – 2015)

Since 1968, I dedicated more than three decades at Nagoya to education, research, and administration, before being conferred the status of University Professor in 2003. In the same year, I received an invitation from the government to become the President of RIKEN, Japan's flagship research institution, established in 1917. It seemed a good time to

Fig. 9.66.
Tomoya Ogawa.

Fig. 9.67.
Yoshiharu Doi.

Fig. 9.68.
Koji Kaya.

Fig. 9.69.
Kohei Tamao.

Fig. 9.70.
Maki Kawai.

Fig. 9.71.
Mikiko Sodeoka.

Fig. 9.72.
Yukishige Ito.

Fig. 9.73.
Zhaomin Hou.

Figs. 9.66.–9.73.
Senior chemists who supported the author's presidency.

serve society more widely, because the commitment to national administrative tasks had already become unavoidable. Although I was still very active in research, I seriously suffered from shortage of time to focus on my individual interests. Thus, upon accepting this offer, I moved to Tokyo and began to commute to Nagoya. In my new role, I took initiatives to make RIKEN a more visible world-class institution. RIKEN had about ten Research Centers led by highly reputed scientists and accommodated some 3,000 researchers, covering research areas ranging from physical to biological sciences and also certain fundamental technologies.

RIKEN's culture was highly international and interdisciplinary. I enjoyed the encounter with the most eminent, well-experienced senior scientists who led the national strategic research centers and numerous researchers of high caliber. During my tenure as the President, thanks to the effort of talented individuals, RIKEN disseminated numerous seminal scientific achievements in basic science and its applications and also acted as a core of brain circulation in Japan and worldwide. Some unforgettable events include the synthesis of nihonium (Nh), which is the first element created in Asia, and the launch of the first clinical research using induced pluripotent stem cells (iPS) cells. In addition, RIKEN established the world's then fastest, 10-petaflops K computer (now renewed by the Fugaku computer) and the X-ray free-electron laser (XFEL), SACLA, both of which were large-scale infrastructures designated as national core technology and subjected successfully to the share operation systems in 2012.

My presidential management was greatly supported by the RIKEN Advisor Council consisting of world-renowned scientists and institutional leaders. RIKEN relies on external visionary figures who have provided the institute with extraordinary guidance, and therefore the title of Honorary RIKEN Fellow has been conferred to the Nobel Laureates, such as Leo Esaki, Yuan-Tseh Lee, David Baltimore, and E. J. Corey in addition to the Malaysian Prime Minister Mahathir Mohamad. Upon completing my tenure in 2015, I was honored to be named a RIKEN Fellow as well. This title had been given only once before to Susumu Tonegawa (1987 Laureate in Physiology or Medicine), who displayed an outstanding leadership as Director of the RIKEN Brain Science Institute.

I am pleased that RIKEN's open, autonomous governance was highly trusted despite the disgraceful research misconduct on "stimulus-triggered acquisition of pluripotency (STAP)" published in *Nature* in 2014. This case occurred through a complicated multi-institutional collaboration between the RIKEN Center of Developmental Biology and the Women's Hospital at Harvard Medical School, which made the crucial mutual verification of the experimental findings difficult. This incident seriously damaged RIKEN's glorious tradition founded by sustained efforts of the predecessors.

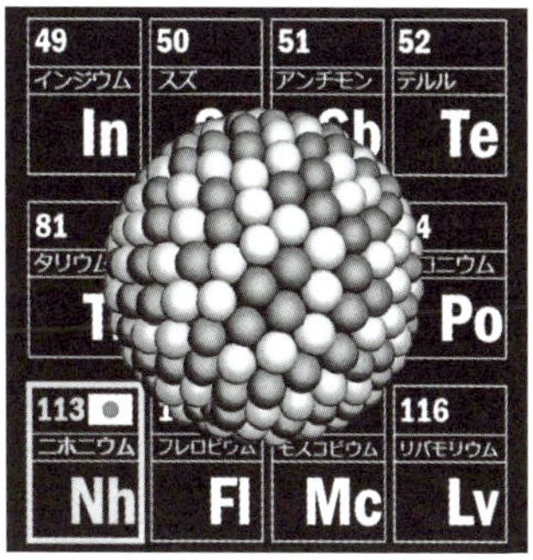

Fig. 9.74. Nihonium.

My memories with Tsung-Dao Lee and Hideki Yukawa

The totally unexpected encounter with Professor Tsung-Dao (T.-D.) Lee (1926–2024, Nobel Laureate in Physics in 1957), who had recently passed away, convinced me that "science is one" and "everything, everybody is connected deeply beyond time and space."

In 2002, I was honored to be selected as a Member of the Pontifical Academy of Sciences by John Paul II of Vatican and, in November of the following year, I participated in the 400th anniversary of the Academy. I was guided through the interior of the Papal Palace and was allowed a special visit to the Basilica of Sistine, where I met T.-D. Lee of Columbia University. Lee was awarded the Nobel Prize at the young age of 31 for his work on the violation of parity law, which is closely related to my work in chemistry on asymmetric synthesis, viz., chemistry-induced violation of molecular symmetry. In fact, I had just been appointed President of RIKEN in October in the same year and learned that T.-D. Lee served enthusiastically as an advisor of RIKEN's Brookhaven National Laboratory (BNL).

The life of a scientist can be long. This first encounter with the brilliant physicist in Vatican occupies a special place in my heart, as this eventually led to bringing back memories of my childhood admiration for Hideki Yukawa, the first Japanese who received a Nobel Prize (**Chapter 1**).

In 2005, as the President of RIKEN, I paid a courtesy visit to T.-D. Lee at Columbia University in New York and asked him for continuous guidance of the RIKEN BNL Research Center. His office that he took over at Columbia in 1953 had been used by the theoretical physicist Hideki Yukawa until immediately before his return to Japan. Out of respect for his predecessor, Lee's office contained furniture from those days, including Yukawa's desk and table. In the middle of our conversation, he suddenly handed me a flat Paulownia box, saying, "this is a present from Yukawa". Inside the box was a Japanese hand cloth that Yukawa had used as a centerpiece on his table. The cloth had been obtained at

Fig. 9.75. Hideki Yukawa's hand cloth presented to the author and his wife by T.-D. Lee at Columbia University in September 2005.

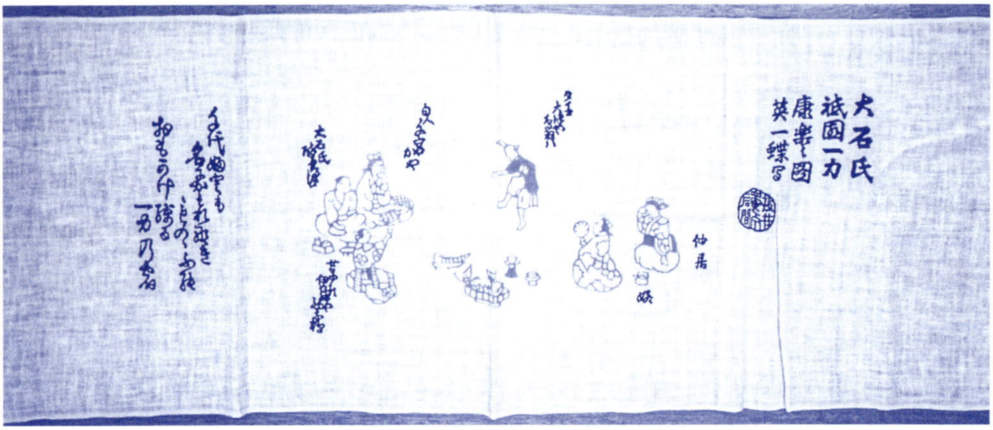

Fig. 9.76. The motif is from "The Loyal 47 Ronin," a famous historical event occurred in 1703 (Edo era) in Japan.

Ichiriki-tei, a 350 year-long famous teahouse in Gion, Kyoto. Thus, from a great physicist in Kyoto, whom I admired as an 11-year-old schoolboy, the cloth crossed the sea and was passed on to a great Chinese-American physicist in 1949, and then, after 56 years, back to me, a Japanese chemist who graduated from Kyoto. Lee, of course, had no way of knowing about the encounter between Yukawa and my parents more than 80 years ago, and this must have been a coincidence. As I could not hide my surprise, I immediately explained the sequence of events regarding my family's connection with Yukawa, and Lee told me that he was delighted that he had given the cloth to the most suitable person. Why and how was I chosen? I could not help but feel that all people around the world are connected beyond time by some invisible thread, and Lee agreed with me saying that outstanding scientists connect culture and history. A mere hand cloth may seem to be totally worthless for many, but to me it is a treasure of my life and I have kept in a safe place at RIKEN (**Figs. 9.75.–9.76.**).

There is more to this story. In September 2021, my friend Tadao Ando, a world-famous architect who received the 1995 Pritzker Prize, took me to visit the former Yukawa residence located in Shimogamo, Kyoto. With the permission of their family, I saw the parlor, where the table was covered with a chic tablecloth, instead of a hand cloth. In his later years, Yukawa had built a magnificent Japanese-style study room for his contemplation of world peace, from where we could also see his beautiful garden. This residence is now used as a guest house of Kyoto University after renovation. On this visit, I was able to express my gratitude to Yukawa's soul for kindling a torch in me, a naïve schoolboy, in 1949. Moreover, I would like to dedicate this beyond-the-age circulatory story to my parents who were acquainted with him earlier in 1939 (**Figs. 9.77.–9.78.**).

Figs. 9.77.–9.78. Author's visit to Hideki Yukawa's residence in September 2021. (Left) From the left: Tadao Ando, Yukiko Yukawa (wife of Yukawa's elder son), and author; (Right) Yukawa residence's garden.

Moving to the Japan Science and Technology Agency

During my experience at RIKEN, I noticed our society was facing numerous imminent problems to be mitigated or solved, and began to assert the responsibility of a science-intensive community for the future of humankind. Towards this end, I felt that we needed to make full use of scientific knowledge and the highest-standard technology, but also create a high-quality science education system. Society must foster and connect the best minds to secure new and diverse leadership. Otherwise, we are unable to eradicate the global problems we face nowadays.

In 2015, I assumed the position of Director-General of the Center of Research and Development Strategy (CRDS) of Japan Science and Technology Agency (JST). The CRDS was established by MEXT in 2003, aiming to navigate science and technology policy in Japan. I had a long relationship with the JST, as it had funded the ERATO Noyori Project (CHAPTER 7). I was interested in how we could properly distribute competitive grants-in-aid for scientific research that promotes science based on the free ideas of researchers. While the Japan Society for the Promotion of Science (JSPS) is the prime funding agency for this process and has actually supported my own research work at university, the JST oversees top-down control of science and technology activities based on the will and strategy of the government. I thought it crucial to create a closer relationship between science and technology innovators and the general public, who enjoy the benefits of scientific knowledge and its applications. To deepen the public's trust in science and technology, I required the research community to disclose more openly what they are doing and why, what they know already, and what they do not know yet. I also wanted to emphasize that scientific progress today is not the result of outstanding individuals, but often requires international co-creation, especially in solving and mitigating global problems such as climate change, energy resources, and infectious diseases. I retired from CRDS of JST in 2024 and became Director-General Emeritus.

9.7 Nothing comes from nothing – thoughts about the future

My parents, despite difficult times after WW II, provided me with excellent educational opportunities. I have an extremely strong sense of affirmation about the schooling, and I am particularly grateful that many of my mentors inspired me to be curious rather than simply imparting facts. Although I had been involved in university education for a long time, I had not contributed to elementary and secondary education. I wanted very much to give back to the schools and also to explore the possibilities of experiencing science outside the classroom.

I am currently the Director of the Science Museum, established in 1964 by the Japan Science Foundation, which is supported by Japanese industry. I am often impressed by the enthusiasm and ability of young boys and girls. I've noticed that children today have very different talents than my generation, and they are growing up in a very different world. I feel that education in Japan tends to be uniform and lacks diversity. It is necessary to fully acquire both tacit knowledge, which is knowledge acquired through cultural and personal experience based on the sense of wonder, and explicit knowledge, which is put together in textbooks and other learning forms. The living environment of childhood is of utmost importance and looking back on my own life, my experience at the *satoyama* (countryside small mountain) through evacuation during WW II was very meaningful in cultivating a physical sense of nature (**CHAPTER 1**). Many children in today's Japan live in artificial, overcrowded living environments in large cities, where they are crammed with only fragmentary explicit knowledge for competitive entrance exams without acquiring basic tacit knowledge on their own.

Science education is essentially designed to provide students with the sustenance they need to live a long and fulfilling life. Scientists cultivate true perspectives of nature and even life by confronting nature and pursuing the truth. Humbleness toward nature is the most important thing, and excessive competitiveness in research and education that ignores it will twist the essence of nature. On the one hand, it is clear that knowledge of the natural sciences forms the basis for modern technological development. New technology, in turn, promotes scientific progress. Thus, science and technology are synergistic, i.e., two wheels on a cart. The quality and quantity of human resources are the decisive factors, which is why not only advanced Western countries but also developing countries are striving to enhance science education. I have no intention to say that science is everything. However, there is no doubt that science is one of the most powerful ways to deal with problems rationally. International cooperation in science and technology is indispensable in this age of global issues. Therefore, as a qualification for "global citizenship," all people today, regardless of their future occupation, should be offered a certain level of universal scientific understanding.

Paradigm change from science and technology (ST) to science, technology and innovation (STI)

Science is a universal endeavor, but scientists cannot pursue it alone. Our individual knowledge is inextricably bound to the combined knowledge of all humanity. So, we must reform the academic system through collaborations between disciplines, sectors, and nations to improve the well-being of humans.

Our intellectual endeavor cannot be clearly categorized. By definition, science pursues the truth in nature through exploring the unknown, leading to discovery, while technology challenges the impossible, resulting in invention. In any event, these activities take place within the context of society. We witness that these scientific, technological, and social impacts are strongly connected or even merged into a single, inseparable entity. Today, science-based technology is meant to enrich our lives and contribute to the security and peaceful sovereignty of our nation and sustain human civilization. Therefore, we must share in the creation of value, seeking out the right time, the right place, and the right people, regardless of geopolitical divisions.

Today, we see that governments of many countries promote science, technology, and innovation (STI) as a source of the nations' competitiveness, public health and welfare, as well as for the mitigation of natural and unnatural disasters, and so forth. Here, innovation is not a mere technological invention, but is meant to be the creation of economic or other public values, which can transform society. In fact, most innovations are founded upon scientific principles and elemental technology invented through basic research, and only came to be after long, arduous collaborative efforts. Solving the most serious social problems is possible only through trusted and fruitful conjunction between research community and other sectors. Recognizing these aspects, the governments must consistently promote both basic science and issue-driven research.

Without ST-based innovations, we could not have realized the affluent, civilized societies we live in today. Thus, the political term STI is a paradigm that is strongly linked with social views and values and is affected by religion, ethics, historical and philosophical views, politics, economy, and others. As such, the way to achieve STI is not straightforward. In this respect, science cannot stand alone but must be properly merged with national or regional cultural heritage. Moreover, I would like to acknowledge industrial collaboration, for it is industry that eventually transforms our basic knowledge into social benefits (CHAPTER 7). Thus, in order to prepare new avenues of value creation, we must connect the best minds worldwide to foster the development of new and diverse leadership in this regard.

Chemists are proud of their profession, because chemistry-based materials are everywhere in this modernized society. The contribution of chemistry to STI has already been enormous, but our community must continue to strive for a peaceful and pleasant world. For instance, synthetic chemists have to pursue artificial photosynthesis to preserve energy resources and elements to overcome the material resource problem while at the same time protecting our environment. Here again, intensive interdisciplinary collaboration is needed.

Our very modest contributions to the SDGs

I have frequently asked youths, "Where are you now? Where are you going? Is your destiny utopia or dystopia?" Seniors like myself are very much aware of our unfortunate roles in the combined effects of various complicated social issues since the Industrial Revolution that have brought modern civilization to a serious state of crisis. While older people must do whatever they can, it is up to the younger generations to fully end the crisis. This is what led the United Nations in 2015 to pass a general assembly resolution titled, "The 2030 Agenda for Sustainable Development" that adds 17 Sustainable Development Goals (SDGs) (**Fig. 9.79.**) to the Millennium Goals (MDGs) already in place. The overall slogan is: "No one should be left behind." The achievement of many of these goals or targets relies on the advancement of science and technology. All Asian countries, together with the rest of the world, must play an active role in achieving these SDGs. We cannot remain passive but must consider this as an opportunity for new development and progress. If we do not change our focus now, there can be no tomorrow. Every one of the SDGs represents a mammoth objective that is impossible for any individual researcher to attain alone. To be quite frank, in my university days six decades ago (**CHAPTER 2**), we did not have the required foresight, and very few of us seriously considered these problems. However, I believe that even I as a tiny individual chemist have contributed – if only a very little bit – toward these goals.

As already described in **CHAPTER 7**, catalysis is the key to achieve Green Chemistry, which corresponds to SDG 12, Responsible Consumption and Production. Our achievements contributed to attaining not only

Fig. 9.79. UN's 17 Sustainable Development Goals.

SDG 12 but also SDG 3, Good Health and Well-being; SDG 9, Industry, Innovation and Infrastructure; and SDG 17, Partnerships for the Goals, in view of our fruitful cooperation with industry.

My very small, but most memorable research event was the 1966 discovery of asymmetric induction caused by a chiral Cu molecular complex catalyst (**CHAPTER 2**) [3]. Honestly, however, I could not think of any practical application at the time and abandoned this line of chemistry. But it turned out that this then overlooked publication became a starting point for a very different journey. A student in Kyoto, Tadatoshi Aratani, who happened to be in the same Nozaki laboratory when we found the first asymmetric carbene reaction, joined Sumitomo Chemical Company, developed an excellent chiral Cu catalyst and consequently established a large-scale synthesis of chrysanthemic esters 1 (**Fig. 9.80.**) and a building block of cilastatin 2, an *in vivo* stabilizer of carbapenem antibiotic at Merck in the US [A]. Such technological development in the 1980s could perhaps be foreseen, and may not be surprising, because it is logical. Thus, the scientific essence did not disappear and, in fact, the reaction was steadily improved in efficiency through the development of new chiral catalysts by many other groups. At the same time that Sumitomo Chemical acknowledged SDGs in its corporate concept, it established Olyset Net technology [B]. Sumitomo's researchers synthesized permethrin 3, a new chiral insecticide with a cyclopropane structure, by the use of our chemical principle, incorporated it into high-density polyethylene fiber to manufacture Olyset Nets. The new material allows the very gradual release of the potent pyrethroid over five years. The Olyset Net technology was provided to Tanzania free of charge to combat malaria. Consequently, more than 8,000 new jobs were created, and school facilities were significantly improved (**Fig. 9.81.**). Since 2003, this new technology has been used in response to the WHO's Roll Back Malaria campaign, initiated in 1998 in Tanzania. Furthermore, I would like to add that this success catalyzed the establishment of the Sakura Girls Secondary School (SGSS) near Kilimanjaro for nurturing female science-intensive leaders. This beneficial activity has since been extended worldwide, and as such, I am inclined to say that my "insignificant" endeavor back in 1966 links to SDG 1, No Poverty; SDG 4, Quality Education; SDG 5, Gender Equality; the aforementioned SDGs 3, 9, and 12; and most importantly, SDG 17, Partnerships for the Goals. I must say all of these are largely societal goals that go beyond simple science and technology.

RBM Partnership to End Malaria: https://endmalaria.org/ (last accessed March. 31, 2025).

More than half a century has passed since we made a tiny scientific finding in a poorly equipped laboratory in Kyoto. First of all, who would have thought that such a very small chemical seed would eventually make such an important contribution to global welfare? What does it teach us about science, technology, and societal innovation? And what

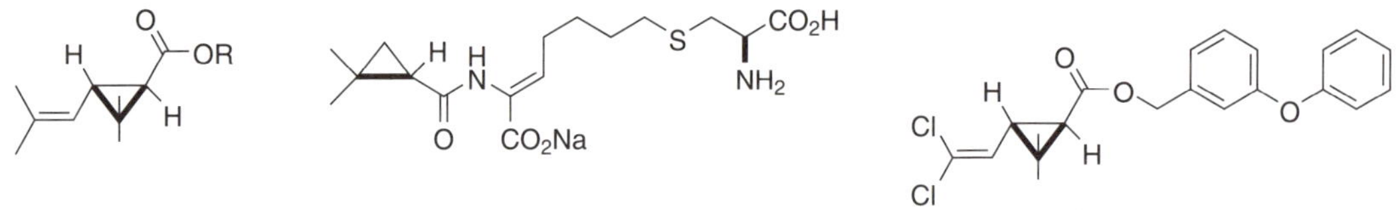

1, chrysanthemates **2**, cilastatin **3**, permethrin

Fig. 9.80. Sumitomo asymmetric synthesis of cyclopropanes.

are the necessary factors to bring this about? The scientific community, and members of industry, government, and many other sectors must carefully rethink these issues. It is clear that we scientists are merely a tiny existence on this planet. I see, however, "fine red threads" connecting all scientific knowledge. Furthermore, in today's global society, and with the contributions of many countries and diverse social sectors, even the smallest research outcomes may contribute to realizing the SDGs. Therefore, I would like to recommend younger generations to think globally, act locally. My own story represents just a small example of how this approach can give rise to great change. Young researchers have within them great potential. I very much hope that they will come to fully realize their capabilities in their future activities in either science or society at large.

Fig. 9.81. The Olyset Net project in Tanzania.

9.8 My message to students and young researchers: Interdisciplinary research and collective knowledge

Knowledge of ignorance.

We should motivate high school or university students in different ways. Basically, science pursues the truth. The essence of science must always be the Socratic "knowledge of ignorance," the ability to recognize that we know nothing. The creation of new knowledge opens up new worlds of the unknown. However, the accumulated scientific discoveries of our forebears have transformed all things of heaven and earth from myth to reality. In order to encourage young science students, I often cite Isaac Newton's: "If I have seen further, it is by standing on the shoulders of giants." In every age, it is the young who venture to explore these frontiers, and we must be patient as they make their way. As such, young researchers are creating new science on the assets founded by their predecessors. The younger generations must be proud that the level of their own study and research is much higher than that of Einstein, Watson and Crick, Ziegler and Natta, or R. B. Woodward. This is very true as Max Weber noticed, "Science is destined to make progress." This is why the younger generation must apply themselves with diligence.

Interdisciplinarity.

Young chemists are responsible for shaping the future of science. Now more than ever, we must perceive science as a single entity, a comprehensive whole. "Science is one." Science is to be regarded as a single entity, because it is based on common natural laws. What we need is not mere interdisciplinary linkage but rather disruptive anti-disciplinary action that will open the way to true scientific progress. Science must continue to progress and expand, but the scientific community must not allow unbridled splitting into narrow categories. Our community will lose its trust if we cannot discard outdated perceptions, and if we allow ourselves to fossilize into a rigid academic system of small, narrow-minded specialist groups.

Scientific research requires diverse and basic elements of knowledge, but there is no rationality in separating science into such conventional fields as physics, chemistry, or biology. Neither does it make much sense to subdivide chemistry into organic, inorganic, physical, or polymer chemistry, as was the case in our young days. Chemistry is in essence a science of substances and materials, and with the aid of the mathematical science and advanced observation and information technology, this is what links all scientific fields. Our younger generation is asked to make chemistry a truly central science.

Collective knowledge.

Discovery, by definition, is difficult to design. Historically, many of the discoveries were made thorough personal serendipity. Still, the capability of individual researchers is limited. Certain geniuses do create frameworks for science, but there is a real need in some fields for the kind of systematic laboratory work that can be done by almost anyone. In this context, even any institution will not be standing alone in their efforts to open new scientific fields. Therefore, we must prepare a research environment or ecosystem that will encourage cooperation, rather than harsh competition (**CHAPTER 9.4**), thereby fostering collective knowledge. The recent digital revolution will accelerate this trend.

Time is rapidly moving. As stated by Blaise Pascal in *Pensées*: "Man is but a reed, the most feeble thing in nature; but he is a thinking reed." I do believe that serendipity will remain the fountainhead of scientific creativity. Unlike our days, however, the chemistry community is currently in the midst of an information revolution. Deep learning derived from real brain functions is becoming the core technology for artificial intelligence (AI) and it is now possible to extract meaningful information and knowledge from so-called big data. People and objects are being connected through the Internet of Things (IoT), and linkage of science, education, medicine, and industry are leading to a wealth of creative innovation. The world of chemistry, spanning both academia and industry, will need to consider, with the utmost care and humility, more creative ways to contribute to human society. If we are to adapt successfully to the new era, we have to implement drastic overhauls of our education and research system.

References for chapter 9

[A] (a) T. Aratani: Catalytic Asymmetric Synthesis of Cyclopropanecarboxylic Acids: An Application of Chiral Copper Carbenoid Reaction. *Pure Appl. Chem.* **1985**, *57*, 1839–1844; (b) T. Aratani: *Comprehensive Asymmetric Catalysis, vol. 3* (eds.: E. N. Jacobsen, A. Pfaltz and H. Yamamoto). Springer, Berlin, **1999**, 1451.

[B] (a) T. Otsubo: *Kagaku-to-Kogyo (Chemistry and Industry)* **2006**, *59*, 780; (b) *Sumitomo Chemical CSR Highlight* **2010**, 8; (c) R. C. Malima, S. M. Magesa, P. K. Tungu, V. Mwingira, F. S. Magogo, W. Sudi, F. W. Mosha, C. F. Curtis, C. Maxwell, and M. Rowland: An Experimental Hut Evaluation of Olyset® Nets against Anopheline Mosquitoes after Seven Years Use in Tanzanian Villages. *Malar. J.* **2008**, *7*, 38.

[C] *Science to enable sustainable plastics.* A white paper from the 8th Chemical Sciences and Society Summit (CS3). London, UK, **2020**.

EPILOGUE
FAMILY
ACKNOWLEDGE-
MENT
AFTERWORD
GLOSSARY
ABBREVIATIONS

 DOI.ORG/10.47261/1564-A

Epilogue – Facts are the enemy of truth

Science pursues the truth. We scientists must show our unending gratitude for the many beautiful truths and wonderful blessings that science yields by always being mindful of how we can advance our research activities in ways that will benefit society and bring wellbeing and happiness to people. In addition to developing scientific expertise in our various fields, we hope that we will cultivate within ourselves a strong sense of responsibility to society, coupled with the rich sensibility that is needed to understand and eventually to express the beauty of science.

Any phenomenal hypotheses are based on experimental facts, but still falsifiability is the essential part of science. Looking back on my 40 years of research, particularly on asymmetric hydrogenation, I am reminded of a line from the musical *Man of La Mancha*: "Fact is the enemy of truth". The roots of academic research are purely for the self-exaltation of the individual, but the research results would largely be a tombstone of the soul. What I have published in journals are scientific "facts" but perhaps not the "truth".

A vast number of researchers work daily in the wide world and report various scientific facts to scientific journals. However, these statements are correct only under very limited conditions experienced by the researchers and do not represent universal "truth" in the wider scientific field. In order to challenge the unexplored, a more holistic view on science and even society is needed in addition to the traditional reductionism. We must accept that research can only supply limited answers, while at the same time celebrating that curiosity-driven research is more joyful for us than anything else. Fortunately, some of the dreams of the Don Quixote of Nagoya did come true thanks to quite a few Sancho Panzas mentioned throughout this book [338].

It is not easy to plan a great discovery. Scientific discovery requires intellect, sensibility, and superior skills and technology. But there is one more essential thing we need, and that is serendipity or the ability to attract good luck. Science history teaches us that people find something meaningful when they were actually looking for something else. Likewise, as with the catalytic functions (CHAPTER 6, Fig. 6.25.), original research ideas rarely lead directly to the correct answer. In particular, since many "rational" ideas are based on existing knowledge, they are often ordinary, and even if they work, they do not yield truly significant results. However, we should remain optimistic. Research sometimes develops beyond intentions. I was very fortunate to have collaborators with a rich sense of serendipity who transformed my rather foolish idea (pre-catalyst) into a new concept or a powerful principle or method (true catalyst) through their keen experimental observation skills. However, synthetic chemistry faces more than just an intellectual challenge.

Resolving a range of scientific and technological problems reveals that superficial novelty does not make any sense. Because of the ever-increasing significance of our endeavors, we academics must pursue "practical elegance" (CHAPTER 7).

I strongly believe that basic science is the fountainhead of human knowledge and as such has eternal cultural value. The technology that we gain from our scientific knowledge is the foundation of civilized society. Moreover, creative innovation derived from science and technology is what makes us internationally competitive and assures our continued survival as independent nations. In order to invent superior technology, researchers and engineers must consolidate basic knowledge through interdisciplinary or transdisciplinary cooperation and be convinced of its success. The advancement of ethical values that may result from scientific or technological innovation is in fact more important than the discoveries alone, requiring not just knowledge but also wisdom. Social values, in the broadest sense, germinate from the small seeds of our cooperative endeavors and are, in fact, essential to the continued survival of humanity. Research achievements are also affected in part by other threads in the social fabric, such things as the era, region, and culture in which individuals are working. The social elements of Japan are very different from those in Europe, the Americas, Arab, or African countries, for example. However, 50 years from now will certainly be very different. We must remember that we cannot walk alone; our individual knowledge is bound to the combined knowledge of all humanity. We all have a role to play in caring for our planet Earth.

I became interested in science at an early age and eventually chose the path of a chemist. Due to the difficult circumstances during my youth, I have always wished to be the "only one," since it was difficult to become the "number one." Fortunately, I have been able to obtain some new and meaningful results, but they form only a small part of the vast body of knowledge generated by people all over the world. Moreover, science is destined to make progress, and the facts I have discovered are becoming fossilized, buried under new knowledge. However, they must remain recorded somewhere as publicly available common assets and always be available when needed. Before this book, I had not shared my life story in such detail with the international chemical community. Although it is quite unique for a chemist, its significance is subjective, but it may bring some insights to future generations living around the globe. I sincerely hope the young readers of this monograph will also enjoy their own fruitful journeys in science.

Author's Family:

Promotion to Professor and Marriage (1972)

In August 1972, at the age of 33, I was promoted to Full Professor of Nagoya University, and in December of the same year I married Hiroko Oshima who was studying the immunology of cancer at a research institute in Tokyo. She is a daughter of Masamitsu, a Professor of Medicine at the University of Tokyo, and Fusako. She is the mother of two sons, Eiji and Koji, and now has four granddaughters. After half a century together, I owe almost all of my private life to her.

When it was announced that I would receive the Roger Adams Award from the American Chemical Society in 2001, I did a long interview with Maureen Rouhi of *Chemical & Engineering News*. I must say that this was not easy for me. The following month's issue featured a four-page article. The last paragraph quotes me as saying, "I try to select very general and fundamental problems. Perhaps simplicity is the hallmark of my work. Complicated subjects are beyond my ability." I also mentioned in the interview that if I were an American, I should thank my family, especially my wife Hiroko, but since I still followed an old Japanese custom, I would refrain from doing so in public. This comment was not covered in the article, maybe because the interviewer felt somewhat strange about this custom. I feel that this point may need further explanation to younger readers, especially those in the West. Hiroko grew up in a family of Japanese scholars, learning how to behave not only from her mother but also from her grandmother and many aunts. When our two sons were older, Hiroko began to accompany me on many of my

"Organometallic Adventures". *C&EN*, **2001**, *79*, July 23, 33–36.

Fig. A.1.
The author's wedding to Hiroko Oshima in 1972.

Fig. A.2.
The author (40), Hiroko (34), Eiji (5), and Koji (newborn) in 1978.

Fig. A.3.
The family in 1992.

overseas trips and had the opportunity to share in some of the more formal honors. However, she stubbornly resisted being praised for her support of me, even as a courtesy. "As a scholar's wife, I have only done what I learned from my mothers," she said. "I have only supported to the best of my ability the life of my husband who has devoted his life to chemistry from his youth, and all the support towards research has come from his collaborators and students. I must not let them misunderstand where the credit should go to," she continued. She is right, and actually, this may be the pride of a traditional Japanese housewife who has taken good care of our family. Hiroko continued to give me the utmost freedom for my research life. I would like to thank everyone involved for their dedication.

Fig. A.4. The growing family in 2022.

Acknowledgement

In this book, I have been able to reflect on my entire life, spanning more than 80 years. I feel that my life has been enriched not only by my research and teaching activities, but also by the many encounters I have had with people around the world. In writing this book, I would like to thank all of those who gave me encouragement and useful suggestions. Dr. Shogo Mori (Assistant Professor at Nagoya University), a young chemist, was amazed at the research environment of our time, but he made clear drawings based on his understanding my old research. Yuriko Nakamura (Nagoya University), Mirei Takizawa (JST/CRDS), and Hiroki Kadowaki (RIKEN) supported me by accurately selecting meaningful photos from my vast and scattered collection of photographs. Dr. Ayako Miyazaki's dedicated help is also greatly appreciated. She is a talented chemist with a broad knowledge not only in the field of science but also in international cultures. Her sharp questions, one after another, brought back memories that had been buried in the corner of my mind. Without her language skills and education, it would have been impossible for her to understand my sentiments and lifestyle as a Japanese scientist. I am also grateful to my wife Hiroko, for her patience with my writing over a long period of time. Finally, I would like to thank the History of Chemistry Division of the German Chemical Society and its Advisory Board members of the "Lives in Chemistry" series (see page 3). I thank Peter Gölitz, Armin de Meijere, and Martin Oestreich for careful reading of the manuscript.

Afterword by Eiji Noyori –
The art of hexagons: Memories with my father

My illustrated reference book cut into pieces

On my bookshelf at home, there remains an illustrated insect reference book that my father bought me during my childhood. Since it has been around for about 40 years, the cover has become old, and the entire book is worn out. Moreover, what is characteristic about this book is that when you turn the pages, there are some large quadrangle holes in them. Is this peculiar book, which looks just like Eric Carle's "The Very Hungry Caterpillar," the work of my young romping daughters? Actually, it is not. The caterpillar of my reference book is the main character of this autobiography, that is, my father, Ryoji Noyori.

What my father cut out from this book were pictures of butterflies with symmetrical, beautiful wings. As an elementary school kid at that time, I had complained that he had ruined my precious book, but while he was soothing me by saying that he would buy me another one, he was using scissors and a box cutter to precisely cut the butterflies out.

Soon after, the butterflies were regenerated as lecture slides. It was one of my father's plots to explain the development of right- and left-handed molecules (enantiomers), and asymmetric catalysts that could freely create both types of isomers, in an easy-to-understand manner. When I shone the light through the completed slides, the beautiful wings appeared. I remember feeling proud as a child because I felt like I had contributed to my father's work.

Creativity to draw on a blank sheet of paper

At the time, my father's laboratory was situated on the corner of the third floor of the building of School of Science at Nagoya University. In his office, there were many molecular models lined up, and the whiteboard was covered with marks of writing and erasing of chemical structures that combine hexagonal molecules (such as benzene), which looked like turtle shells.

There must have been loads of work that my father had to do beyond normal working hours. On Sundays, he would take me to the lab and devote himself to writing papers and other work. While he was busy, I would sit on the carpeted floor in his room, taking apart and putting together molecular models like toy blocks. I had also banged away randomly on the typewriter and enjoyed the touch of it. The lab was my playground.

On my father's desk was a stack of exam answer sheets, which appeared unique to me as a boy. Unlike the test papers at my school, the sheets had no questions written on them and no boxes to write in the answers. On such white sheets of paper, there were some answer sheets filled with chemical formulas written by devoted students, while one line on another answer sheet simply stated, "I didn't study. I will take the supplementary exam," indicating that the student had given up.

It is difficult to respond to open-ended questions during an exam, but it is especially challenging to fill a blank sheet of paper with original ideas. It is easy to color a coloring book, in which the positions and colors are already designated. However, to freely draw a picture on a blank sheet of paper, one must be able to conceive of the idea and then realize it. According to my father, this process is true creativity.

It appears that the genes of my father, who pursued creativity through synthetic organic chemistry, were expressed in different ways in his two sons. Eventually, I became a journalist for a newspaper, writing articles on a blank sheet of paper, while my brother Koji, five years younger than me, chose to make a living by bringing oil paints to life on a brand new canvas.

Encountering the demon sergeant for once

My father was feared as the "demon sergeant" for his intense work ethic and strict teaching. Hearing him thundering down on students over the phone, as a young child, I apparently said to my mother, "I don't want to be taught by a teacher like him." Nevertheless, once he was back in the household, he never gave us that impression, and he valued meals and conversations with his family more than anything else.

There was one time my father scolded me severely as the "demon sergeant," which I will never forget. It was a Sunday afternoon when I was in second grade. My grandfather had given me an unusual sheet of stamps from a foreign country as a gift, as I was into stamp collecting at that time. However, I then traded them with a friend in the neighborhood without any hesitation. "Go and get it back immediately. Do you not care about people's considerations?" My father ran up the stairs to the second floor where I was and scolded me in a burning rage. I was too shocked to cry out loud. All I could do was to go to my friend's house and ask him to return the stamps with an earnest apology. From the perspective of my father, who had spent his childhood in poverty, he must have not been able to forgive his son's foolishness in neglecting things and people's thoughtfulness.

Although my father allowed us, his sons, to pursue whatever field interested us, he made sure we received a practical education. On holidays, he took us to parks rich in nature so that we could experience being among living things. He also took us to *sumo* training grounds so that we could watch *sumo* wrestlers (*rikishi*) up close as they clashed with each other, covered in sweat and mud. It was because my father that I gained valuable life experience that could by no means be obtained while sitting at a desk. In 1984, he took me out of school for a month and a half so that I could accompany him on a lecture trip to France. Thanks to this wise decision, I had enough courage to go out on simple errands alone in the unfamiliar city center of Paris during the latter half of our long stay. Now, as a journalist, I must go wherever curiosity takes me. From the depths of Siberia, where frozen mammoths sleep at 30 degrees below zero, to the world's most powerful radio telescope, ALMA, in the Chilean desert, 5,000 meters above sea level. Perhaps it was the adventurous spirit that I had developed as a child that allowed me to travel to all these places without any reluctance.

A father-son press conference

When I became a newspaper journalist in 1997 after graduating from university, my boss assigned me a mission every October. He told me to "keep a close eye on my father's movements." By the late 1990s, my father had received numerous international awards and was considered a strong candidate for the Nobel Prize. My boss had figured that if my father received the prize, we might get a better scoop than other newspapers if his own son was reporting. The first four years ended fruitlessly. Then came the fifth year. It was past 6:30 p. m. on October 10, 2001, and I was waiting in the room of the Nagoya University press office when I heard a university official's voice coming from the next room, saying, "It looks like he has been selected." I probably had difficulty in understanding the situation, as my colleague said that my face had turned pale, and I did not move as if I had been sucked into the chair. It must have been about 20 to 30 minutes later. Among countless camera flashes of light, my father appeared at the press conference. He said, "It is an honor beyond my capacity, but I really feel that I am in a very special situation." I could see that he could not hide his nervousness. Next, it was the journalists' turn to ask questions. By some chance, I was the first one up. After taking a breath, I stood up and said, "Congratulations. My name is Noyori from *Yomiuri Newspaper*. What do you think of the significance of this prize?" Looking at my father's face, I saw his mouth loosen into a relaxed expression. I will never forget the smile on his face. This is a scene from a father-son press conference that is probably unprecedented in more than 100 years of the Nobel Prize history.

By a fortunate coincidence, I later became a member of the Science News Department of Yomiuri Newspaper and began working on a variety of scientific articles. When the articles appeared in the paper, my father sometimes "graded" or commented on them. After all, he was a Nobel Prize Laureate in chemistry. He was a tough person to deal with and served as a good measure of my maturity as a science journalist. Seven years ago, on the 250th anniversary of the discovery of hydrogen by the British chemist, Henry Cavendish, I wrote an article on the subject, summarizing its potential as a clean energy source and the challenges of its storage. My father provided his feedback on my article saying,

"The article is well written and scores 99 points." It was a good enough score, but I would have liked to have received a perfect score of 100. When I asked him, "What's missing to score one more point?" He grinned and said, "The article did not mention that a chemist named Noyori had succeeded in developing an asymmetric hydrogenation reaction." I could only laugh at the fact that I had missed the finishing touch.

To end with

This is the first time in more than 20 years that I have had the opportunity to collaborate with my father on a book, and I decided to write about the real image of my father, who is often idolized (especially in Japan), and my small memories of him, from the perspective of his eldest son and a newspaper journalist. My father's name is Ryoji and mine is Eiji. In Japanese characters, "Ryo" means good, and "Ei" means excellent. If there is one way in which I am superior to him beyond my name, it is that I observe and record people's activities in minute detail and portray their essence in very simple terms.

This scientist, who has devoted half his life to the art of hexagons, loves and is loved by my mother, who has been with him for 50 years, nearly as long as he has been with chemistry, his two sons (and their wives), and four grandchildren. If I can convey even a small part of the richness of his humanity to the readers, I feel that it would be the greatest sign of my filial deference to him.

Well, Father, what is the score of this afterword? I sincerely hope it scores the pass mark, and with this, I would like to finish.

Glossary

Achiral: A molecule that is superposable on its mirror image, i. e., symmetric.

Asymmetric induction: The preferential formation in a chemical reaction of one enantiomer or diastereoisomer over the other due to the influence of a chiral feature present in the substrate, reagent, catalyst, or environment.

Atropisomers: A subclass of conformers (one of a set of stereoisomers, each of which is characterized by a conformation) that can be isolated as a separate chemical species and that arise from restricted rotation about a single bond.

Chirality: The geometric property of a rigid object (or spatial arrangement of points or atoms) that makes it non-superposable on its mirror image.

Diastereo-isomers: Stereoisomers not related as mirror images.

Enantiomeric excess (ee): A way of measuring how optically pure a mixture is by calculating the purity of the major enantiomer. The excess of one enantiomer over the other in a mixture of enantiomers. Expressed mathematically as enantiomeric excess = % of major enantiomer – % of minor enantiomer.

Enantiomers: One of a pair of molecular entities that are mirror images of each other and are non-superposable.

Enantioselectivity: The preferential formation in a chemical reaction of one enantiomer over another.

HPLC: High Performance Liquid Chromatography. An analytical technique to separate, identify, and quantify components dissolved in a mixture.

Ligand: The atoms or groups joined to the central atom.

Optical activity: A sample of material able to rotate the plane of polarization of a beam of transmitted plane-polarized light.

Optical resolution: A process by which a racemic mixture is separated into its two constituent enantiomers.

Optical yield: The ratio of the optical purity of the product to that of the precursor, reactant, or catalyst in a chemical reaction involving chiral reactants and products.

Periselectivity: The preferential generation of one of a group of products having different symmetries during a reaction. This is often, though not always, controlled by molecular orbital symmetry.

Racemate (racemic mixture): An equimolar mixture of a pair of enantiomers, which does not exhibit optical activity.

Regioselectivity: The preferential formation of one regioisomer where one direction of bond making or breaking occurs preferentially over all other possible directions.

Stereogenic center (stereocenter): An atom that is a focus of stereoisomerism, such that an interchange of two groups attached to this atom leads to a stereoisomer.

Stereoisomers: Isomers that possess identical constitution, but which differ in the arrangement of their atoms in space.

Stereoselectivity: The preferential formation in a chemical reaction of one stereoisomer over another.

TOF: Turn Over Frequency. TON in a certain period of time.

TON: Turn Over Number. The number of the molecules produced per catalytic site before deactivation under a given set of reaction conditions.

Abbreviations

Ag: Silver
Al: Aluminum
B: Boron
Ba: Barium
BINAL-H: Lithium Aluminum Hydride-2,2'-Dihydroxy-1,1'-binaphthyl
b. p.: Boiling point
Bu: Butyl
Ce: Cerium
CHIRAPHOS: *rel*-[(2*R*,3*R*)-Butane-2,3-diyl] bis(diphenylphosphane)
Cl: Chlorine / Chloride
Co: Cobalt
Cu: Copper
D: Deuterium
DAIPEN: 1,1-Bis(4-methoxyphenyl)-3-methyl-1,2-butanediamine
DIOP: *O*-Isopropylidene-2,3-dihydroxy-1,4-bis(diphenylphosphino)butane
DIPAMP: (Ethane-1,2-diyl)bis[(2-methoxyphenyl)(phenyl)phosphane]
DMF: Dimethylformamide
DMSO: Dimethylsulfoxide
ESI-TOF-MS: Electrospray ionization time-of-flight mass spectrometry
Et: Ethyl
F: Fluorine / Fluoride
Fe: Iron
H: Hydrogen / Hydride
HCl: Hydrogen chloride
Hg: Mercury
I: Iodine / Iodide
Ir: Iridium
Li: Lithium
K: Potassium
Me: Methyl
MeCN: Acetonitrile
Mg: Magnesium
m. p.: Melting point
N: Nitrogen
Na: Sodium
Ni: Nickel
O: Oxygen
OAc: Acetate
P: Phosphorus
Pd: Palladium
Pd / C: Palladium on Carbon
Ph: Phenyl
Pr: Propyl
Pt: Platinum
Rh: Rhodium
Ru: Ruthenium
S: Sulfur
Si: Silicon
Solv: Solvent
THF: Tetrahydrofuran
Ti: Titanium
TolBINAP: 2,2'-Bis(di-*p*-tolylphosphino)-1,1'-binaphthyl
W: Tungsten
XylBINAP: 2,2'-Bis[di(3,5-xylyl)phosphino]-1,1'-binaphthyl
Zn: Zinc

VITA CHEMISTREE HOLOGRAPHS

Vita

1938 Born on Sept. 3, Kobe, Japan; first of three sons and one daughter of Suzuko and Kaneki Noyori, engineer

1945 Elementary school linked to Kobe Univ.

1951 Nada Junior and Senior High School, Kobe

1957 Kyoto Univ., Faculty of Eng., Dept. of Industrial Chem.

1961 Bachelor degree, Kyoto Univ., supervisor K. Sisido

1963 Master degree, Kyoto Univ., supervisor K. Sisido; Instructor in H. Nozaki's group

1967 Dr. Eng., Kyoto Univ., supervisor H. Nozaki

1968 Assoc. Prof., Nagoya Univ., Dept. of Chemistry

1969 Postdoc with E. J. Corey, Harvard Univ.

1972 Prof., Nagoya Univ.; marriage to Hiroko Oshima

1973 Birth of Eiji Noyori, first son

1978 Birth of Koji Noyori, second son

1981 Karl Pfister Visiting Prof., MIT

1982 Pacific Coast Lectureship, USA and Canada

1984 Victor J. Chambers Memorial Lecture, Univ. of Rochester

1985 The Chemical Society of Japan (CSJ) Award

1988 Lemieux Lecture, University of Ottawa

1989 The Centenary Medal, RSC; Merck-Schuchardt Lecture, FRG

1990 Arthur Birch Lecture, Austr. Natl. Univ.; Baker Lecture, Cornell University

1991 Director, ERATO Noyori Molecular Catalysis Project; Morris S. Kharasch Visiting Prof., Univ. of Chicago; J. G. Kirkwood Award, Yale Univ.

1993 Asahi Prize, Japan; Tetrahedron Prize

1994 Merck Lecture, Univ. of Cambridge; Max Tishler Lecture, Harvard Univ.; Karl Ziegler Prof., Max Planck Institute for Coal Research; Publication of "Asymmetric Catalysis in Organic Synthesis," Wiley

1995 Dr. h. c., Techn. Univ. Munich; Japan Acad. Prize

1996 Reuben Benjamin Sandin Lecture, Univ. of Alberta

1997 Arthur C. Cope Scholar Award, ACS; Dean, Graduate School of Science, Nagoya Univ.; George Kenner Award, Univ. of Liverpool

1998 Person of Cultural Merit (Japan Government); Melvin Calvin Lecture, UC Berkeley

1999 Dr. h. c. Univ. of Rennes; King Faisal Prize for Science (with D. Seebach), Saudi Arabia; Bohlmann Lecture, Techn. Univ. of Berlin; Georg Wittig Lecture, Univ. Heidelberg

2000 Distinguished Visiting Prof., Hong Kong Polytech. Univ.; The Order of Culture from the Emperor and Japan Government; Kurt Alder Lecture, Univ. of Cologne; Werner E. Bachmann Lecture, Univ. of Michigan

2001 Board of Directors Member (External), Takasago Int. Corp., Japan; Foreign Member, Am. Acad. of Arts and Sciences; Wolf Prize in

Chemistry (with H. B. Kagan and K. B. Sharpless); Roger Adams Award, ACS; Nobel Prize in Chemistry (with W. S. Knowles and K. B. Sharpless); Georg Olah Lecture, Univ. of Southern California

President of CSJ; Member of Pontifical Acad. of Sci. and Japan Acad.; Dr. h. c., Univ. of Bologna; Luigi Sacconi Medal, Ital. Chem. Soc.; Arnold O. Beckman Lecture, Caltech 2002

President of RIKEN; Univ. Prof., Nagoya Univ.; Foreign Menber, US Natl. Acad. Sci. and Russian Acad. Sci.; Dr. h. c. from Univ. of Alicante, Uppsala, Ottawa, and Chicago. 2003

Chair of Sci. and Technol. Council of MEXT, Japan; Honorary Foreign Member of the Korean Natl. Acad. Sci., the Royal Society, London and Polish Acad. Sci.; Dr. h. c., RWTH Aachen; New Year's Ceremonial Lecture at the Imperial Court; Chulabhorn Distinguished Lecture, Bangkok 2005

Chair of Education Council in the Abe Cabinet, Japan; Amedeo Avogadro Medal, Ital. Chem. Soc.; Morris Travers Memorial Lecture, Indian Inst. Sci.; Asia Science Lecture at Indian Natl. Sci. Acad. 2006

Head of the Official Suite for the Visit of the Emperor and Empress to Europe; Dr. h. c., Nanjing Univ. Sci. Technol. and Hong Kong Polytechnic Univ.; Bernal-Castejon Chair Lecture, Univ. Zaragoza 2007

Dr. h. c., Hanyang Univ. 2008

125th Anniv. Lecture of Roy. Soc. of Canada 2009

Honorary Academician, Academia Sinica; Foreign Member, Chin. Acad. Sci; Lomonosov Large Gold Medal, Russ. Acad. Sci; Sir Derek Barton Medal, RSC; Julius Bredt Lecture, RWTH Aachen; Acad. Sinica Lecture 2010

Dr. h. c., Univ. Sci. Malaysia 2011

Dr. h. c., Kazan Federal Univ. 2012

Dr. h. c., Saint-Petersburg State Univ. and Liverpool Univ.; Linus Pauling Lecture, Jawaharlal Nehru Center for Adv. Sci. Res., Bangalore; K. T. Wang Bio-Organic Lecture, Acad. Sinica 2013

Director-General, Center R&D Strategy (CRDS), Japan Sci. and Tech. Agency (JST); Director, Science Museum, Japan Sci. Found.; Member (External), Board of Directors of Toray Industries; RIKEN Fellow 2015

Dr. h. c., Pohang Univ. Sci. Technol. 2016

Dr. h. c., Inst. Chem. Technol., India 2018

Honorary Member, Uzbekistan Acad. Sci. 2019

Citation for Chemical Breakthrough Award (Hist. Chem. Div., ACS) to Nagoya Univ., Inst. of Mol. Sci., and Takasago Int. Corp. 2021

Vice President, Japan Academy 2022

Autobiography "Research Should be Fresh, Simple and Clear" in the Series "Lives in Chemistry" 2025

Bachman, Werner Emmanuel
(1901 – 1951)
Ann Arbor, MI, USA

Corey, Elias James
(* 1928) Cambridge, MA, USA

Crooks, William
(1832 – 1919)
London, UK

Divers, Edward
(1837 – 1912)
Tokyo, Japan

Doi, Hisashi
(* 1971) Osaka, Japan

Doucet, Henri
(* 1967) Rennes, France

Fukui, Kenichi
(1918 – 1998) Kyoto, Japan

Gao, Jing-Xing
Xiamen, China

Gomberg, Moses
(1866 – 1947)
Ann Arbor, MI, USA

Hashimoto, Shunichi
(* 1949) Hokkaido, Japan

Hayakawa, Yoshihiro
(* 1945) Nagoya, Japan

Hayashi, Masahiko
(* 1960) Kobe, Japan

Hilt, Gerhard
(* 1968) Marburg, Germany

Hirata, Yoshimasa
(1915 – 2000)
Nagoya, Japan

Hofmann, August Wilhelm von
(1818 – 1892) London, UK

Ikariya, Takao
(1948 – 2017) Tokyo, Japan

Itami, Kenichiro
(* 1971)
Nagoya, Wako, Japan

Iwasa, Seiji
Toyohashi, Japan

Jaisankar, Parasuraman
(* 1965) Kolkata, India

Jessop, Philip G.
Kingston, Ontario, Canada

Karrer, Paul
(1889 – 1971)
Zurich, Switzerland

Kawakita, Michitada
(1853 – 1925)
Tokyo, Japan

Kita, Gen-itsu
(1883 – 1952)
Kyoto, Japan

Kitamura, Masato
(* 1956) Nagoya, Japan

Kodama, Shinjiro
(1906 – 1996)
Kyoto, Japan

Kumamoto, Takuya
(* 1968) Hiroshima, Japan

Liebig, Justus von
(1803 – 1873)
Giessen, Germany

Lu, Yixin
(* 1970) Singapore

Lubell, William
(* 1962) Montreal, Canada

Meinwald, Jerrold
(1927-2018)
Ithaca, NY, USA

Morita, Yasushi
(* 1960) Aichi, Japan

Muniz, Kilian
(1970 – 2020)
Tarragona, Spain

Murata, Shizuaki
(* 1953) Nagoya, Japan

Nishizawa, Mugio
(1945 – 2010)
Tokushima, Japan

Noyori, Ryoji
(* 1938) Nagoya, Japan

Nozaki, Hitosi
(1922 – 2019)
Kyoto, Japan

Ohkuma, Takeshi
(* 1962) Hokkaido, Japan

Saito, Susumu
(* 1969) Nagoya, Japan

Sakurada, Ichiro
(1904 – 1986)
Kyoto, Japan

Sato, Kazuhiko
(* 1963) Tsukuba, Japan

Sato, Tsuneo
(* 1952) Okayama, Japan

Sheehan, John C.
(1915 – 1992)
Cambridge, MA, USA

Sisido, Keiiti
(1908 – 1995)
Kyoto, Japan

Suga, Seiji
(* 1963) Okayama, Japan

Suzuki, Masaaki
(* 1947) Gifu, Japan

Takaya, Hidemasa
(1940 – 1995)
Kyoto, Japan

Tokunaga, Makoto
(* 1966) Fukuoka, Japan

Widhalm, Michael
(* 1954) Vienna, Austria

Woodward, Robert Burns
(1917 – 1979)
Cambridge, MA, USA

Xiao, Jianliang
Liverpool, UK

Yamakawa, Masashi
(* 1950)
Nagoya, Japan

Yamamoto, Hisashi
(* 1943)
Nagoya, Japan

Yanagisawa, Akira
(* 1961)
Chiba, Japan

Yoshimura, Masahiro
(* 1973)
Nagoya, Japan

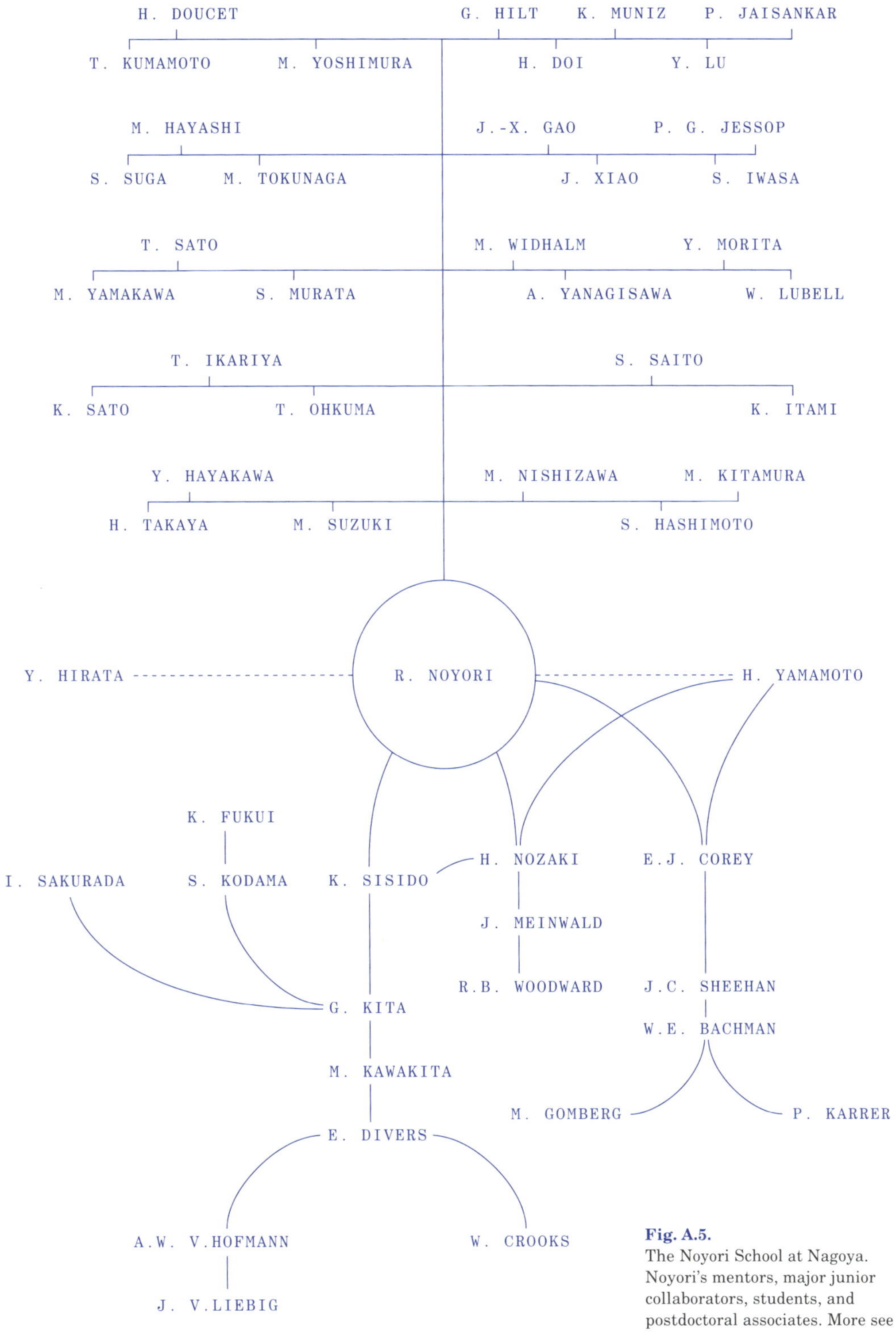

Fig. A.5.
The Noyori School at Nagoya. Noyori's mentors, major junior collaborators, students, and postdoctoral associates. More see supporting material ›Links [6].

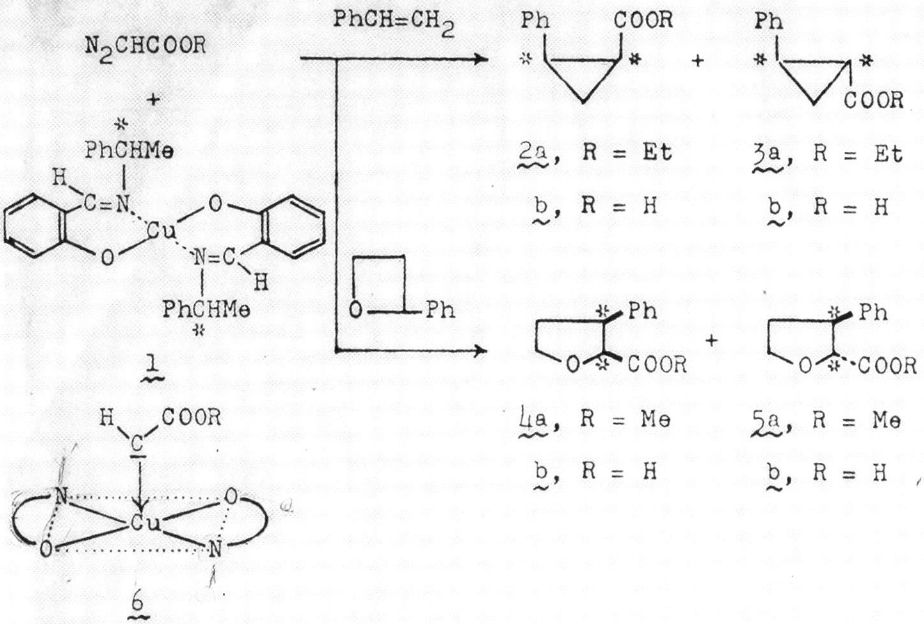

of the chelate 1 with either olefinic or ethereal substrate. Strong coordination of bases such as pyridine[4] to the chelate 1 destroyed the catalytic activity. The reasonable account for the dissymmetric alkoxycarbonylcarbene reactions must be given by assuming the formation of a reactive complex which should result from nucleophilic attack by diazoacetate at the open coordination site of the chelate 1 and following elimination of nitrogen.[10,11] In this complex the carbene

(10) Analogous reaction paths and transient complexes were postulated: (a) F. D. Mango and I. Dvoretzky, J. Am. Chem. Soc., 88, 1654 (1966); (b) R. K. Armstrong, J. Org. Chem., 31, 618 (1966); (c) C. Rüchardt and G. N. Schrauzer, Chem. Ber., 93, 1840 (1960). Rigorously, however, we can not rule out the less likely alternative that a complex initially formed from diazoacetate and 1 reacts directly with the substrates, as this has been suggested in ref. 10b.

(11) Stable complexes of carbene with wolfram and chromium were obtained recently, and their structures were established by X-ray

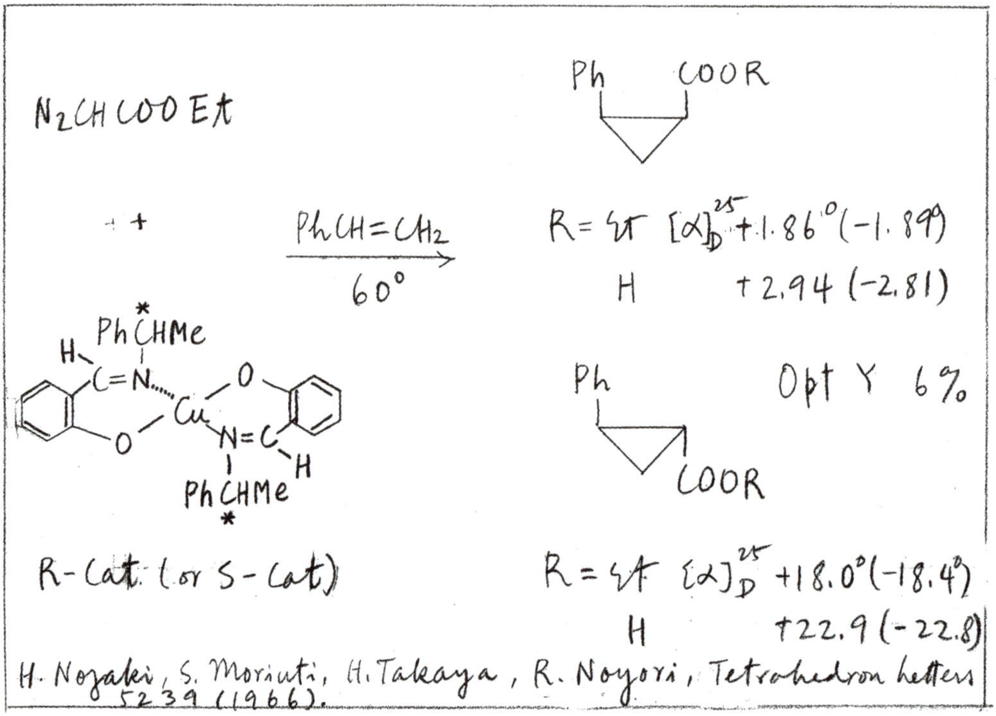

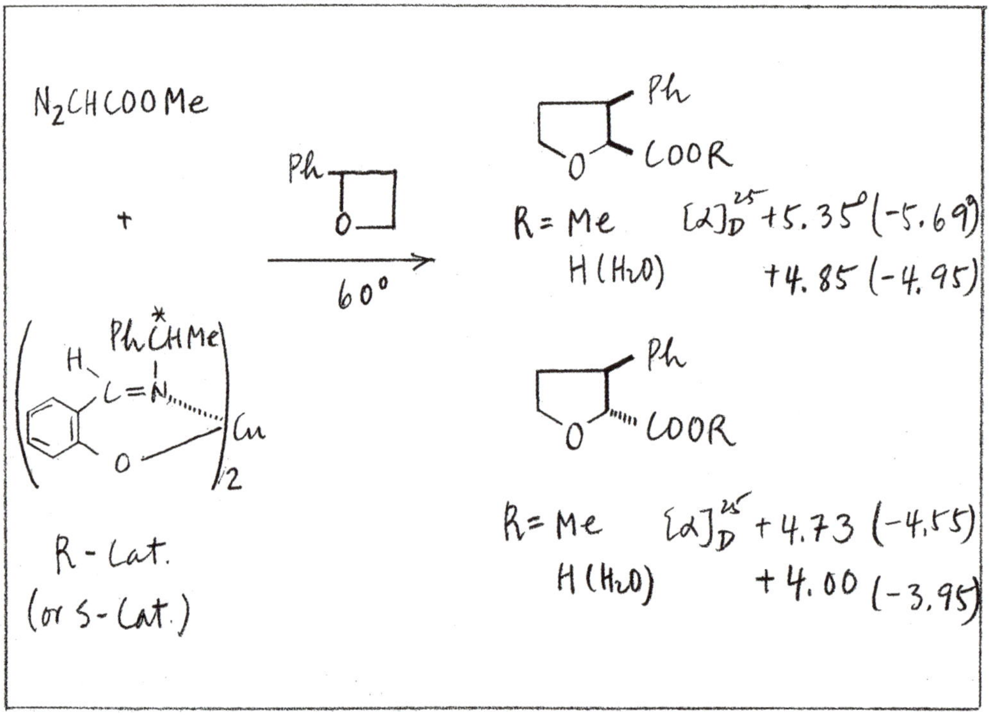

Figs. A.6. –A.8. The birth of asymmetric organometallic molecular catalysis in 1966 with some crude drawings for a lecture in Japan. Part of the manuscript which was submitted to the *Journal of the American Chemical Society* but rejected for publication. For additional holographs, see the supporting material ›Links [6].

PUBLICATIONS

Publications

[1] K. Sisido, R. Noyori, and H. Nozaki: The Stereochemistry of 9,10-Dihydro-9,10-*o*-xylyleneanthracene Derivatives. *J. Am. Chem. Soc.* **1962**, *84*, 3562 – 3566.

[2] H. Nozaki, R. Noyori, and K. Sisido: Chemistry of Xylylenes and Related Compounds. *Nippon Kagaku Zassi* **1966**, *87*, 641 – 651.

[3] H. Nozaki, S. Moriuti, H. Takaya, and R. Noyori: Asymmetric Induction in Carbenoid Reaction by Means of a Dissymmetric Copper Chelate. *Tetrahedron Lett.* **1966**, *7*, 5239 – 5244.

[4] H. Nozaki, H. Takaya, S. Moriuti, and R. Noyori: Homogeneous Catalysis in the Decomposition of Diazo Compounds by Copper Chelates. Asymmetric Carbenoid Reactions. *Tetrahedron* **1968**, *24*, 3655 – 3669.

[5] R. Noyori, H. Takaya, Y. Nakanisi, and H. Nozaki: Partial Asymmetric Synthesis of Methylenecyclopropanes and Spiropentanes. *Can. J. Chem.* **1969**, *47*, 1242 – 1245.

[6] H. Nozaki, T. Aratani, T. Toraya, and R. Noyori: Asymmetric Syntheses by Means of (–)-Sparteine Modified Organometallic Reagents. *Tetrahedron* **1971**, *27*, 905 – 913.

[7] H. Nozaki, M. Kurita, and R. Noyori: Photo-Induced Polar Addition of Protic Solvents to 2-Cycloheptenone. *Tetrahedron Lett.* **1968**, *9*, 2025 – 2027.

[8] R. Noyori, A. Watanabe, and M. Katô: The Reactive Intermediate in the Photoinduced Alcohol Addition of *cis*-2-Cyclooctenone. *Tetrahedron Lett.* **1968**, *9*, 5443 – 5444.

[9] R. Noyori, H. Inoue, and M. Katô: Divergence of the Photochemical Reaction of Cyclo-octa-2,6-dienone. *J. Chem. Soc. D: Chem. Commun.* **1970**, 1695 – 1696.

[10] R. Noyori and M. Katô: Photo-Induced Polar Addition of Protic Solvents to Cycloalkenones. Evidence for the Ground-State *trans* Isomers as Chemically-Reactive Intermediates. *Bull. Chem. Soc. Jpn.* **1974**, *47*, 1460 – 1466.

[11] R. Noyori, H. Inoue, and M. Katô: Photolysis of 1-Acetylcyclooctene. Direct Observation of Dienol Intermediate in Photochemical Deconjugation of α,β-Unsaturated Ketone. *J. Am. Chem. Soc.* **1970**, *92*, 6699 – 6700.

[12] R. Noyori, H. Inoue, and M. Katô: Direct Observation of Dienol Intermediates in Photochemical Deconjugation of an α,β-Unsaturated Ketone. Photoisomerization of 1-Acetylcyclooctene. *Bull. Chem. Soc. Jpn.* **1976**, *49*, 3673 – 3678.

[13] K. Yokoyama, M. Katô, and R. Noyori: Photo-Induced Cycloaddition of 1-Nitrocyclooctene and Cyclopentadiene. *Bull. Chem. Soc. Jpn.* **1977**, *50*, 2201 – 2202.

[14] H. Nozaki, M. Katô, R. Noyori, and M. Kawanisi: Photochemical Alkylation of Nitrogen Heteroaromatics by Carboxylic Acids under Decarboxylation. *Tetrahedron Lett.* **1967**, *8*, 4259 – 4260.

[15] R. Noyori, M. Katô, M. Kawanisi, and H. Nozaki: Photochemical Reaction of Benzopyridines with Alkanoic Acids: Novel Reductive Alkylation of Acridine, Quinoline, and Isoquinoline under Decarboxylation. *Tetrahedron* **1969**, *25*, 1125 – 1136.

[16] H. Nozaki, M. Kurita, and R. Noyori: Photochemical Reactions of 2,6-Cycloheptadienone. *Tetrahedron Lett.* **1968**, *9*, 3635 – 3638.

[17] R. Noyori and M. Katô: Photochemical Reaction of 2,7-Cyclooctadienone in Protic Solvents. *Tetrahedron Lett.* **1968**, *9*, 5075 – 5077.

[18] R. Noyori, Y. Ohnishi, and M. Katô: Photo-Induced Cyclization of *cis,cis*-2,8-Cyclononadienone *via* the *cis,trans* Isomer. *Tetrahedron Lett.* **1971**, *12*, 1515 – 1518.

[19] R. Noyori, Y. Ohnishi, and M. Katô: Photochemistry of 2,6-Cycloheptadienones in Strong Acid. A Protonated Version of Cycloheptadienone-Oxyheptatrienyl Transformation. *J. Am. Chem. Soc.* **1972**, *94*, 5105 – 5106.

[20] R. Noyori, Y. Ohnishi, and M. Katô: Novel Photorearrangements of Cyclic Cross-Conjugated Dienones in Sulfuric Acid. Transformations Involving a Stereospecific Symmetry-Forbidden Process. *J. Am. Chem. Soc.* **1975**, *97*, 928 – 929.

[21] R. Noyori, T. Nishimura, and H. Takaya: Reaction of Methylenecyclopropanes with Enneacarbonyldi-iron: A New Route Tricarbonyltrimethylenemethaneiron Complexes. *J. Chem. Soc. D: Chem. Commun.* **1969**, 89 – 89.

[22] R. Noyori, Y. Hayakawa, S. Makino, N. Hayakawa, and H. Takaya: Carbon-Carbon Bond Formations Promoted by Transition Metal Carbonyls. VIII. Iron Carbonyl Promoted Reaction of α,α'-Dibromo Ketones and Carboxamides. A Convenient Route to Muscarines *via* Furanones. *J. Am. Chem. Soc.* **1973**, *95*, 4103 – 4105.

[23] R. Noyori and H. Takaya: Reaction of Methylenecyclopropanes with Palladium Chloride. *J. Chem. Soc. D: Chem. Commun.* **1969**, 525 – 525.

[24] R. Noyori, T. Odagi, and H. Takaya: Nickel-Catalyzed Reactions Involving Strained σ Bonds. I. Nickel(0)-Catalyzed Reaction of Methylenecyclopropanes with Olefins. A Novel $[_{\sigma}2 + _{\pi}2]$ Cycloaddition. *J. Am. Chem. Soc.* **1970**, *92*, 5780 – 5781.

[25] R. Noyori, Y. Kumagai, I. Umeda, and H. Takaya: Nickel-Catalyzed Reactions Involving Strained σ Bonds. IV. Nickel(0)-Catalyzed Reaction of Methylenecyclopropane with Olefins. Orientation and Stereochemistry. *J. Am. Chem. Soc.* **1972**, *94*, 4018 – 4020.

[26] H. Takaya, N. Hayashi, T. Ishigami, and R. Noyori: Nickel-Catalyzed Reactions Involving Strained σ Bonds. IX. The Nickel-Catalyzed Isomerization of Methylenecyclopropane to Butadiene. *Chem. Lett.* **1973**, *2*, 813 – 816.

[27] R. Noyori, M. Yamakawa, and H. Takaya: Nickel Catalyzed Reactions Involving Strained Bonds. 15. Evidence for a Trimethylenemethane Complex Intermediate in the Nickel(0) Catalyzed Reaction of Methylenecyclopropane. *Tetrahedron Lett.* **1978**, *19*, 4823 – 4826.

[28] E. J. Corey and R. Noyori: A Total Synthesis of Prostaglandin $F_{2\alpha}$ (*dl*) from 2-Oxabicyclo[3.3.0]oct-6-en-3-one. *Tetrahedron Lett.* **1970**, *11*, 311 – 313.

[29] E. J. Corey, B. W. Erickson, and R. Noyori: A New Synthesis of α,β-Unsaturated Aldehydes Using 1,3-Bis(methylthio)allyllithium. *J. Am. Chem. Soc.* **1971**, *93*, 1724 – 1729.

[30] E. J. Corey, R. Noyori, and T. K. Schaaf: Total Synthesis of Prostaglandins $F_{1\alpha}$, E_1, $F_{2\alpha}$, and E_2 (Natural Forms) from a Common Synthetic Intermediate. *J. Am. Chem. Soc.* **1970**, *92*, 2586 – 2587.

[31] R. Noyori, S. Makino, and H. Takaya: Carbon-Carbon Bond Formations Promoted by Transition Metal Carbonyls. I. Reaction of α,α'-Dibromo Ketones with Iron Carbonyls in the Presence of 1,3-Dienes. A New Route to Troponoid Compounds. *J. Am. Chem. Soc.* **1971**, *93*, 1272 – 1273.

[32] R. Noyori, Y. Baba, S. Makino, and H. Takaya: Carbon-Carbon Bond Formations Promoted by Transition Metal Carbonyls. V. Reaction of α,α'-Dibromo Ketones with Iron Carbonyls in the Presence of Five-Membered Heterocycles. *Tetrahedron Lett.* **1973**, *14*, 1741 – 1744.

[33] R Noyori, S. Makino, and H. Takaya: Carbon-Carbon Bond Formations Promoted by Transition Metal Carbonyls. VI. A Convenient Route to Troponoids. *Tetrahedron Lett.* **1973**, *14*, 1745 – 1746.

[34] R. Noyori, Y. Hayakawa, S. Makino, and H. Takaya: Carbon-Carbon Bond Formations Promoted by Transition Metal Carbonyls. IV. A Convenient Route to 4,5-Homotropones. *Chem. Lett.* **1973**, *2*, 3 – 4.

[35] H. Takaya, S. Makino, Y. Hayakawa, and R. Noyori: Carbon-Carbon Bond Formations Promoted by Transition Metal Carbonyls. 17. Reactions of Polybromo Ketones with 1,3-Dienes in the Presence of Iron Carbonyls. New 3 + 4 → 7 Cyclocoupling Reaction Forming 4-Cycloheptenones. *J. Am. Chem. Soc.* **1978**, *100*, 1765 – 1777.

[36] H. Takaya, Y. Hayakawa, S. Makino, and R. Noyori: Carbon-Carbon Bond Formations Promoted by Transition Metal Carbonyls. 18. New Synthesis of Troponoid Compounds via the Iron Carbonyl Promoted Cyclocoupling between Polybromo Ketones and 1,3-Dienes. *J. Am. Chem. Soc.* **1978**, *100*, 1778 – 1785.

[37] R. Noyori, Y. Hayakawa, M. Funakura, H. Takaya, S. Murai, R. Kobayashi, and S. Tsutsumi: Carbon-Carbon Bond Formations Promoted by Transition Metal Carbonyls. III. Mechanistic Aspects of the Reaction of α,α'-Dibromo Ketones and Iron Carbonyl. Reductive Rearrangements of Dibromo Ketones. *J. Am. Chem. Soc.* **1972**, *94*, 7202 – 7204.

[38] R. Noyori: Iron Carbonyls in Organic Synthesis. *Ann. New York Acad. Sci.* **1977**, *295*, 225 – 238.

[39] M. Nishizawa and R. Noyori: Carbon-Carbon Bond Formation Promoted by Transition Metal Carbonyls. 30. A Cationic [3,4] Sigmatropic Rearrangement. *Bull. Chem. Soc. Jpn.* **1981**, *54*, 2233 – 2234.

[40] R. Noyori, K. Yokoyama, and Y. Hayakawa: Carbon-Carbon Bond Formations Promoted by Transition Metal Carbonyls. VII. Reaction of α,α'-Dibromo Ketones and Aromatic Olefins Promoted by Iron Carbonyl. A Cationic 3 + 2 → 5 Cycloaddition. *J. Am. Chem. Soc.* **1973**, *95*, 2722 – 2723.

[41] Y. Hayakawa, K. Yokoyama, and R. Noyori: Carbon-Carbon Bond Formation Promoted by Transition Metal Carbonyls. 20. Iron Carbonyl Promoted Reaction of α,α'-Dibromo Ketones and Aromatic Olefins Leading to 3-Arylcyclopentanones. The [3 + 2] Cycloaddition Involving an Allylic Cation. *J. Am. Chem. Soc.* **1978**, *100*, 1791 – 1799.

[42] R. Noyori, K. Yokoyama, S. Makino, and Y. Hayakawa: Carbon-Carbon Bond Formations Promoted by Transition Metal Carbonyls. II. Reaction of α,α'-Dibromo Ketones and Enamines with the Aid of Iron Carbonyls. A Novel Cyclopentenone Synthesis. *J. Am. Chem. Soc.* **1972**, *94*, 1772 – 1773.

[43] Y. Hayakawa, K. Yokoyama, and R. Noyori: Carbon-Carbon Bond Formation Promoted by Transition Metal Carbonyls. 21. Novel Cyclopentenone Synthesis via the Iron Carbonyl Aided Cyclocoupling between α,α'-Dibromo Ketones and Enamines. *J. Am. Chem. Soc.* **1978**, *100*, 1799 – 1806.

[44] R. Noyori, K. Yokoyama, and Y. Hayakawa: Cyclopentenones from α,α'-Dibromoketones and Enamines: 2,5-Dimethyl-3-phenyl-2-cyclopenten-1-one. *Org. Synth.* **1978**, *58*, 56 – 63.

[45] Y. Hayakawa, H. Takaya, S. Makino, N. Hayakawa, and R. Noyori: Carbon-Carbon Bond Formations Promoted by Transition Metal Carbonyls. XXIII. Synthesis of 3(2*H*)-Furanones by the Iron Carbonyl-Promoted Cyclocoupling Reaction of α,α'-Dibromo Ketones and

Carboxamides. A Convenient Route to Muscarines. *Bull. Chem. Soc. Jpn.* **1977**, *50*, 1990–1998.

[46] R. Noyori, F. Shimizu, K. Fukuta, H. Takaya, and Y. Hayakawa: Carbon-Carbon Bond Formations Promoted by Transition Metal Carbonyls. 15. Regioselectivity of the Iron Carbonyl Promoted Cyclocoupling Reaction of α,α'-Dibromo Ketones with Olefins and Dienes. *J. Am. Chem. Soc.* **1977**, *99*, 5196–5198.

[47] Y. Hayakawa, K. Yokoyama, and R. Noyori: Carbon-Carbon Bond Formations Promoted by Transition Metal Carbonyls. XIV. Stereochemistry of the Electrophilic Olefinic Substitution of 2-Oxyallyl-Iron(II) Species. Inspection of the Conformation of the Zwitterionic Intermediates. *Tetrahedron Lett.* **1976**, *17*, 4347–4350.

[48] R. Noyori, F. Shimizu, and Y. Hayakawa: Carbon-Carbon Bond Formation Promoted by Transition Metal Carbonyls. 27. Iron Carbonyl Promoted Reaction of α,α'-Dibromo Ketones and Isobutylene, an Ene Reaction Involving Allylic Cations. *Tetrahedron Lett.* **1978**, *19*, 2091–2094.

[49] R. Noyori: Carbon-Carbon Bond Formation Promoted by Transition Metal Carbonyls. 28. Organic Syntheses via the Polybromo Ketone-Iron Carbonyl Reaction. *Acc. Chem. Res.* **1979**, *12*, 61–66.

[50] R. Noyori and Y. Hayakawa: Natural Product Synthesis via the Polybromo Ketone-Iron Carbonyl Reaction. (Tetrahedron Symposium-In-Print, Ed.: M. F. Semmelhack), *Tetrahedron* **1985**, *41*, 5879–5886.

[51] Y. Hayakawa, M. Sakai, and R. Noyori: Carbon-Carbon Bond Formations Promoted by Transition Metal Carbonyls. XIII. A New Route to Nezukone. *Chem. Lett.* **1975**, *4*, 509–510.

[52] R. Noyori, T. Souchi, and Y. Hayakawa: Carbon-Carbon Bond Formations Promoted by Transition Metal Carbonyls. XII. Synthesis of Carbocamphenilone and Its 6,7-Dehydro Derivative. *J. Org. Chem.* **1975**, *40*, 2681–2682.

[53] Y. Hayakawa, F. Shimizu, and R. Noyori: Carbon-Carbon Bond Formation Promoted by Transition Metal Carbonyls. 25. A Single-Step Synthesis of (±)-α-Cuparenone. *Tetrahedron Lett.* **1978**, *19*, 993–994.

[54] R. Noyori, M. Nishizawa, F. Shimizu, Y. Hayakawa, K. Maruoka, S. Hashimoto, H. Yamamoto, and H. Nozaki: Carbon-Carbon Bond Formation Promoted by Transition Metal Carbonyls. 29. Intramolecular Dibromo Ketone-Iron Carbonyl Reaction in Terpene Synthesis. *J. Am. Chem. Soc.* **1979**, *101*, 220–222.

[55] R. Noyori, Y. Baba, and Y. Hayakawa: Carbon-Carbon Bond Formations Promoted by Transition Metal Carbonyls. X. A New, General Synthesis of Tropane Alkaloids. *J. Am. Chem. Soc.* **1974**, *96*, 3336–3338.

[56] Y. Hayakawa and R. Noyori: The Stereoselective Reduction of Tropinone to Tropine. *Bull. Chem. Soc. Jpn.* **1974**, *47*, 2617.

[57] Y. Hayakawa, Y. Baba, S. Makino, and R. Noyori: Carbon-Carbon Bond Formation Promoted by Transition Metal Carbonyls. 19. General Synthesis of Tropane Alkaloids via the Polybromo Ketone-Iron Carbonyl Reaction. *J. Am. Chem. Soc.* **1978**, *100*, 1786–1791.

[58] R. Noyori, S. Makino, T. Okita, and Y. Hayakawa: Carbon-Carbon Bond Formations Promoted by Transition Metal Carbonyls. XI. A Method for the Generation of a Synthetic Equivalent of Unsubstituted Oxyallyl via the Bromo Ketone-Iron Carbonyl Reaction. A New Route to Thujaplicins. *J. Org. Chem.* **1975**, *40*, 806–807.

[59] R. Noyori, T. Sato, and Y. Hayakawa: Carbon-Carbon Bond Formation Promoted by Transition Metal Carbonyls. 24. *C*-Nucleoside Synthesis. I. A Stereocontrolled General Synthesis of *C*-Nucleosides. *J. Am. Chem. Soc.* **1978**, *100*, 2561–2563.

[60] T. Sato, R. Ito, Y. Hayakawa, and R. Noyori: Carbon-Carbon Bond Formation Promoted by Transition Metal Carbonyls. 26. *C*-Nucleoside Synthesis. II. Stereocontrolled Synthesis of Showdomycin and 6-Azapseudouridines. *Tetrahedron Lett.* **1978**, *19*, 1829–1832.

[61] T. Sato, Y. Hayakawa, and R. Noyori: *C*-Nucleoside Synthesis. 17. A Stereocontrolled Total Synthesis of *C*-Nucleosides. *Bull. Chem. Soc. Jpn.* **1984**, *57*, 2515–2525.

[62] T. Sato, M. Watanabe, and R. Noyori: *C*-Nucleoside Synthesis. IV. Synthesis of Pseudouridines Modified at the C-5' Position. *Tetrahedron Lett.* **1978**, *19*, 4403–4406.

[63] T. Sato, M. Watanabe, and R. Noyori: *C*-Nucleoside Synthesis. V. A Convenient Route to 5'-Modified Pseudoisocytidines and 2-Thiopseudouridines. *Chem. Lett.* **1978**, *7*, 1297–1300.

[64] T. Sato, M. Watanabe, and R. Noyori: *C*-Nucleoside Synthesis. VI. A Stereocontrolled Synthesis of C-4' Alkylated Pyrimidine *C*-Nucleosides. *Tetrahedron Lett.* **1979**, *20*, 2897–2900.

[65] T. Sato, K. Marunouchi, and R. Noyori: *C*-Nucleoside Synthesis. VII. A General, Stereocontrolled Entry to Pyrimidine Homo-*C*-Nucleosides. *Tetrahedron Lett.* **1979**, *20*, 3669–3672.

[66] T. Sato and R. Noyori: *C*-Nucleoside Synthesis. VIII. Synthesis of Homoshowdomycin and Homopyrazomycin. *Heterocycles* **1979**, *13*, 141–144.

[67] T. Sato and R. Noyori: *C*-Nucleoside Synthesis. IX. Synthesis of 3-Carbamoyl-4-[(β-D-ribofuranosyl)methyl]pyrazole, a Pyrazole Homo-*C*-nucleoside. *Bull. Chem. Soc. Jpn.* **1980**, *53*, 1195–1196.

[68] T. Sato, M. Watanabe, and R. Noyori: *C*-Nucleoside Synthesis. 11. Synthesis of 4'-Phenylated Pyrimidine *C*-Nucleosides. *Heterocycles* **1980**, *14*, 761 – 764.

[69] T. Sato, M. Watanabe, and R. Noyori: *C*-Nucleoside Synthesis. 12. Stereocontrolled Synthesis of 1',4'-Dialkylated Pyrimidine Ribo-*C*-nucleosides. *Chem. Lett.* **1980**, *9*, 679 – 682.

[70] T. Sato, H. Kobayashi, and R. Noyori: *C*-Nucleoside Synthesis. 10. Synthesis of 2'-Methylated Pyrimidine *C*-Nucleosides. *Tetrahedron Lett.* **1980**, *21*, 1971 – 1974.

[71] T. Sato and R. Noyori: *C*-Nucleoside Synthesis. 13. Synthesis of 4'-Hydroxymethylated Pyrimidine Ribo-*C*-nucleosides. *Tetrahedron Lett.* **1980**, *21*, 2535 – 2538.

[72] T. Sato, H. Kobayashi, and R. Noyori: *C*-Nucleoside Synthesis. 16. Stereocontrolled Entry to Pyrimidine Hamamelo-*C*-nucleosides. *Heterocycles* **1981**, *15*, 321 – 324.

[73] T. Sato and R. Noyori: *C*-Nucleoside Synthesis. 20. General Synthesis of Homo-*C*-nucleosides. *Bull. Chem. Soc. Jpn.* **1983**, *56*, 2700 – 2705.

[74] R. Noyori, Y. Hayakawa, H. Takaya, S. Murai, R. Kobayashi, and N. Sonoda: Carbon-Carbon Bond Formations Promoted by Transition Metal Carbonyls. 16. Reaction of α,α'-Dibromo Ketones and Iron Carbonyls. Mechanistic Aspects. *J. Am. Chem. Soc.* **1978**, *100*, 1759 – 1765.

[75] R. Noyori and Y. Hayakawa: *Reductive Dehalogenation of Polyhalo Ketones with Low-Valent Metals and Related Reducing Agents in Organic Reactions* (ed: W. G. Dauben), Vol. 29, Chapter 2. John Wiley & Sons, New York, **1983**.

[76] T. Sato and R. Noyori: *C*-Nucleoside Synthesis. III. The Use of Zinc / Silver Couple in the Cyclocoupling Reaction of Polybromo Ketones and Furan. *Bull. Chem. Soc. Jpn.* **1978**, *51*, 2745 – 2746.

[77] R. Noyori, T. Odagi, and H. Takaya: Nickel-Catalyzed Reactions Involving Strained σ Bonds. I. Nickel(0)-Catalyzed Reaction of Methylenecyclopropanes with Olefins. A Novel [$_{\sigma}2 + _{\pi}2$] Cycloaddition. *J. Am. Chem. Soc.* **1970**, *92*, 5780 – 5781.

[78] R. Noyori, T. Suzuki, Y. Kumagai, and H. Takaya: Nickel-Catalyzed Reactions Involving Strained σ Bonds. II. Nickel(0)-Catalyzed Reaction of Bicyclo[1.1.0]butanes with Olefins. *J. Am. Chem. Soc.* **1971**, *93*, 5894 – 5896.

[79] R. Noyori, T. Suzuki, and H. Takaya: Nickel-Catalyzed Reactions Involving Strained σ Bonds. III. Nickel(0)-Catalyzed Reaction of Bicyclo[2.1.0]pentane with Olefins. *J. Am. Chem. Soc.* **1971**, *93*, 5896 – 5897.

[80] R. Noyori, Y. Kumagai, I. Umeda, and H. Takaya: Nickel-Catalyzed Reactions Involving Strained σ Bonds. IV. Nickel(0)-Catalyzed Reaction of Methylenecyclopropane with Olefins. Orientation and Stereochemistry. *J. Am. Chem. Soc.* **1972**, *94*, 4018 – 4020.

[81] R. Noyori, T. Ishigami, N. Hayashi, and H. Takaya: Nickel-Catalyzed Reactions Involving Strained s Bonds. VII. Transition Metal Catalyzed [2 + 2] Cross-Addition of Olefins. Nickel(0)-Catalyzed Cycloaddition of Norbornadiene and Methylenecyclopropane. *J. Am. Chem. Soc.* **1973**, *95*, 1674 – 1676.

[82] R. Noyori: Nickel-Catalyzed Reactions Involving Strained s Bonds. VI. On the Nature of Carbenoids Generated from Bicyclo[1.1.0]butanes and Transition Metal Complexes. *Tetrahedron Lett.* **1973**, *14*, 1691 – 1696.

[83] H. Takaya, M. Yamakawa, and R. Noyori: Nickel-Catalyzed Reactions Involving Strained s Bonds. VIII. Nickel(0)-Catalyzed Valence Isomerization of the 1,1'-Bishomocubane System. *Chem. Lett.* **1973**, *2*, 781 – 782.

[84] R. Noyori, Y. Kumagai, and H. Takaya: Nickel Catalyzed Reactions Involving Strained Bonds. X. Nickel(0) Catalyzed Cycloaddition of Bicyclo[2.1.0]pentane and Olefins. Contrasting Stereochemistry of the Thermal and Transition Metal Catalyzed Reactions. *J. Am. Chem. Soc.* **1974**, *96*, 634 – 636.

[85] R. Noyori, H. Kawauchi, and H. Takaya: Nickel Catalyzed Reactions Involving Strained Bonds. XI. Stereochemistry of the Nickel(0) Catalyzed Reaction of Bicyclo[1.1.0]butane and Electron-Deficient Olefins. *Tetrahedron Lett.* **1974**, *15*, 1749 – 1752.

[86] R. Noyori, I. Umeda, H. Kawauchi, and H. Takaya: Nickel-Catalyzed Reactions Involving Strained Bond. XII. Nickel(0)-Catalyzed Reaction of Quadricyclane with Electron-Deficient Olefins. *J. Am. Chem. Soc.* **1975**, *97*, 812 – 820.

[87] R. Noyori, M. Yamakawa, and H. Takaya: Nickel Catalyzed Reactions Involving Strained Bonds. XIII. Peri- and Regioselectivities of the Nickel(0)-Catalyzed Valence Isomerization of the 1,8-Bishomocubane System. A Molecular Orbital Consideration. *J. Am. Chem. Soc.* **1976**, *98*, 1471 – 1478.

[88] H. Takaya, T. Suzuki, Y. Kumagai, M. Yamakawa, and R. Noyori: Nickel Catalyzed Reactions Involving Strained Bonds. 16. Nickel(0)-Catalyzed Reactions of Bicyclo[2.1.0]pentane and Electron-Deficient Olefins. *J. Org. Chem.* **1981**, *46*, 2846 – 2854.

[89] H. Takaya, T. Suzuki, Y. Kumagai, M. Hosoya, H. Kawauchi, and R. Noyori: Nickel Catalyzed Reactions Involving Strained Bonds. 17. Nickel(0)-Catalyzed Reactions of Bicyclo[1.1.0]butanes. Geminal Two-Bond Cleavage Reaction and the Stereospecific Olefin Trapping of the Carbenoid Intermediate. *J. Org. Chem.* **1981**, *46*, 2854 – 2861.

[90] H. Takaya, M. Yamakawa, and R. Noyori: Nickel Catalyzed Reactions Involving Strained Bonds.

18. Nickel(0) Catalyzed [2 + 2] Cross-Addition of Bicyclo[2.2.1]heptene Derivatives with Electron-Deficient Olefins. *Bull. Chem. Soc. Jpn.* **1982**, *55*, 852 – 858.

[91] R. Noyori: Nickel(0) Catalyzed Reactions Involving Strained *s* Bonds in *Organotransition-Metal Chemistry* (Eds.: Y. Ishii and M. Tsutsui). Plenum Press, New York, **1975**, 231 – 241.

[92] H. Takaya, M. Yamakawa, and R. Noyori: Nickel(0) Catalyzed Reaction of Bicyclo[1.1.0] butanes with Electron-Deficient Olefins in *Fundamental Research in Homogeneous Catalysis* (Eds.: Y. Ishii and M. Tsutsui), Vol. 2. Plenum Press, New York, **1978**, 221 – 239.

[93] R. Noyori, Nickel(0) Catalyzed Reactions of Strained Ring Systems in Inorganic Compounds with Unusual Properties II (Ed.: R. B. King), (Adv. Chem. Ser. Vol. 173). American Chemical Society, Washington, D. C., **1979**, 307 – 324.

[94] R. Noyori and H. Takaya: Binaphthyls: The Beauty and Chiral Uses. *Chemica Scripta* **1985**, *25*, 83 – 89.

[95] R. Noyori, Binaphthyls as Chiral Elements for Asymmetric Synthesis in *(IUPAC) Stereocontrolled Organic Synthesis* (Ed: B. M. Trost). Blackwell Scientific Publications, Oxford, **1994**, 1.

[96] R. Noyori, I. Tomino, and Y. Tanimoto: Asymmetric Synthesis via Axially Dissymmetric Molecules. 1. Virtually Complete Enantioface Differentiation in Carbonyl Group Reduction by a Complex Aluminum Hydride Reagent. *J. Am. Chem. Soc.* **1979**, *101*, 3129 – 3131.

[97] R. Noyori, I. Tomino, and M. Nishizawa: Asymmetric Synthesis via Axially Dissymmetric Molecules. 2. A Highly Efficient Synthesis of Prostaglandin Intermediates Possessing the 15*S* Configuration. *J. Am. Chem. Soc.* **1979**, *101*, 5843 – 5844.

[98] M. Nishizawa and R. Noyori: Asymmetric Synthesis via Axially Dissymmetric Molecules. 3. Asymmetric Synthesis of Chiral Geraniol-1-*d* and Related Terpenic Alcohols. *Tetrahedron Lett.* **1980**, *21*, 2821 – 2824.

[99] M. Nishizawa, M. Yamada, and R. Noyori: Asymmetric Synthesis via Axially Dissymmetric Molecules. 4. Highly Enantioselective Reduction of Alkynyl Ketones by a Binaphthol-Modified Aluminum Hydride Reagent. Asymmetric Synthesis of Some Insect Pheromones. *Tetrahedron Lett.* **1981**, *22*, 247 – 250.

[100] R. Noyori: Asymmetric Synthesis via Axially Dissymmetric Molecules. 5. A Binaphthol-Modified Complex Aluminum Hydride Reagent Possessing Extremely High Ability of Chiral Recognition. *Pure Appl. Chem.* **1981**, *53*, 2315 – 2322.

[101] R. Noyori, I. Tomino, Y. Tanimoto, and M. Nishizawa: Asymmetric Synthesis via Axially Dissymmetric Molecules. 6. Rational Designing of Efficient Chiral Reducing Agents. Highly Enantioselective Reduction of Aromatic Ketones by Binaphthol-Modified Lithium Aluminum Hydride Reagents. *J. Am. Chem. Soc.* **1984**, *106*, 6709 – 6716.

[102] R. Noyori, I. Tomino, M. Yamada, and M. Nishizawa: Asymmetric Synthesis via Axially Dissymmetric Molecules. 7. Synthetic Applications of the Enantioselective Reduction by Binaphthol-Modified Lithium Aluminum Hydride Reagents. *J. Am. Chem. Soc.* **1984**, *106*, 6717 – 6725.

[103] R. Noyori and M. Suzuki: An Organometallic Way to Prostaglandins: The Three-Component Coupling Synthesis. *Chemtracts-Org. Chem.* **1990**, *3*, 173.

[104] Y. Okamoto, S. Honda, I. Okamoto, H. Yuki, S. Murata, R. Noyori, and H. Takaya: Chromatographic Resolution. 4. Novel Packing Material for Optical Resolution: (+)-Poly(triphenylmethyl methacrylate) Coated on Macroporous Silica Gel. *J. Am. Chem. Soc.* **1981**, *103*, 6971 – 6973.

[105] R. Noyori und M. Suzuki: Prostaglandin Synthesis. 9. Prostaglandin-Synthesen durch Dreikomponenten-Kupplung. *Angew. Chem.* **1984**, *96*, 854 – 882; R. Noyori and M. Suzuki: Prostaglandin Synthesis. 9. Prostaglandin Syntheses by Three-Component Coupling. *Angew. Chem. Int. Ed.* **1984**, *23*, 847 – 876.

[106] M. Suzuki, A. Yanagisawa, and R. Noyori: Prostaglandin Synthesis. 10. An Extremely Short Way to Prostaglandins. *J. Am. Chem. Soc.* **1985**, *107*, 3348 – 3349.

[107] M. Suzuki, A. Yanagisawa and R. Noyori: Prostaglandin Synthesis. 16. The Three-Component Coupling Synthesis of Prostaglandins. *J. Am. Chem. Soc.* **1988**, *110*, 4718 – 4726.

[108] R. Noyori, A. Yanagisawa, H. Koyano, M. Kitamura, M. Nishizawa, and M. Suzuki: Organometallics in Prostaglandin Synthesis. *Phil. Trans. R. Soc. Lond. A* **1988**, *326*, 579 – 585.

[109] R. Noyori: Prostaglandins Made Simple. *Chem. Brit.* **1989**, *25*, 883 – 888.

[110] M. Suzuki, T. Suzuki, T. Kawagishi, and R. Noyori: Vicinal Carba-condensations. 1. Organocopper Conjugate Addition Revisited. *Tetrahedron Lett.* **1980**, *21*, 1247 – 1250.

[111] M. Suzuki, T. Suzuki, T. Kawagishi, Y. Morita, and R. Noyori: Conjugate Addition of Phosphine-Complexed Organocopper Reagents to α,β-Unsaturated Ketones. *Isr. J. Chem.* **1984**, *24*, 118 – 124.

[112] Y. Morita, M. Suzuki, and R. Noyori: An Organozinc Aid in Alkylation and Acylation of Lithium Enolates. *J. Org. Chem.* **1989**, *54*, 1785 – 1787.

[113] M. Suzuki, Y. Morita, H. Koyano, M. Koga, and R. Noyori: Three-Component Coupling Synthesis of Prostaglandins. A Simplified, General Procedure. *Tetrahedron* **1990**, *46*, 4809 – 4822.

[114] M. Suzuki, T. Kawagishi, A. Yanagisawa, T. Suzuki, N. Okamura, and R. Noyori: Prostaglandin Synthesis. 17. Three-Component Coupling Synthesis of Prostaglandins: The Aldol Route. *Bull. Chem. Soc. Jpn.* **1988**, *61*, 1299 – 1312.

[115] M. Kitamura, K. Manabe, R. Noyori, and H. Takaya: Kinetic Resolution of 4-Hydroxy-2-cyclopentenone by Rhodium-Catalyzed Asymmetric Isomerization. *Tetrahedron Lett.* **1987**, *28*, 4719 – 4720.

[116] M. Kitamura, I. Kasahara, K. Manabe, R. Noyori, and H. Takaya: Kinetic Resolution of Racemic Allylic Alcohols by BINAP-Ruthenium(II)-Catalyzed Hydrogenation. *J. Org. Chem.* **1988**, *53*, 708 – 710.

[117] R. Noyori, A. Yanagisawa, H. Koyano, and M. Suzuki, Natural and Unnatural Prostaglandins via the Three-Component Coupling Synthesis in *Advances in Prostaglandin, Thromboxane, and Leukotriene Research* (Eds.: B. Samuelsson, P. Y.-K. Wong, and F. F. Sun), Vol. 19. Raven Press, New York, **1989**, 631 – 634.

[118] M. Suzuki, A. Yanagisawa, and R. Noyori: Prostaglandin Synthesis. 6. Synthesis of Prostaglandin D_1 and D_2 via the Three-Component Coupling Process. *Tetrahedron Lett.* **1984**, *25*, 1383 – 1386.

[119] M. Suzuki, A. Yanagisawa, and R. Noyori: Prostaglandin Synthesis. 4. A Short Way to Prostacyclin. *Tetrahedron Lett.* **1983**, *24*, 1187 – 1188.

[120] M. Suzuki, I.-H. Son, R. Noyori, and H. Masuda: Synthesis, Molecular Structures, and Solution-Phase Behavior of New Anionic Pentacoordinate Triorganotin(IV) Compounds: Tris(dimethylamino)sulfonium Dichlorotriorganostannates, Bis(2,6-dimethylphenoxy)triorganostannates, and Chloro(2,6-dimethylphenoxy) triorganostannates. *Organometallics* **1990**, *9*, 3043 – 3053.

[121] M. Suzuki, T. Kawagishi, and R. Noyori: Trialkylsilyl Triflates in Organic Synthesis. 9. Vicinal Carba-condensations. 2. A New Procedure for α-Alkoxyalkylation of α,β-Unsaturated Ketones. *Tetrahedron Lett.* **1981**, *22*, 1809 – 1812.

[122] M. Suzuki, A. Yanagisawa and R. Noyori: Regiospecific α-Dialkoxymethylation of Preformed Enolates. *Tetrahedron Lett.* **1982**, *23*, 3595 – 3596.

[123] M. Kitamura, T. Miki, K. Nakano, and R. Noyori: Conjugate Addition of Diorganozincs to α,β-Unsaturated Ketones Catalyzed by a Copper(I)-Sulfonamide Combined System. *Tetrahedron Lett.* **1996**, *37*, 5141 – 5144.

[124] M. Suzuki, T. Kawagishi, T. Suzuki, and R. Noyori: Vicinal Carba-condensations. 3. Prostaglandin Synthesis. 1. A Facile Synthesis of (–)-Prostaglandin E_1 via a Three-Component Coupling Process. *Tetrahedron Lett.* **1982**, *23*, 4057 – 4060.

[125] M. Suzuki, T. Kawagishi, and R. Noyori: Prostaglandin Synthesis. 3. A General Synthesis of Primary Prostaglandins. *Tetrahedron Lett.* **1982**, *23*, 5563 – 5566.

[126] R. Noyori, H. Koyano, M. Mori, R. Hirata, Y. Shiga, T. Kokura, and M. Suzuki: Unnatural Prostaglandins of Biochemical and Physiological Significance. *Pure Appl. Chem.* **1994**, *66*, 1999 – 2005.

[127] M. Suzuki, H. Koyano, and R. Noyori: Prostaglandin Synthesis. 14. A Controlled Synthesis of Isocarbacyclin. *J. Org. Chem.* **1987**, *52*, 5583 – 5588.

[128] K. Bannai, T. Tanaka, N. Okamura, A. Hazato, S. Sugiura, K. Manabe, K. Tomimori, Y. Kato, S. Kurozumi, and R. Noyori: Syntheses of Isocarbacyclin by Highly Regioselective Alkylation of Allylic Alcohols. *Tetrahedron* **1990**, *46*, 6689 – 6704.

[129] M. Suzuki, H. Koyano, R. Noyori, H. Hashimoto, M. Negishi, A. Ichikawa, and S. Ito: An Azido-Functionalized Isocarbacyclin Analogue Acting as an Efficient Photoaffinity Probe for a Prostacyclin Receptor. *Tetrahedron* **1992**, *48*, 2635 – 2658.

[130] S. Ito, H. Hashimoto, M. Negishi, M. Suzuki, H. Koyano, R. Noyori, and A. Ichikawa: Identification of the Prostacyclin Receptor by Use of [15-$^3H_1$]19-(3-Azidophenyl)-20-norisocarbacyclin, an Irreversible Specific Photoaffinity Probe. *J. Biol. Chem.* **1992**, *267*, 20326 – 20330.

[131] M. Suzuki, K. Kato, R. Noyori, Y. Watanabe, H. Takechi, K. Matsumura, B. Långström, und Y. Watanabe: (15*R*)-16-*m*-Tolyl-17,18,19,20-tetranorisocarbacyclin: ein stabiler, hochselektiver Ligand mit hoher Bindungsaffinität für einen Prostacyclin-Rezeptor im zentralen Nervensystem. *Angew. Chem.* **1996**, *108*, 366 – 369; M. Suzuki, K. Kato, R. Noyori, Y. Watanabe, H. Takechi, K. Matsumura, B. Långström, and Y. Watanabe: (15*R*)-16-*m*-Tolyl-17,18,19,20-tetranorisocarbacyclin: A Stable Ligand with High Binding Affinity and Selectivity for a Prostacyclin Receptor in the Central Nervous System. *Angew. Chem. Int. Ed.* **1996**, *35*, 334 – 336.

[132] H. Takechi, K. Matsumura, Y. Watanabe, K. Kato, R. Noyori, M. Suzuki, and Y. Watanabe: A Novel Subtype of the Prostacyclin Receptor Expressed in the Central Nervous System. *J. Biol. Chem.* **1996**, *271*, 5901 – 5906.

[133] M. Suzuki, K. Kato, Y. Watanabe, T. Satoh, K. Matsumura, Y. Watanabe, and R. Noyori: 15-Deoxy-16-(*m*-tolyl)-17,18,19,20-tetranorisocarbacyclin: A Simple TIC Derivative with Potent Anti-apoptotic Activity for Neuronal Cells. *Chem. Commun.* **1999**, 307 – 308.

[134] M. Suzuki, R. Noyori, B. Långström, and Y. Watanabe: Molecular Design of Prostaglandin Probes in Brain Research: High, Specific Binding to a Novel Prostacyclin Receptor in the Central Nervous System. *Bull. Chem. Soc. Jpn.* **2000**, *73*, 1053 – 1070.

[135] M. Suzuki, H. Doi, M. Björkman, Y. Andersson, B. Långström, Y. Watanabe, and R. Noyori: Rapid Coupling of Methyl Iodide with Aryltributylstannanes Mediated by Palladium(0) Complexes: A General Protocol for the Synthesis of $^{11}CH_3$-Labeled PET Tracers. *Chem. Eur. J.* **1997**, *3*, 2039 – 2042.

[136] M. Suzuki, H. Doi, K. Kato, M. Björkman, B. Långström, Y. Watanabe, and R. Noyori: Rapid Methylation for the Synthesis of a ^{11}C-Labeled Tolylisocarbacyclin Imaging the IP_2 Receptor in a Living Human Brain. *Tetrahedron* **2000**, *56*, 8263 – 8273.

[137] M. Björkman, H. Doi, B. Resul, M. Suzuki, R. Noyori, Y. Watanabe, and B. Långström: Synthesis of a ^{11}C-Labelled Prostaglandin F_{2a} Analogue Using an Improved Method for Stille Reactions with [^{11}C]Methyl Iodide. *J. Labelled Cpd. Radiopharm.* **2000**, *43*, 1327 – 1334.

[138] T. Kato, M. Fukushima, S. Kurozumi, and R. Noyori: Antitumor Activity of Δ^7-Prostaglandin A_1 and Δ^{12}-Prostaglandin J_2 *in Vitro* and *in Vivo*. *Cancer Res.* **1986**, *46*, 3538 – 3542.

[139] R. Noyori and M. Suzuki: Organic Synthesis of Prostaglandins: Advancing Biology. *Science* **1993**, *259*, 44 – 45.

[140] K. Akimaru, M. T. Kuo, K. Furuta, M. Suzuki, R. Noyori, and T. Ishikawa: Induction of *MRP/GS-X* Pump and Cellular Resistance to Anticancer Prostaglandins. *Cytotechnology* **1996**, *19*, 221 – 227.

[141] M. Suzuki, M. Mori, T. Niwa, R. Hirata, K. Furuta, T. Ishikawa, and R. Noyori: Chemical Implications for Antitumor and Antiviral Prostaglandins: Reaction of Δ^7-Prostaglandin A_1 and Prostaglandin A_1 Methyl Esters with Thiols. *J. Am. Chem. Soc.* **1997**, *119*, 2376 – 2385.

[142] M. Suzuki, T. Kiho, K. Furuta, S. Fukushima, Y. Takeuchi, and R. Noyori: Δ^7-Prostaglandin C_1: A Primary Metabolite of Antitumor Δ^7-Prostaglandin A_1 in the Sera. *Tetrahedron* **1997**, *53*, 17009 – 17014.

[143] M. Suzuki, T. Kiho, K. Tomokiyo, K. Furuta, S. Fukushima, Y. Takeuchi, M. Nakanishi, and R. Noyori: Rational Design of Antitumor Prostaglandins with High Biological Stability. *J. Med. Chem.* **1998**, *41*, 3084 – 3090.

[144] M. Suzuki, Y. Morita, A. Yanagisawa, R. Noyori, B. J. Baker, and P. J. Scheuer: Prostaglandin Synthesis. 12. Synthesis of (7*E*)- and (7*Z*)-Punaglandin 4. Structural Revision. *J. Am. Chem. Soc.* **1986**, *108*, 5021 – 5022.

[145] M. Suzuki, Y. Morita, A. Yanagisawa, B. J. Baker, P. J. Scheuer, and R. Noyori: Prostaglandin Synthesis. 15. Synthesis and Structural Revision of (7*E*)- and (7*Z*)-Punaglandin 4. *J. Org. Chem.* **1988**, *53*, 286 – 295.

[146] M. Suzuki, R. Noyori, and N. Hamanaka: Ruthenium(II)-Catalyzed Reaction of 1,4-Epiperoxides. *J. Am. Chem. Soc.* **1982**, *104*, 2024 – 2025.

[147] M. Suzuki, H. Ohtake, Y. Kameya, N. Hamanaka, and R. Noyori: Ruthenium(II)-Catalyzed Reactions of 1,4-Epiperoxides. *J. Org. Chem.* **1989**, *54*, 5292 – 5302.

[148] M. Suzuki, Y. Oda, N. Hamanaka, and R. Noyori: Palladium(0) Catalyzed Reactions of 1,4-Epiperoxides. *Heterocycles* **1990**, *30*, 517 – 535.

[149] M. Suzuki, R. Noyori, and N. Hamanaka: Palladium(0) Catalyzed Reaction of 1,4-Epiperoxides. Conversion of a Prostaglandin Endoperoxide to Primary Prostaglandins. *J. Am. Chem. Soc.* **1981**, *103*, 5606 – 5607.

[150] M. Suzuki, Y. Oda, and R. Noyori: Palladium(0) Catalyzed Reaction of 1,3-Diene Epoxides. A Useful Method for the Site-Specific Oxygenation of 1,3-Dienes. *J. Am. Chem. Soc.* **1979**, *101*, 1623 – 1625.

[151] M. Suzuki, A. Watanabe, and R. Noyori: Palladium(0)-Catalyzed Reaction of α,β-Epoxy Ketones Leading to β-Diketones. *J. Am. Chem. Soc.* **1980**, *102*, 2095 – 2096.

[152] M. Suzuki, A. Watanabe, and R. Noyori: Palladium(0)-Catalyzed Isomerization of α,β-Epoxy Ketones to β-Diketones. *Recl. Trav. Chim. Pays-Bas* **1988**, *107*, 230 – 236.

[153] R. Noyori, K. Yokoyama, J. Sakata, I. Kuwajima, E. Nakamura, and M. Shimizu: Fluoride Ion Catalyzed Aldol Reaction between Enol Silyl Ethers and Carbonyl Compounds. *J. Am. Chem. Soc.* **1977**, *99*, 1265 – 1267.

[154] E. Nakamura, M. Shimizu, I. Kuwajima, J. Sakata, K. Yokoyama, and R. Noyori: Fluoride Ion Catalyzed Aldol Reaction between Enol Silyl Ethers and Carbonyl Compounds. *J. Org. Chem.* **1983**, *48*, 932 – 945.

[155] R. Noyori, I. Nishida, J. Sakata, and M. Nishizawa: Superanions. I. Tris(dialkylamino) sulfonium Enolates. *J. Am. Chem. Soc.* **1980**, *102*, 1223 – 1225.

[156] R. Noyori, I. Nishida, and J. Sakata: Superanions. II. Alkylation via Tris(dialkylamino) sulfonium Enolates. *Tetrahedron Lett.* **1980**, *21*, 2085 – 2088.

[157] R. Noyori, I. Nishida, and J. Sakata: Superanions. 5. Tris(dialkylamino)sulfonium Enolates.

Synthesis, Structure, and Reactions. *J. Am. Chem. Soc.* **1983**, *105*, 1598 – 1608.

[158] R. Noyori: Tris(dialkylamino)sulfonium Enolates and Phenoxide in *Selectivity – A Goal for Synthetic Efficiency* (Eds.: W. Bartmann and B. M. Trost), Verlag Chemie, Weinheim, **1984**, 121.

[159] R. Noyori, I. Nishida, and J. Sakata: Superanions. 4. A Tris(dialkylamino)sulfonium Phenoxide. *Tetrahedron Lett.* **1981**, *22*, 3993 – 3996.

[160] R. Noyori, I. Nishida, and J. Sakata: Superanions. 3. Erythro-Selective Aldol Reaction via Tris(dialkylamino)sulfonium Enolates. *J. Am. Chem. Soc.* **1981**, *103*, 2106 – 2108.

[161] S. Murata, M. Suzuki, and R Noyori: Trialkylsilyl Triflates. 5. A Stereoselective Aldol-Type Condensation of Enol Silyl Ethers and Acetals Catalyzed by Trimethylsilyl Trifluoromethanesulfonate. *J. Am. Chem. Soc.* **1980**, *102*, 3248 – 3249.

[162] S. Murata, M. Suzuki, and R. Noyori: Trimethylsilyl Triflate Catalyzed Aldol-Type Reaction of Enol Silyl Ethers and Acetals or Related Compounds. *Tetrahedron* **1988**, *44*, 4259 – 4275.

[163] M. Kitamura, K. Nakano, T. Miki, M. Okada, and R. Noyori: Stereochemistry of Aldols: Configuration and Conformation of Aldols Derived from Cycloalkanones and Aldehydes. *J. Am. Chem. Soc.* **2001**, *123*, 8939 – 8950.

[164] S. Murata, M. Suzuki, and R. Noyori: Trialkylsilyl Triflates in Organic Synthesis. VII. Condensation of Enol Silyl Ethers and Dialkoxymethanes Catalyzed by Trimethylsilyl Trifluoromethanesulfonate. Regiospecific Synthesis of α-Alkoxymethyl Ketones. *Tetrahedron Lett.* **1980**, *21,* 2527 – 2528.

[165] S. Murata and R. Noyori: Trialkylsilyl Triflates in Organic Synthesis. 14. Condensation of Enol Silyl Ethers with 2-Acetoxytetrahydrofuran and -tetrahydropyrans. *Tetrahedron Lett.* **1982**, *23*, 2601 – 2602.

[166] T. Tsunoda, M. Suzuki, and R. Noyori: Trialkylsilyl Triflates. II. Reaction of Acetals and Trialkylsilanes Catalyzed by Trimethylsilyl Trifluoromethanesulfonate. A Simple Method for Conversion of Acetals to Ethers. *Tetrahedron Lett.* **1979**, *20*, 4679 – 4680.

[167] T. Tsunoda, M. Suzuki, and R. Noyori: Trialkylsilyl Triflates. III. Trimethylsilyl Trifluoromethanesulfonate as a Catalyst of the Reaction of Allyltrimethylsilane and Acetals. *Tetrahedron Lett.* **1980**, *21*, 71 – 74.

[168] M. Suzuki, H. Takada, and R. Noyori: Trialkylsilyl Triflates in Organic Synthesis. 13. Bis(trimethylsilyl) Peroxide for the Baeyer-Villiger Type Oxidation. *J. Org. Chem.* **1982**, *47*, 902 – 904.

[169] R. Noyori, S. Murata, and M. Suzuki: Trialkylsilyl Triflates in Organic Synthesis. 11. Trimethysilyl Triflate in Organic Synthesis. (Tetrahedron Symposium-In-Print, Ed.: E. J. Corey), *Tetrahedron* **1981**, *37*, 3899 – 3910.

[170] T. Tsunoda, M. Suzuki, and R. Noyori: Trialkylsilyl Triflates. VI. A Facile Procedure for Acetalization under Aprotic Conditions. *Tetrahedron Lett.* **1980**, *21,* 1357 – 1358.

[171] S. Murata and R. Noyori: Trialkylsilyl Triflates. IV. Silylation with a Perfluorinated Resinsulfonic Acid Trimethylsilyl Ester. *Tetrahedron Lett.* **1980**, *21,* 767 – 768.

[172] S. Murata, M. Suzuki, and R. Noyori: Trialkylsilyl Triflates. I. A New Method for Converting Oxiranes to Allylic Alcohols by an Organosilicon Reagent. *J. Am. Chem. Soc.* **1979**, *101*, 2738 – 2739.

[173] H. Shirahama, K. Hayano, Y. Kanemoto, S. Misumi, T. Ohtsuka, N. Hashiba, A. Furusaki, S. Murata, R. Noyori, and T. Matsumoto: Trialkylsilyl Triflates in organic synthesis. 8. Conformationally Selective Transannular Cyclizations of Humulene 9,10-Epoxide. Synthesis of the Two Skeletally Different Cyclohumulanoids: DL-Bicyclohumulenone and DL-Africanol. *Tetrahedron Lett.* **1980**, *21,* 4835 – 4838.

[174] S. Murata, M. Suzuki, and R. Noyori: Trialkylsilyl Triflates in Organic Synthesis. 12. Ring Opening of Oxiranes by Trimethylsilyl Trifluoromethanesulfonate. *Bull. Chem. Soc. Jpn.* **1982**, *55*, 247 – 254.

[175] H. Shirahama, S. Murata, T. Fujita, B. R. Chhabra, R. Noyori, and T. Matsumoto: Trimethylsilyl Triflate Induced Reaction of Humulene 6,7-Epoxide. Cyclization to 5-Hydroxy-4,8,11,11-tetramethyltricyclo[6.3.0.0^{2,4}]undec-9-ene. *Bull. Chem. Soc. Jpn.* **1982**, *55*, 2691 – 2692.

[176] R. Noyori, M. Hayashi, and S. Hashimoto: Selectivity in Silicon-Mediated Organic Reactions: Origin and Application in *Organosilicon and Bioorganosilicon Chemistry: Structure, Bonding, Reactivity and Synthetic Application* (Ed.: H. Sakurai). Ellis Horwood, West Sussex, **1985**, 213.

[177] S. Hashimoto, M. Hayashi, and R. Noyori: An Easy Preparation of Triphenylmethyl Carboxylates. *Bull. Chem. Soc. Jpn.* **1984**, *57*, 1431 – 1432.

[178] S. Murata and R. Noyori: Trialkylsilyl Triflates in Organic Synthesis. 10. A Facile Procedure for *O*-Tritylation. *Tetrahedron Lett.* **1981**, *22*, 2107 – 2108.

[179] M. Hayashi, S. Hashimoto, and R. Noyori: Simple Synthesis of Glycosyl Fluorides. *Chem. Lett.* **1984**, *13*, 1747 – 1750.

[180] S. Hashimoto, M. Hayashi, and R. Noyori: Glycosylation Using Glucopyranosyl Fluorides and Silicon-Based Catalysts. Solvent Dependency of the Stereoselection. *Tetrahedron Lett.* **1984**, *25*, 1379 – 1382.

[181] R. Noyori and M. Hayashi: Condensation of 1-Fluorofuranoses and Silylated Nucleobases Catalyzed by Tetrafluorosilane. *Chem. Lett.* **1987**, *16*, 57 – 60.

[182] S. Hashimoto, I. Kurimoto, Y. Fujii, and R. Noyori: Novel Nucleophilic Substitution Reaction by Radical Cation Intermediates. Photosensitized Transacetalization via $S_{ON}1$ Mechanism. *J. Am. Chem. Soc.* **1985**, *107*, 1427 – 1429.

[183] R. Noyori and I. Kurimoto: Photoinduced Transacetalization using a Tris(bipyridine) ruthenium(II)-Methyl Viologen Cosensitizing System. *J. Chem. Soc. Chem. Commun.* **1986**, 1425 – 1426.

[184] R. Noyori and I. Kurimoto: Electrochemical Glycosylation Method. *J. Org. Chem.* **1986**, *51*, 4320 – 4322.

[185] Y. Hayakawa, Y. Aso, M. Uchiyama, and R. Noyori: Chemoselective Phosphorylation of *N*-Unprotected Nucleosides via Aluminum Alkoxides. *Tetrahedron Lett.* **1983**, *24*, 5641 – 5644.

[186] M. Uchiyama, Y. Aso, R. Noyori, and Y. Hayakawa: O-Selective Phosphorylation of Nucleosides without N-Protection. *J. Org. Chem.* **1993**, *58*, 373 – 379.

[187] Y. Hayakawa, M. Uchiyama, and R. Noyori: A Convenient Method for the Formation of Internucleotide Linkage. *Tetrahedron Lett.* **1984**, *25*, 4003 – 4006.

[188] Y. Hayakawa, M. Uchiyama, H. Kato, and R. Noyori: Allyl Protection of Internucleotide Linkage. *Tetrahedron Lett.* **1985**, *26*, 6505 – 6508.

[189] Y. Hayakawa, M. Hirose, and R. Noyori: *O*-Allyl Protection of Guanine and Thymine Residues in Oligodeoxyribonucleosides. *J. Org. Chem.* **1993**, *58*, 5551 – 5555.

[190] Y. Hayakawa, H. Kato, M. Uchiyama, H. Kajino, and R. Noyori: Allyloxycarbonyl Group: A Versatile Blocking Group for Nucleotide Synthesis. *J. Org. Chem.* **1986**, *51*, 2400 – 2402.

[191] Y. Hayakawa, S. Wakabayashi, T. Nobori, and R. Noyori: A General Approach to Nucleoside 3'- and 5'-Monophosphates. *Tetrahedron Lett.* **1987**, *28*, 2259 – 2262.

[192] Y. Hayakawa, M. Uchiyama, and R. Noyori: Nonaqueous Oxidation of Nucleoside Phosphites to the Phosphates. *Tetrahedron Lett.* **1986**, *27*, 4191 – 4194.

[193] Y. Hayakawa, M. Kataoka, and R. Noyori: Benzimidazolium Triflate as an Efficient Promoter for Nucleotide Synthesis via the Phosphoramidite Method. *J. Org. Chem.* **1996**, *61*, 7996 – 7997.

[194] Y. Hayakawa, M. Uchiyama, T. Nobori, and R. Noyori: A Convenient Synthesis of 2'-5' Linked Oligoribonucleotides. *Tetrahedron Lett.* **1985**, *26*, 761 – 764.

[195] A. Tominaga, S. Saito, S. Kohno, K. Sakurai, Y. Hayakawa, and R. Noyori: Antiviral Effects of 2', 5'Oligoadenylates (2-5As) and Related Compounds. *Microbiol. Immunol.* **1990**, *34*, 737 – 747.

[196] R. Noyori, M. Uchiyama, T. Nobori, M. Hirose, and Y. Hayakawa: Practical Synthesis of 2'-5'-Linked Oligoadenylates (2-5A Oligomers). *Aust. J. Chem.* **1992**, *45*, 205 – 225.

[197] Y. Hayakawa, T. Nobori, R. Noyori, and J. Imai: Synthesis of 2'-5',3'-5' Linked Triadenylates. *Tetrahedron Lett.* **1987**, *28*, 2623 – 2626.

[198] Y. Hayakawa, M. Hirose, and R. Noyori: General, Regiocontrolled Synthesis of Unsymmetrically Linked Dinucleoside Adenosine 2',3'-Bis(phosphate)s., *Nucleic Acids Symp. Ser.* **1989**, *21*, 103.

[199] Y. Hayakawa, M. Hirose, M. Hayakawa, and R. Noyori: General Synthesis and Binding Affinity of Position-Selective Phosphonodiester- and Phosphotriester-Incorporated Oligodeoxyribonucleotides. *J. Org. Chem.* **1995**, *60*, 925 – 930.

[200] Y. Hayakawa, M. Uchiyama, and R. Noyori: Solid-Phase Synthesis of Oligodeoxyribonucleotides Using the Bis(trimethylsilyl)peroxide Oxidation of Phosphites. *Tetrahedron Lett.* **1986**, *27*, 4195 – 4196.

[201] Y. Hayakawa, S. Wakabayashi and R. Noyori: Allylic Protecting Groups in Solid-Phase DNA Synthesis. *Nucleic Acids Symp. Ser.* **1988**, *20*, 75

[202] Y. Hayakawa, S. Wakabayashi, H. Kato, and R. Noyori: The Allylic Protection Method in Solid-Phase Oligonucleotide Synthesis. An Efficient Preparation of Solid-Anchored DNA Oligomers. *J. Am. Chem. Soc.* **1990**, *112*, 1691 – 1696.

[203] R. Noyori, M. Uchiyama, H. Kato, S. Wakabayashi, and Y. Hayakawa: Organometallic Methodologies for Nucleic Acid Synthesis. *Pure Appl. Chem.* **1990**, *62*, 613 – 622.

[204] Y. Hayakawa, H. Harada, M. Hirose, R. Noyori, S. Wakabayashi, K. Miyazaki, Y. Kawase, and I. Kato: Direct Synthesis of Solid-Anchored DNA Oligomers. *Nucleic Acids Symp. Ser.* **1991**, *25*, 63.

[205] R. Noyori, S. Suga, K. Kawai, S. Okada, and M. Kitamura: Enantioselective Alkylation of Carbonyl Compounds. From Stoichiometric to Catalytic Asymmetric Induction. *Pure Appl. Chem.* **1988**, *60*, 1597 – 1606.

[206] R. Noyori: Centenary Lecture. Chemical Multiplication of Chirality: Science and Applications. *Chem. Soc. Rev.* **1989**, *18*, 187 – 208.

[207] R. Noyori: Organometallic Ways for the Multiplication of Chirality. *Tetrahedron* **1994**, *50*, 4259 – 4292.

[208] M. Kitamura, S. Suga, K. Kawai, and R. Noyori: Catalytic Asymmetric Induction. Highly Enantioselective Addition of Dialkylzincs to Aldehydes. *J. Am. Chem. Soc.* **1986**, *108*, 6071 – 6072.

[209] M. Kitamura, H. Oka, S. Suga, and R. Noyori: Catalytic Enantioselective Addition of Dialkylzincs to Aldehydes Using (2*S*)-(–)-3-*exo*-(Dimethylamino)isoborneol [(2*S*)-DAIB]: (*S*)-1-Phenyl-1-propanol. *Org. Synth.* **2004**, *10*, 635.

[210] R. Noyori, S. Suga, K. Kawai, S. Okada, M. Kitamura, N. Oguni, M. Hayashi, T. Kaneko, and Y. Matsuda: Enantioselective Addition of Diorganozincs to Aldehydes Catalyzed by β-Amino Alcohols. *J. Organomet. Chem.* **1990**, *382*, 19 – 37.

[211] M. Yamakawa and R. Noyori: An *Ab Initio* Molecular Orbital Study on the Amino Alcohol-Promoted Reaction of Dialkylzincs and Aldehydes. *J. Am. Chem. Soc.* **1995**, *117*, 6327 – 6335.

[212] M. Yamakawa and R. Noyori: Asymmetric Addition of Dimethylzinc to Benzaldehyde Catalyzed by (2*S*)-3-*exo*-(Dimethylamino) isoborneol. A Theoretical Study on the Origin of Enantioselection. *Organometallics* **1999**, *18*, 128 – 133.

[213] M. Kitamura, S. Okada, S. Suga, and R. Noyori: Enantioselective Addition of Dialkylzincs to Aldehydes Promoted by Chiral Amino Alcohols. Mechanism and Nonlinear Effect. *J. Am. Chem. Soc.* **1989**, *111*, 4028 – 4036.

[214] R. Noyori und M. Kitamura: Enantioselektive Addition von Organometallreagentien an Carbonylverbindungen: Übertragung, Vervielfältigung und Verstärkung der Chiralität. *Angew. Chem.* **1991**, *103*, 34 – 35; R. Noyori and M. Kitamura: Enantioselective Addition of Organometallic Reagents to Carbonyl Compounds: Chirality Transfer, Multiplication, and Amplification. *Angew. Chem. Int. Ed.* **1991**, *30*, 49 – 69.

[215] R. Noyori, S. Suga, H. Oka, and M. Kitamura: Self and Nonself Recognition of Chiral Catalysts: The Origin of Nonlinear Effects in the Amino-Alcohol Catalyzed Asymmetric Addition of Diorganozincs to Aldehydes. *Chem. Rec.* **2001**, *1*, 85 – 100.

[216] M. Kitamura, M. Yamakawa, H. Oka, S. Suga, and R. Noyori: Homochiral and Heterochiral Dimers of the Methylzinc Alkoxide Formed from Dimethylzinc and Enantiomeric 3-*exo*-(Dimethylamino)isoborneol – Origin of the Distinct Differences in Solution-Phase Behavior and Crystal Structures. *Chem. Eur. J.* **1996**, *2*, 1173 – 1181.

[217] M. Kitamura, S. Suga, H. Oka, and R. Noyori: Quantitative Analysis of the Chiral Amplification in the Amino Alcohol-Promoted Asymmetric Alkylation of Aldehydes with Dialkylzincs. *J. Am. Chem. Soc.* **1998**, *120*, 9800 – 9809.

[218] M. Kitamura, H. Oka, and R. Noyori: Asymmetric Addition of Dialkylzincs to Benzaldehyde Derivatives Catalyzed by Chiral β-Amino Alcohols. Evidence for the Monomeric Alkylzinc Aminoalkoxide as Catalyst. *Tetrahedron* **1999**, *55*, 3605 – 3614.

[219] M. Kitamura, S. Suga, M. Niwa, R. Noyori, Z.-X. Zhai, and H. Suga: Enantiomer Recognition of Asymmetric Catalysts. Thermodynamic Properties of Homochiral and Heterochiral Dimers of the Methylzinc Alkoxide Formed from Dimethylzinc and Enantiomeric 3-*exo*-(Dimethylamino)isoborneol. *J. Phys. Chem.* **1994**, *98*, 12776 – 12781.

[220] M. Kitamura, S. Suga, M. Niwa, and R. Noyori: Self and Nonself Recognition of Asymmetric Catalysts. Nonlinear Effects in the Amino Alcohol-Promoted Enantioselective Addition of Dialkylzincs to Aldehydes. *J. Am. Chem. Soc.* **1995**, *117*, 4832 – 4842.

[221] A. Miyashita, A. Yasuda, H. Takaya, K. Toriumi, T. Ito, T. Souchi, and R. Noyori: Synthesis of 2,2'-Bis(diphenylphosphino)-1,1'-binaphthyl (BINAP), an Atropisomeric Chiral Bis(triaryl) phosphine, and Its Use in the Rhodium(I)-Catalyzed Asymmetric Hydrogenation of α-(Acylamino)acrylic Acids. *J. Am. Chem. Soc.* **1980**, *102*, 7932 – 7934.

[222] A. Miyashita, H. Takaya, T. Souchi, and R. Noyori: 2,2'-Bis(diphenylphosphino)-1,1'-binaphthyl (BINAP), A New Atropisomeric Bis(triaryl)phosphine. Synthesis and Its Use in the Rh(I)-Catalyzed Asymmetric Hydrogenation of α-(Acylamino)acrylic Acids. (Tetrahedron Symposium-In-Print, Ed.: A. I. Meyers), *Tetrahedron* **1984**, *40*, 1245 – 1253.

[223] H. Takaya, S. Akutagawa, and R. Noyori: (*R*)-(+)- and (*S*)-(–)-2,2'-Bis(diphenylphosphino)-1,1'-binaphthyl (BINAP). *Org. Synth.* **1989**, *67*, 20 – 32.

[224] H. Takaya, K. Mashima, K. Koyano, M. Yagi, H. Kumobayashi, T. Taketomi, S. Akutagawa, and R. Noyori: Practical Synthesis of (*R*)- or (*S*)-2,2'-Bis(diarylphosphino)-1,1'-binaphthyls (BINAPs). *J. Org. Chem.* **1986**, *51*, 629 – 635.

[225] K. Tani, T. Yamagata, S. Otsuka, S. Akutagawa, H. Kumobayashi, T. Taketomi, H. Takaya, A. Miyashita, and R. Noyori: Cationic Rhodium(I) Complex-Catalysed Asymmetric Isomerisation of Allylamines to Optically Active Enamines. *J. Chem. Soc. Chem. Commun.* **1982**, 600 – 601.

[226] K. Tani, T. Yamagata, S. Akutagawa, H. Kumobayashi, T. Taketomi, H. Takaya, A. Miyashita, R. Noyori, and S. Otsuka: Highly Enantioselective Isomerization of Prochiral Allylamines Catalyzed by Chiral Diphosphine Rhodium(I) Complexes. Preparation of Optically Active Enamines. *J. Am. Chem. Soc.* **1984**, *106*, 5208 – 5217.

[227] M. Yamakawa and R. Noyori: Ab Initio Molecular Orbital Study on Rhodium(I)-Catalyzed Isomerization of Allylic Amines to Enamines. *Organometallics* **1992**, *11*, 3167 – 3169.

[228] R. Noyori: *Asymmetric Catalysis in Organic Synthesis*, John Wiley & Sons, New York, **1994**.

[229] R. Noyori: Asymmetrische Katalyse: Kenntnisstand und Perspektiven (Nobel-Vortrag). *Angew. Chem.* **2002**, *114*, 2108 – 2123; R. Noyori: Aymmetric Catalysis: Science and Opportunities (Nobel Lecture). *Angew. Chem. Int. Ed.* **2002**, *41*, 2008 – 2022.

[230] R. Noyori: Chiral Metal Complexes as Discriminating Molecular Catalysts. *Science*, **1990**, *248*, 1194 – 1199.

[231] R. Noyori and H. Takaya: BINAP: An Efficient Chiral Element for Asymmetric Catalysis. *Acc. Chem. Res.* **1990**, *23*, 345 – 350.

[232] T. Ohta, H. Takaya, and R. Noyori: BINAP-Ruthenium(II) Dicarboxylate Complexes: New, Highly Efficient Catalysts for Asymmetric Hydrogenations. *Inorg. Chem.* **1988**, *27*, 566 – 569.

[233] M. Kitamura, M. Tokunaga, and R. Noyori: Practical Synthesis of BINAP-Ruthenium(II) Dicarboxylate Complexes. *J. Org. Chem.* **1992**, *57*, 4053 – 4054.

[234] R. Noyori: Asymmetric Catalysis by Chiral Metal-Complexes. *CHEMTECH* **1992**, *26*, 360 – 367.

[235] R. Noyori, M. Ohta, Y. Hsiao, M. Kitamura, T. Ohta, and H. Takaya: Asymmetric Synthesis of Isoquinoline Alkaloids by Homogeneous Catalysis. *J. Am. Chem. Soc.* **1986**, *108*, 7117 – 7119.

[236] M. Kitamura, Y. Hsiao, M. Ohta, M. Tsukamoto, T. Ohta, H. Takaya, and R. Noyori: General Asymmetric Synthesis of Isoquinoline Alkaloids. Enantioselective Hydrogenation of Enamides Catalyzed by BINAP-Ruthenium(II) Complexes. *J. Org. Chem.* **1994**, *59*, 297 – 310.

[237] M. Kitamura, Y. Hsiao, R. Noyori, and H. Takaya: General Asymmetric Synthesis of Benzomorphans and Morphinans via Enantioselective Hydrogenation. *Tetrahedron Lett.* **1987**, *28*, 4829 – 4832.

[238] M. Kitamura, M. Tsukamoto, H. Takaya, and R. Noyori: Conformational Study on 2-Acyl-1-alkylidene-1,2,3,4-tetrahydroisoquinolines. *Bull. Chem. Soc. Jpn.* **1996**, *69*, 1695 – 1700.

[239] M. Kitamura, M. Tsukamoto, Y. Bessho, M. Yoshimura, U. Kobs, M. Widhalm, and R. Noyori: Mechanism of Asymmetric Hydrogenation of α-(Acylamimo)acrylic Esters Catalyzed by BINAP-Ruthenium(II) Diacetate. *J. Am. Chem. Soc.* **2002**, *124*, 6649 – 6667.

[240] W. D. Lubell, M. Kitamura, and R. Noyori: Enantioselective Synthesis of β-Amino Acids Based on BINAP-Ruthenium(II) Catalyzed Hydrogenation. *Tetrahedron: Asymmetry* **1991**, *2*, 543 – 554.

[241] M. Kitamura, M. Yoshimura, M. Tsukamoto, and R. Noyori: Synthesis of α-Amino Phosphonic Acids by Asymmetric Hydrogenation. *Enantiomer* **1996**, *1*, 281 – 303.

[242] H. Takaya, T. Ohta, N. Sayo, H. Kumobayashi, S. Akutagawa, S. Inoue, I. Kasahara, and R. Noyori: Enantioselective Hydrogenation of Allylic and Homoallylic Alcohols. *J. Am. Chem. Soc.* **1987**, *109*, 1596 – 1597.

[243] H. Takaya, T. Ohta, S. Inoue, M. Tokunaga, M. Kitamura, and R. Noyori: Asymmetric Hydrogenation of Allylic Alcohols Using BINAP-Ruthenium Complexes: (*S*)-(–)-Citronellol. *Org. Synth.* **1995**, *72*, 74 – 85.

[244] S. Inoue, M. Osada, K. Koyano, H. Takaya, and R. Noyori: Asymmetric Hydrogenation of Geraniol and Nerol Catalyzed by BINAP-Rhodium(I) Complexes. *Chem. Lett.* **1985**, *14*, 1007 – 1008.

[245] M. Kitamura, K. Nagai, Y. Hsiao, and R. Noyori: Stereoselective Synthesis of a Precursor of 1β-Methylcarbapenems. *Tetrahedron Lett.* **1990**, *31*, 549 – 552.

[246] T. Ohta, H. Takaya, M. Kitamura, K. Nagai, and R. Noyori: Asymmetric Hydrogenation of Unsaturated Carboxylic Acids Catalyzed by BINAP-Ruthenium(II) Complexes. *J. Org. Chem.* **1987**, *52*, 3174 – 3176.

[247] T. Ohta, H. Takaya, and R. Noyori: Stereochemistry and Mechanism of the Asymmetric Hydrogenation of Unsaturated Carboxylic Acids Catalyzed by BINAP-Ruthenium(II) Dicarboxylate Complexes. *Tetrahedron Lett.* **1990**, *31*, 7189 – 7192.

[248] R. Noyori, T. Ohkuma, M. Kitamura, H. Takaya, N. Sayo, H. Kumobayashi, and S. Akutagawa: Asymmetric Hydrogenation of β-Keto Carboxylic Ester. A Practical, Purely Chemical Access to β-Hydroxy Esters in High Enantiomeric Purity. *J. Am. Chem. Soc.* **1987**, *109*, 5856 – 5858.

[249] M. Kitamura, M. Tokunaga, T. Ohkuma, and R. Noyori: Asymmetric Hydrogenation of 3-Oxo Carboxylates Using BINAP-Ruthenium Complexes: (*R*)-(–)-Methyl 3-Hydroxybutanoate. *Org. Synth.* **1993**, *71*, 1 – 13.

[250] M. Kitamura, T. Ohkuma, S. Inoue, N. Sayo, H. Kumobayashi, S. Akutagawa, T. Ohta, H. Takaya, and R. Noyori: Homogeneous Asymmetric Hydrogenation of Functionalized Ketones. *J. Am. Chem. Soc.* **1988**, *110*, 629 – 631.

[251] M. Kitamura, M. Tokunaga, T. Ohkuma, and R. Noyori: Convenient Preparation of BINAP-Ruthenium(II) Complexes Catalyzing Asymmetric Hydrogenation of Functionalized

Ketones. *Tetrahedron Lett.* **1991**, *32*, 4163–4166.

[252] K. Mashima, K. Kusano, T. Ohta, R. Noyori, and H. Takaya: Synthesis of New Cationic BINAP-Ruthenium(II) Complexes and their Use in Asymmetric Hydrogenation [BINAP = 2,2'-bis(diphenylphosphino)-1,1'-binaphthyl]. *J. Chem. Soc. Chem. Commun.* **1989**, 1208–1210.

[253] R. Noyori und T. Ohkuma: Asymmetrische Katalyse mit hinsichtlich Struktur und Funktion gezielt entworfenen Molekülen: die chemo- und stereoselektive Hydrierung von Ketonen. *Angew. Chem.* **2001**, *113*, 40–75; R. Noyori and T. Ohkuma: Asymmetric Catalysis by Architectural and Functional Molecular Engineering: Practical Chemo- and Stereoselective Hydrogenation of Ketones. *Angew. Chem. Int. Ed.* **2001**, *40*, 40–73.

[254] M. Kitamura, M. Yoshimura, N. Kanda, and R. Noyori: Asymmetric Synthesis of β-Hydroxy Sulfonic Acids by BINAP / Ru-Catalyzed Hydrogenation. *Tetrahedron* **1999**, *55*, 8769–8785.

[255] T. Ohkuma and R. Noyori: Carbonyl Hydrogenations in *Transition Metals for Organic Synthesis* (Eds.: M. Beller and C. Bolm), Vol. 2. Wiley-VCH, Weinheim, **1998**, 1.1.3.

[256] T. Ohkuma and R. Noyori, Hydrogenation of Carbonyl Groups in *Comprehensive Asymmetric Catalysis* (Eds.: E. N. Jacobsen, A. Pfaltz, H. Yamamoto), Vol. 1. Springer Berlin, Heidelberg, **1999**, Chapter 6.1.

[257] T. Ohkuma, M. Kitamura, and R. Noyori: Asymmetric Hydrogenation in *Catalytic Asymmetric Synthesis* (Ed: I. Ojima), 2nd Ed. Wiley-VCH, New York, **2000**, Chapter 1.

[258] M. Kitamura, T. Ohkuma, H. Takaya, and R. Noyori: A Practical Asymmetric Synthesis of Carnitine. *Tetrahedron Lett.* **1988**, *29*, 1555–1556.

[259] T. Nishi, M. Kitamura, T. Ohkuma, and R. Noyori: Synthesis of Statine and its Analogues by Homogeneous Asymmetric Hydrogenation. *Tetrahedron Lett.* **1988**, *29*, 6327–6330.

[260] R. Noyori, T. Ikeda, T. Ohkuma, M. Widhalm, M. Kitamura, H. Takaya, S. Akutagawa, N. Sayo, T. Saito, T. Taketomi, and H. Kumobayashi: Stereoselective Hydrogenation via Dynamic Kinetic Resolution. *J. Am. Chem. Soc.* **1989**, *111*, 9134–9135.

[261] M. Kitamura, T. Ohkuma, M. Tokunaga, and R. Noyori: Dynamic Kinetic Resolution in BINAP-Ruthenium(II) Catalyzed Hydrogenation of 2-Substituted 3-Oxo Carboxylic Esters. *Tetrahedron: Asymmetry* **1990**, *1*, 1–4.

[262] R. Noyori, M. Tokunaga, and M. Kitamura: Stereoselective Organic Synthesis via Dynamic Kinetic Resolution. *Bull. Chem. Soc. Jpn.* **1995**, *68*, 36–56.

[263] M. Kitamura, M. Tokunaga, and R. Noyori: Asymmetric Hydrogenation of β-Keto Phosphonates: A Practical Way to Fosfomycin. *J. Am. Chem. Soc.* **1995**, *117*, 2931–2932.

[264] M. Kitamura, M. Tokunaga, T. Pham, W. D. Lubell, and R. Noyori: Asymmetric Synthesis of α-Amino β-Hydroxy Phosphonic Acids via BINAP-Ruthenium Catalyzed Hydrogenation. *Tetrahedron Lett.* **1995**, *36*, 5769–5772.

[265] M. Kitamura, M. Tokunaga, and R. Noyori: Quantitative Expression of Dynamic Kinetic Resolution of Chirally Labile Enantiomers: Stereoselective Hydrogenation of 2-Substituted 3-Oxo Carboxylic Esters Catalyzed by BINAP-Ruthenium(II) Complexes. *J. Am. Chem. Soc.* **1993**, *115*, 144–152.

[266] Y. Kishimoto, P. Eckerle, T. Miyatake, T. Ikariya, and R. Noyori: Living Polymerization of Phenylacetylenes Initiated by $Rh(C{\equiv}CC_6H_5)(2,5\text{-norbornadiene})[P(C_6H_5)_3]_2$. *J. Am. Chem. Soc.* **1994**, *116*, 12131–12132.

[267] Y. Kishimoto, M. Itou, T. Miyatake, T. Ikariya, and R. Noyori: Polymerization of Monosubstituted Acetylenes with a Zwitterionic Rhodium(I) Complex, $Rh^+(2,5\text{-norbornadiene})[(\eta^6\text{-}C_6H_5)B^-(C_6H_5)_3]$. *Macromolecules* **1995**, *28*, 6662–6666.

[268] Y. Kishimoto, T. Miyatake, T. Ikariya, and R. Noyori: An Efficient Rhodium(I) Initiator for Stereospecific Living Polymerization of Phenylacetylenes. *Macromolecules* **1996**, *29*, 5054–5055.

[269] Y. Kishimoto, P. Eckerle, T. Miyatake, M. Kainosho, A. Ono, T. Ikariya, and R. Noyori: Well-Controlled Polymerization of Phenylacetylenes with Organorhodium(I) Complexes: Mechanism and Structure of the Polyenes. *J. Am. Chem. Soc.* **1999**, *121*, 12035–12044.

[270] R. Noyori and S. Hashiguchi: Asymmetric Transfer Hydrogenation Catalyzed by Chiral Ruthenium Complexes. *Acc. Chem. Res.* **1997**, *30*, 97–102.

[271] R. Noyori, M. Yamakawa, and S. Hashiguchi: Metal-Ligand Bifunctional Catalysis: A Nonclassical Mechanism for Asymmetric Hydrogen Transfer between Alcohols and Carbonyl Compounds. *J. Org. Chem.* **2001**, *66*, 7931–7944.

[272] S. Hashiguchi, A. Fujii, J. Takehara, T. Ikariya, and R. Noyori: Asymmetric Transfer Hydrogenation of Aromatic Ketones Catalyzed by Chiral Ruthenium(II) Complexes. *J. Am. Chem. Soc.* **1995**, *117*, 7562–7563.

[273] J. Takehara, S. Hashiguchi, A. Fujii, S. Inoue, T. Ikariya, and R. Noyori: Amino Alcohol Effects on the Ruthenium(II)-Catalysed Asymmetric Transfer Hydrogenation of Ketones in Propan-2-ol. *Chem. Commun.* **1996**, 233–234.

[274] J.-X. Gao, T. Ikariya, and R. Noyori: A Ruthenium(II) Complex with a C_2-Symmetric Diphosphine / Diamine Tetradentate Ligand for Asymmetric Transfer Hydrogenation of Aromatic Ketones. *Organometallics* **1996**, *15*, 1087 – 1089.

[275] S. Inoue, K. Nomura, S. Hashiguchi, R. Noyori, and Y. Izawa: Asymmetric Transfer Hydrogenation Catalyzed by Diamine-Iridium(I) Complexes. *Chem. Lett.* **1997**, *26*, 957 – 958.

[276] K. Murata, T. Ikaraya, and R. Noyori: New Chiral Rhodium and Iridium Complexes with Chiral Diamine Ligands for Asymmetric Transfer Hydrogenation of Aromatic Ketones. *J. Org. Chem.* **1999**, *64*, 2186 – 2187.

[277] I. Yamada and R. Noyori: Asymmetric Transfer Hydrogenation of Benzaldehydes. *Org. Lett.* **2000**, *2*, 3425 – 3427.

[278] K. Matsumura, S. Hashiguchi, T. Ikariya, and R. Noyori: Asymmetric Transfer Hydrogenation of α,β-Acetylenic Ketones. *J. Am. Chem. Soc.* **1997**, *119*, 8738 – 8739.

[279] K.-J. Haack, S. Hashiguchi, A. Fujii, T. Ikariya und R. Noyori: Katalysatorvorstufe, Katalysator und Zwischenstufe des Ru^{II}-katalysierten, asymmetrischen Wasserstofftransfers zwischen Alkoholen und Ketonen. *Angew. Chem.* **1997**, *109*, 297 – 300; K.-J. Haack, S. Hashiguchi, A. Fujii, T. Ikariya, and R. Noyori: The Catalyst Precursor, Catalyst, and Intermediate in the Ru^{II}-Promoted Asymmetric Hydrogen Transfer between Alcohols and Ketones. *Angew. Chem. Int. Ed.* **1997**, *36*, 285 – 288.

[280] M. Yamakawa, I. Yamada, and R. Noyori: CH / p Attraction: The Origin of Enantioselectivity in Transfer Hydrogenation of Aromatic Carbonyl Compounds Catalyzed by Chiral η^6-Arene-Ruthenium(II) Complexes. *Angew. Chem. Int. Ed.* **2001**, *40*, 2818 – 2821.

[281] A. Matsuoka, C. A. Sandoval, M. Uchiyama, R. Noyori, and H. Naka: Why *p*-Cymene? Conformational Effect in Asymmetric Hydrogenation of Aromatic Ketones with a η^6-Arene / Ruthenium(II) Catalyst. *Chem. Asian J.* **2015**, *10*, 112-115.

[282] A. Fujii, S. Hashiguchi, N. Uematsu, T. Ikariya, and R. Noyori: Ruthenium(II)-Catalyzed Asymmetric Transfer Hydrogenation of Ketones Using a Formic Acid-Triethylamine Mixture. *J. Am. Chem. Soc.* **1996**, *118*, 2521 – 2522.

[283] S. Hashiguchi, A. Fujii, K.-J. Haack, K. Matsumura, T. Ikariya und R. Noyori: Kinetische Racematspaltung sekundärer Alkohole durch Ruthenium(II)-katalysierte Transferhydrierung. *Angew. Chem.* **1997**, *109*, 300 – 303; S. Hashiguchi, A. Fujii, K.-J. Haack, K. Matsumura, T. Ikariya, and R. Noyori: Kinetic Resolution of Racemic Secondary Alcohols by Ru^{II}-Catalyzed Hydrogen Transfer. *Angew. Chem. Int. Ed.* **1997**, *36*, 288 – 290.

[284] K. Murata, K. Okano, M. Miyagi, H. Iwane, R. Noyori, and T. Ikariya: A Practical Stereoselective Synthesis of Chiral Hydrobenzoins via Asymmetric Transfer Hydrogenation of Benzils. *Org. Lett.* **1999**, *1*, 1119 – 1121.

[285] T. Ikariya, S. Hashiguchi, K. Murata, and R. Noyori: Preparation of Optically Active (*R*,*R*)-Hydrobenzoin from Benzoin or Benzil [1,2-Ethanediol, 1,2-Diphenyl-, (1*R*,2*R*)-]. *Org. Synth.* **2005**, *82*, 10 – 17.

[286] N. Uematsu, A. Fujii, S. Hashiguchi, T. Ikariya, and R. Noyori: Asymmetric Transfer Hydrogenation of Imines. *J. Am. Chem. Soc.* **1996**, *118*, 4916 – 4917.

[287] T. Ohkuma, N. Utsumi, K. Tsutsumi, K. Murata, C. Sandoval, and R. Noyori: The Hydrogenation / Transfer Hydrogenation Network: Asymmetric Hydrogenation of Ketones with Chiral η^6-Arene / *N*-Tosylethylenediamine-Ruthenium(II) Catalysts. *J. Am. Chem. Soc.* **2006**, *128*, 8724 – 8725.

[288] C. A. Sandoval,, T. Ohkuma, N. Utsumi, K. Tsutsumi, K. Murata, and R. Noyori: Mechanism of Asymmetric Hydrogenation of Acetophenone Catalyzed by Chiral η^6-Arene-*N*-Tosylethylenediamine-Ruthenium(II) Complexes. *Chem. Asian. J.* **2006**, *1-2*, 102 – 110.

[289] T. Ohkuma, K. Tsutsumi, N. Utsumi, N. Arai, R. Noyori, and K. Murata: Asymmetric Hydrogenation of α-Chloro Aromatic Ketones Catalyzed by η^6-Arene / TsDPEN-Ruthenium(II) Complexes. *Org. Lett.* **2007**, *9*, 255 – 257.

[290] R. Noyori: Asymmetric Hydrogenation. *Acta Chem. Scand.* **1996**, *50*, 380 – 390.

[291] T. Ohkuma, H. Ooka, S. Hashiguchi, T. Ikariya, and R. Noyori: Practical Enantioselective Hydrogenation of Aromatic Ketones. *J. Am. Chem. Soc.* **1995**, *117*, 2675 – 2676.

[292] T. Ohkuma. H. Takeno, Y. Honda, and R. Noyori: Asymmetric Hydrogenation of Ketones with Polymer-Bound BINAP / Diamine Ruthenium Catalysts. *Adv. Synth. Catal.* **2001**, *343*, 369 – 375.

[293] T. Ohkuma, M. Koizumi, H. Doucet, T. Pham, M. Kozawa, K. Murata, E. Katayama, T. Yokozawa, T. Ikariya, and R. Noyori: Asymmetric Hydrogenation of Alkenyl, Cyclopropyl, and Aryl Ketones. $RuCl_2$(xylbinap)(1,2-diamine) as a Precatalyst Exhibiting a Wide Scope. *J. Am. Chem. Soc.* **1998**, *120*, 13529 – 13530.

[294] T. Ohkuma, M. Koizumi, H. Ikehira, T. Yokozawa, and R. Noyori: Selective Hydrogenation of Benzophenones to Benzhydrols. Asymmetric Synthesis of Unsymmetrical Diarylmethanols. *Org. Lett.* **2000**, *2*, 659 – 662.

[295] T. Ohkuma, M. Koizumi, M. Yoshida, and R. Noyori: General Asymmetric Hydrogenation of Hetero-aromatic Ketones. *Org. Lett.* **2000**, *2*, 1749 – 1751.

[296] T. Ohkuma, D. Ishii, H. Takeno, and R. Noyori: Asymmetric Hydrogenation of Amino Ketones Using Chiral $RuCl_2$(diphophine)(1,2-diamine) Complexes. *J. Am. Chem. Soc.* **2000**, *122*, 6510 – 6511.

[297] T. Ohkuma, M. Koizumi, K. Muñiz, G. Hilt, C. Kabuto, and R. Noyori: *trans*-RuH(η^1-BH_4)(binap)(1,2-diamine): A Catalyst for Asymmetric Hydrogenation of Simple Ketones under Base-Free Conditions. *J. Am. Chem. Soc.* **2002**, *124*, 6508 – 6509.

[298] M. Yamakawa, H. Ito, and R. Noyori: The Metal-Ligand Bifunctional Catalysis: A Theoretical Study on the Ruthenium(II)-Catalyzed Hydrogen Transfer between Alcohols and Carbonyl Compounds. *J. Am. Chem. Soc.* **2000**, *122*, 1466 – 1478.

[299] C. A. Sandoval, Y. Yamaguchi, T. Ohkuma, K. Kato, and R. Noyori: Solution Structures and Behavior of *trans*-Ru-H(η^1-BH_4)(binap)(1,2-diamine) Complexes. *Magn. Reson. Chem.* **2006**, *44*, 66 – 75.

[300] C. A. Sandoval, T. Ohkuma, K. Muñiz, and R. Noyori: Mechanism of Asymmetric Hydrogenation of Ketones Catalyzed by BINAP / 1,2-Diamine-Ruthenium(II) Complexes. *J. Am. Chem. Soc.* **2003**, *125*, 13490 – 13503.

[301] R. Noyori, C. A. Sandoval, K. Muñiz, and T. Ohkuma: Metal-Ligand Bifunctional Catalysis for Asymmetric Hydrogenation. *Phil. Trans. R. Soc. A* **2005**, *363*, 901 – 912.

[302] H. Doucet, T. Ohkuma, K. Murata, T. Yokozawa, M. Kozawa, E. Katayama, A. F. England, T. Ikariya, und R. Noyori: *trans*-[$RuCl_2$(phosphan)$_2$(1,2-diamin)]- und chirale *trans*-[Ru-Cl_2(diphosphan)(1,2-diamin)]-Komplexe: lagerstabile Katalysatorvorstufen für die schnelle, produktive und stereoselektive Hydrierung von Ketonen. *Angew. Chem.* **1998**, *110*, 1792 – 1796; H. Doucet, T. Ohkuma, K. Murata, T. Yokozawa, M. Kozawa, E. Katayama, A. F. England, T. Ikariya, and R. Noyori: *trans*-[$RuCl_2$(phosphane)$_2$(1,2-diamine)] and Chiral *trans*-[$RuCl_2$(diphosphane)(1,2-diamine): Shelf-Stable Precatalysts for the Rapid, Productive, and Stereoselective Hydrogenation of Ketones. *Angew. Chem. Int. Ed.* **1998**, *37*, 1703 – 1707.

[303] C. A. Sandoval, Q. Shi, S. Liu, and R. Noyori: NH / π Attraction: A Role in Asymmetric Hydrogenation of Aromatic Ketones with Binap / 1,2-Diamine-Ruthenium(II) Complexes. *Chem. Asian J.* **2009**, *4*, 1221 – 1224.

[304] T. Ohkuma, H. Ikehira, T. Ikariya, and R. Noyori: Asymmetric Hydrogenation of Cyclic α,β-Unsaturated Ketones to Chiral Allylic Alcohols. *Synlett* **1997**, 467 – 468.

[305] T. Ohkuma, H. Doucet, T. Pham, K. Mikami, T. Korenaga, M. Terada, and R. Noyori: Asymmetric Activation of Racemic Ruthenium(II) Complexes for Enantioselective Hydrogenation. *J. Am. Chem. Soc.* **1998**, *120*, 1086 – 1087.

[306] K. Mikami, T. Korenaga, M. Terada, T. Ohkuma, T. Pham und R. Noyori: Konformativ flexible Biphenyl-Phosphan-liganden für Ru-katalysierte enantioselektive Hydrierungen. *Angew. Chem.* **1999**, *111*, 517 – 519; K. Mikami, T. Korenaga, M. Terada, T. Ohkuma, T. Pham, and R. Noyori: Conformationally Flexible Biphenyl-phosphane Ligands for Ru-Catalyzed Enantioselective Hydrogenation. *Angew. Chem. Int. Ed.* **1999**, *38*, 495 – 497.

[307] K. Mikami, T. Korenaga, T. Ohkuma, and R. Noyori: Asymmetric Activation / Deactivation of Racemic Ru Catalysts for Highly Enantioselective Hydrogenation of Ketonic Substrates. *Angew. Chem. Int. Ed.* **2000**, *39*, 3707 – 3710.

[308] T. Ohkuma, H. Ooka, M. Yamakawa, T. Ikariya, and R. Noyori: Stereoselective Hydrogenation of Simple Ketones Catalyzed by Ruthenium(II) Complexes. *J. Org. Chem.* **1996**, *61*, 4872 – 4873.

[309] T. Ohkuma, J. Li, and R. Noyori: Asymmetric Hydrogenation of 2-Arylated Cycloalkanones through Dynamic Kinetic Resolution. *Synlett* **2004**, *8*, 1383 – 1386.

[310] T. Ohkuma, T. Hattori, H. Ooka, T. Inoue, and R. Noyori: BINAP / 1,4-Diamine-Ruthenium(II) Complexes for Efficient Asymmetric Hydrogenation of 1-Tetralones and Analogues. *Org. Lett.* **2004**, *6*, 2681 – 2683.

[311] T. Ohkuma, C. A. Sandoval, R. Srinivasan, Q. Lin, Y. Wei, K. Muñiz, and R. Noyori: Asymmetric Hydrogenation of *tert*-Alkyl Ketones. *J. Am. Chem. Soc.* **2005**, *127*, 8288 – 8289.

[312] Ryoji Noyori (Profile), *Green Chem.* **2003**, *5*, G37 – G39.

[313] R. Noyori: Organic Synthesis: Current Status in Japan. *Science* **1992**, *258*, 584 – 585.

[314] T. Ohkuma, H. Ooka, T. Ikariya, and R. Noyori: Preferential Hydrogenation of Aldehydes and Ketones. *J. Am. Chem. Soc.* **1995**, *117*, 10417 – 10418.

[315] P. G. Jessop, T. Ikariya, and R. Noyori: Homogeneous Hydrogenation of Carbon Dioxide. *Chem. Rev.* **1995**, *95*, 259 – 272.

[316] P. G. Jessop, T. Ikariya, and R. Noyori: Homogeneous Catalysis in Supercritical Fluids. *Science* **1995**, *269*, 1065 – 1069.

[317] P. G. Jessop, T. Ikariya, and R. Noyori: Homogeneous Catalytic Hydrogenation of Supercritical Carbon Dioxide. *Nature* **1994**, *368*, 231 – 233.

[318] P. G. Jessop, Y. Hsiao, T. Ikariya, and R. Noyori: Methyl Formate Synthesis by Hydrogenation of Supercritical Carbon Dioxide in the Presence of Methanol. *J. Chem. Soc. Chem. Commun.* **1995**, 707 – 708.

[319] P. G. Jessop, Y. Hsiao, T. Ikariya, and R. Noyori: Catalytic Production of Dimethylformamide from Supercritical Carbon Dioxide. *J. Am. Chem. Soc.* **1994**, *116*, 8851 – 8852.

[320] P. G. Jessop, Y. Hsiao, T. Ikariya, and R. Noyori: Homogeneous Catalysis in Supercritical Fluids: Hydrogenation of Supercritical Carbon Dioxide to Formic Acid, Alkyl Formates, and Formamides. *J. Am. Chem. Soc.* **1996**, *118*, 344 – 355.

[321] R. Noyori: Supercritical Fluids: Introduction. *Chem. Rev.* **1999**, *99*, 353 – 354.

[322] T. Ikariya and R. Noyori: Organic Reactions in Supercritical Fluids in *Transition Metal Catalysed Reactions,* IUPAC Chemistry for the 21st Century Ser. (Eds.: S.-I. Murahashi and S. G. Davies). Blackwell Science, **1999**, 1 – 28.

[323] J. Xiao, S. C. A. Nefkens, P. G. Jessop, T. Ikariya, and R. Noyori: Asymmetric Hydrogenation of α,β-Unsaturated Carboxylic Acids in Supercritical Carbon Dioxide. *Tetrahedron Lett.* **1996**, *37*, 2813 – 2816.

[324] P. G. Jessop, T. Ikariya, and R. Noyori: Selectivity for Hydrogenation or Hydroformylation of Olefins by Hydridopentacarbonylmanganese(I) in Supercritical Carbon Dioxide. *Organometallics* **1995**, *14*, 1510 – 1513.

[325] Y. Kayaki, Y. Noguchi, S. Iwasa, T. Ikariya, and R. Noyori: An Efficient Carbonylation of Aryl Halides Catalysed by Palladium Complexes with Phosphite Ligands in Supercritical Carbon Dioxide. *Chem. Commun.* **1999**, 1235 – 1236.

[326] P. G. Jessop, T. Ikariya, and R. Noyori: Homogenous Catalysis in Supercritical Fluids. *Chem. Rev.* **1999**, *99*, 475 – 493.

[327] R. Noyori and T. Ikariya: Supercritical Fluids for Organic Synthesis in *Stimulating Concepts in Chemistry* (Eds: F. Vögtle, J. F. Stoddart, and M. Shibasaki). Wiley-VCH, New York, **2000**, 13 – 24.

[328] R. Noyori, M. Aoki, and K. Sato: Green Oxidation with Aqueous Hydrogen Peroxide. *Chem. Commun.* **2003**, 1977 – 1986.

[329] K. Sato, M. Aoki, J. Takagi, and R. Noyori: Organic Solvent- and Halide-Free Oxidation of Alcohols with Aqueous Hydrogen Peroxide. *J. Am. Chem. Soc.* **1997**, *119*, 12386 – 12387.

[330] K. Sato, J. Takagi, M. Aoki, and R. Noyori: Hydrogen Peroxide Oxidation of Benzylic Alcohols to Benzaldehydes and Benzoic Acids Under Halide-Free Conditions. *Tetrahedron Lett.* **1998**, *39*, 7549 – 7552.

[331] K. Sato, M. Aoki, J. Takagi, K. Zimmermann, and R. Noyori: A Practical Method for Alcohol Oxidation with Aqueous Hydrogen Peroxide under Organic Solvent- and Halide-Free Conditions. *Bull. Chem. Soc. Jpn.* **1999**, *72*, 2287 – 2306.

[332] K. Sato, M. Hyodo, J. Takagi, M. Aoki, and R. Noyori: Hydrogen Peroxide Oxidation of Aldehydes to Carboxylic Acids: An Organic Solvent-, Halide- and Metal-Free Procedure. *Tetrahedron Lett.* **2000**, *41*, 1439 – 1442.

[333] K. Sato, M. Hyodo, M. Aoki, X.-Q. Zheng, and R. Noyori: Oxidation of Sulfides to Sulfoxides and Sulfones with 30% Hydrogen Peroxide Under Organic Solvent- and Halogen-Free Conditions. *Tetrahedron* **2001**, *57*, 2469 – 2576.

[334] K. Sato, M. Aoki, M. Ogawa, T. Hashimoto, and R. Noyori: A Practical Method for Epoxidation of Terminal Olefins with 30% Hydrogen Peroxide under Halide-Free Conditions. *J. Org. Chem.* **1996**, *61*, 8310 – 8311.

[335] K. Sato, M. Aoki, M. Ogawa, T. Hashimoto, D. Panyella, and R. Noyori: A Halide-Free Method for Olefin Epoxidation with 30% Hydrogen Peroxide. *Bull. Chem. Soc. Jpn.* **1997**, *70*, 905 – 915.

[336] K. Sato, M. Aoki, and R. Noyori: A "Green" Route to Adipic Acid: Direct Oxidation of Cyclohexenes with 30 percent Hydrogen Peroxide. *Science* **1998**, *281*, 1646 – 1647.

[337] W. S. Knowles and R. Noyori: Pioneering Perspectives on Asymmetric Hydrogenation. *Acc. Chem. Res.* **2007**, *40*, 1238 – 1239.

[338] R. Noyori: Fakten sind der Feind der Wahrheit – Betrachtungen über Zufallsentdeckungen und ungeahnte Entwicklungen der asymmetrischen Katalyse. *Angew. Chem.* **2013**, *125*, 83 – 98; R. Noyori: Facts are the Enemy of Truth – Reflections on Serendipitous Discovery and Unforeseen Developments in Asymmetric Catalysis. *Angew. Chem. Int. Ed.* **2013**, *52*, 79 – 92.

[339] R. Noyori: *Research Should Be Fresh, Simple, and Clear.* (Lives in Chemistry.) GNT Publishing, Berlin, **2025**.

TABLE OF CONTENTS

Table of contents

LINKS
SOURCES
INDEX

 DOI.ORG/10.47261/1564-A

Links and literature

With this section of the appendix we wish to stimulate interactive reading and additional browsing. We have compiled a set of links to authors, people, organizations, awards, and platforms that provide additional information. They should inspire thoughts and satisfy your curiosity: Who has also been awarded these prizes, how do I submit a proposal, what are the missions and activities of these organizations, how can I contact the author, can I become a member of an organization, etc.?

1. Book Series & Feedback

[1] For questions and feedback to the author, please use this address
rnoyori@l-i-c.org

[2] For questions, suggestions, and feedback to the Advisory Board and publisher, please use
https://l-i-c.org/1135#feedback

[3] For further information and supporting information on this book, and all other books of this series, please look at the homepage
https://l-i-c.org/

[4] As well as the homepage of the Division of History of Chemistry of the GDCh
https://en.gdch.de/network-structures/gdch-structures/history-of-chemistry.html

[5] For more lives in chemistry then and now, please follow us on social media
https://bsky.app/profile/livesinchem.bsky.social
https://www.linkedin.com/in/livesinchem/
https://x.com/livesinchem

[6] Supporting Material
https://l-i-c.org/1135#supporting

2. Knowledge

[7] Yasu Furukawa: Exploring the History of Chemistry in Japan. *Ambix* **2021**, *28*, 302 – 317.
https://doi.org/10.1080/00026980.2021.1930675

[8] ACS "Voices of Inorganic Chemistry" interview: Richard R. Schrock (**2011**)
https://youtu.be/pdNFgXBMNm4

[9] The New Kyoto School
https://medium.com/@masatoshihonda/the-new-kyoto-school-bridging-tradition-and-modernity-in-japanese-thought-8872db5a3d78

3. Persons

[10] Elias James Corey (* 1928)
https://www.nobelprize.org/prizes/chemistry/1990/corey/facts/

[11] Kenichi Fukui (1918 – 1998)
https://www.nobelprize.org/prizes/chemistry/1981/fukui/facts/

[12] Gen-itsu Kita (1883 – 1952)
https://doi.org/10.1142/9781786344557_0007

[13] Hitosi Nozaki (1922 – 2019)
https://en.wikipedia.org/wiki/Hitoshi_Nozaki

[14] Hisashi Yamamoto (* 1943)
https://web.archive.org/web/20091114081542/http://yamamotogroup.uchicago.edu/

4. Institutions & Companies

[15] ERATO – Exploratory Research for Advanced Technology research funding program
https://www.jst.go.jp/erato/en/

[16] OMCOS, Organometallic Catalysis Directed Towards Organic Synthesis
https://omcos22.org/

[17] Kyoto University
https://www.kyoto-u.ac.jp/en

[18] RIKEN
https://www.riken.jp/en/

[19] The Japan Academy
https://www.japan-acad.go.jp/en/

5. Scientific Societies

[20] American Chemical Society (ACS)
https://www.acs.org/

[21] German Chemical Society (GDCh)
https://www.gdch.de/

[22] Royal Society of Chemistry (RSC)
https://www.rsc.org/

[23] The Chemical Society of Japan (CSJ)
https://www.chemistry.or.jp/en/

[24] The Society of Synthetic Organic Chemistry, Japan (SSOCJ)
https://www.ssocj.jp/en/

6. Awards

[25] Nobel Prize in Chemistry 2001
https://www.nobelprize.org/prizes/chemistry/2001/summary/

[26] Ryoji Noyori Prize
https://www.ssocj.jp/en/award/ryoji-noyori-prize/

[27] Ryoij Noyori ACES Award
https://aces.onlinelibrary.wiley.com/hub/journal/1861471x/noyori-aces-award-call

[28] Nagoya Gold Medal
https://www.msd-life-science-foundation.or.jp/en/symp/nagoya

7. Publications

[29] ACS Publications
https://pubs.acs.org/

[30] CSJ Publications
https://www.csj.jp/csj-en/journals/publications.html

[31] GDCh Publications
https://en.gdch.de/publications.html

[32] RSC Publications
https://pubs.rsc.org/

8. From Idea to Success

Karl Reuter initiated and supported the series "Lives in Chemistry". From 1974 to 1981 he studied chemistry at the Giessen University, joined Günther Maier's group, and received his PhD in 1985. He started his career in the chemical industry first and soon founded his own company (Reuter Chemischer Apparatebau e. K.) in Freiburg, Germany, in 1995. Karl Reuter has always been convinced that creation of knowledge demands sound research and intense lab work as well as bold ideas. The Advisory Board is greatly indebted to him and would like to thank him for his trust, leadership, and financial support.

[33] Reuter Chemische Apparatebau e.K., Dr. Karl Reuter, Freiburg i. Br.
http://www.rca-separations.com/

[34] Günther Maier, the starting point of LiC
https://l-i-c.org/1125

Image sources

American Chemical Society: Figs. 4.5., 4.9., 4.10.
Fukui Institute for Fundamental Chemistry, Kyoto University: Fig. 2.7.
Gesellschaft Deutscher Chemiker (GDCh), Bildarchiv: Fig. 8.7.
GNT Publishing: Fig. A.5. (Idea: Eva E. Wille; editing: Eva E. Wille with Ryoji Noyori; design: Andreas Töpfer).
Lawrence Berkeley Laboratory, University of California: Fig. 6.2.
Masaaki Suzuki: Fig. 5.35.
Masashi Yamakawa: Fig. 6.22. (b).
Nagoya University: Figs. 3.1., 9.12. – 9.31.
Noyori Private Archive: Figs. 1.1. – 1.8., 2.1. – 2.6., 2.8. – 2.10., 2.12. – 2.15., 3.2. – 3.8., 4.2. – 4.4., 4.6., 4.7., 4.8., 5.1. – 5.34., 5.36. – 5.62., 5.64. – 5.74., 6.5. – 6.22. (a), 6.23. – 6.45., 7.1. – 7.48., 7.50., 7.51., 8.1., 8.4., 8.5., 8.8., 8.9., 9.1. – 9.11., 9.32. – 9.36., 9.60. – 9.65., 9.77., 9.78., 9.80., A.1. – A.4., A.6. – A.8.
Ono Pharmaceutical Co., Ltd.: Fig. 4.1.
Pressens Bild, Stockholm: Figs. 8.2., 8.6.
RIKEN: Figs. 9.66. – 9.76.
Scripta Scientium / Ralf Hahn: Figs. 6.1., 6.3., 7.49.
Sumitomo Chemical Company, Limited: 9.81.
The Society of Synthetic Organic Chemistry, Japan: Figs. 9.37. – 9.59.
The Yomiuri Shimbun: Fig. 8.3.
United Nations: Fig. 9.79.
Wikimedia Commons: Figs. 2.11., 5.63., 6.4.

Unfortunately, the rights holders of some images could not be determined.
Possible rights holders should contact the publisher.

Index

H

I

J

K

L

M

N

REACTIONS
THE SERIES

Reactions from our readers

These books are most impressive: From the material of the cover to the typographical detail each of the autobiographies is a true marvel that do not only appeal to the expert readers.

Silvia Werfel, Leipzig, in *bindereport*

The Maier book came. It feels good in the hand, and falls open to read, The typography, color, and space are beautifully done, and the coordination of the elements visible on every point. Chemical formulas are very legible and appropriate in contrast. Congratulations to Andreas Töpfer for a beautiful design job.

Roald Hoffmann, Ithaca, USA

The autobiography of Gerhard Ertl is a page-turner; it makes excellent reading for everybody interested in the history of science and certainly for those who want to know more about this Nobel Laureate.

Melanie Schnell, Kiel, in *Angewandte Chemie*

It's a delight to read Henri Brunner's autobiography. I had only briefly met him before, and this book gives me deep insight into his achievements and personality. Furthermore I learned a lot about the chemistry departments of the famous Munich universities, thrilling.

Sonja Herres-Pawlis, Aachen, in an e-mail to Eva E. Wille, Weinheim

Dieter Oesterhelt had discovered early in his career something fundamentally new: Bacteriorhodopsin (BR) that allows bacteria to use sunlight for energy conversion. BR plays a central role in Oesterhelt's autobiography that is part of the wonderful series "Lives in Chemistry"; this autobiography is presented in interview form with Mathias Grote as interviewer.

Ferdinand Hucho, Berlin, in *BIOspektrum*

Steve Kent's autobiography provides an insight into the achievements of this protein chemistry hero and it unveils his motivation for a lifelong "search for truth". The writing is vivid and lucid, the book is a wonderful addition to the bibliophilic designed series "Lives in Chemistry".

Ferdinand Hucho, Berlin, in *BIOspektrum*

Titles in this series

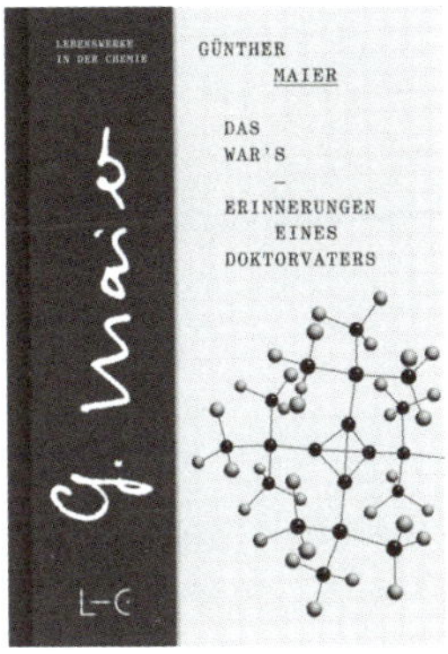

312 PP.
170 FIGS.
39,80 €
L-I-C.ORG/1125

172 PP.
85 FIGS.
34,80 €
L-I-C.ORG/1126

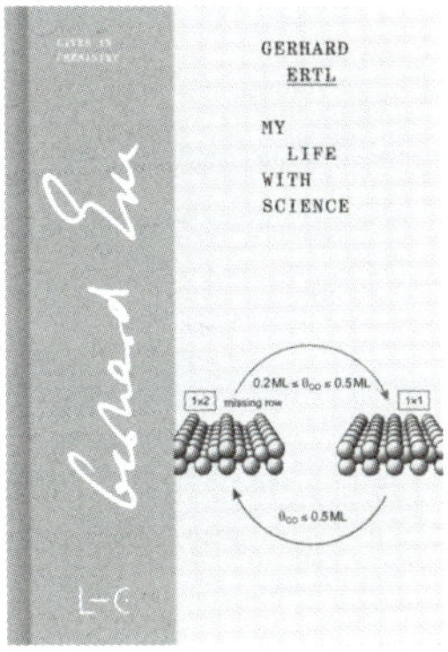

196 PP.
98 FIGS.
39,80 €
L-I-C.ORG/1131

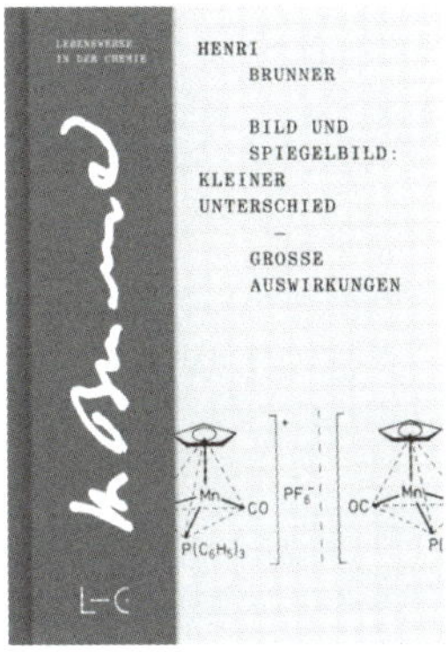

296 PP.
261 FIGS.
39,80 €
L-I-C.ORG/1127

288 PP.
82 FIGS.
39,80 €
L-I-C.ORG/1128

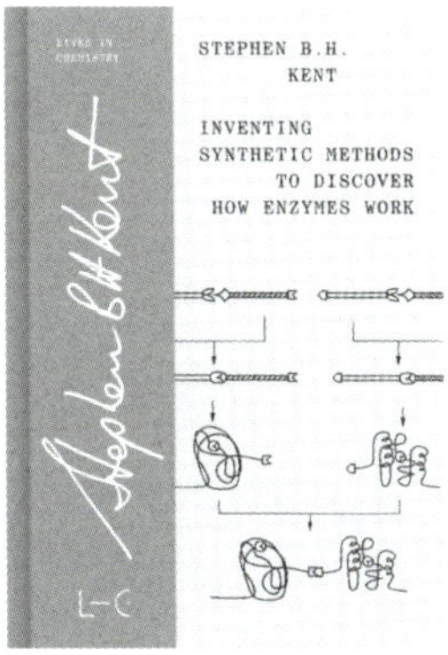

336 PP.
267 FIGS.
39,80 €
L-I-C.ORG/1129

268 PP.
231 FIGS.
39,80 €
L-I-C.ORG/1130

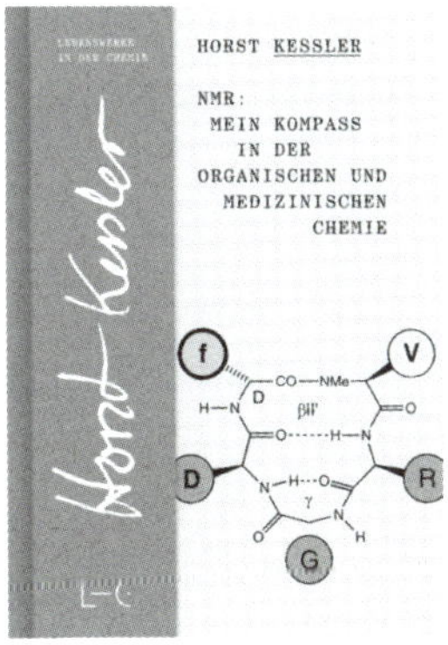

300 PP.
239 FIGS.
39,80 €
L-I-C.ORG/1132

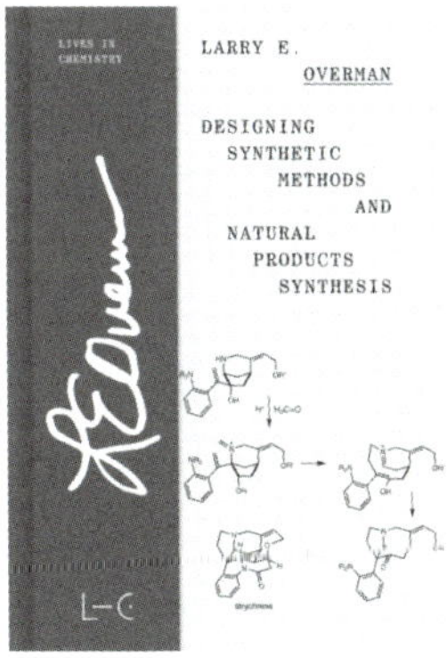

247 PP.
225 FIGS.
39,80 €
L-I-C.ORG/1133

The books in the series "Lives in Chemistry" impress with a clear structure, an attractive design, and a reader-friendly approach. Practical relevance is a common theme in Franz Effenberger's research and other activities. He combined the pleasures of chemical synthesis with a keen interest in fundamental structures and mechanisms.

Dieter Jahn, Hamburg, in *Nachr. Chem.*

Conveying specialized knowledge to non-specialists is a balancing act between precision and comprehensibility, and Horst Kessler has mastered this perfectly. The book is a recommended reading for everyone interested in chemical research in the broadest sense. It has great potential to inspire current and future generations of scientists to pursue bold and independent research.

Thorsten Berg, Leipzig, in *Nachr. Chem.*

This remarkable autobiography serves as a prime example that progress in the chemical and biochemical sciences can only be made if one applies new methodological developments to relevant problems. Horst Kessler's autobiography is recommended for PhD students as well as for all organic and medicinal chemists and biochemists. For the chemical science community this is a book filled with a wealth of information from the period when chemistry and biochemistry reconverged.

Martin Engelhard, Dortmund, in *BIOspektrum*

Larry Overman's autobiography is a must-read for its timeless insights on building and leading a research group. This book is both inspiring and highly relevant for contemporary professionals.

Vera Koester in *ChemistryViews*

The Katharina Kohse-Höinghaus book is a great biography that is a real pleasure to read and can be highly recommended to all aspiring female – but not only female – scientists.

Stefanie Dehnen, Karlsruhe, in her *first reader reaction*

Lives in Chemistry: Rethinking again and again!

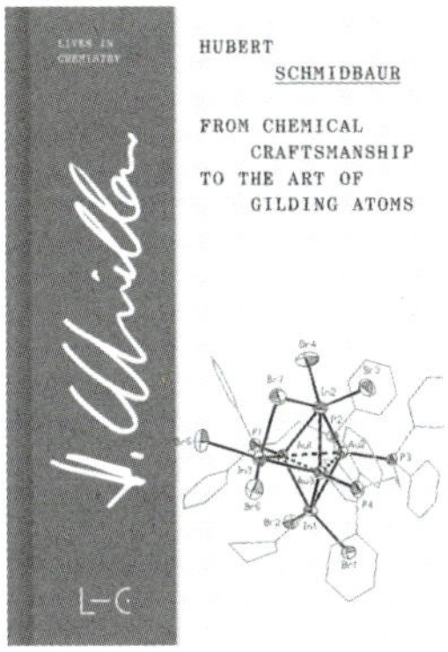

282 PP.
266 FIGS.
39,80 €
L-I-C.ORG/1134

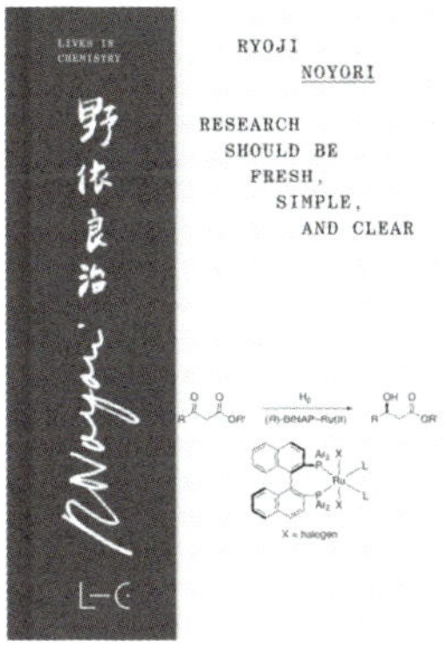

346 PP.,
316 FIGS.
39,80 €
L-I-C.ORG/1135

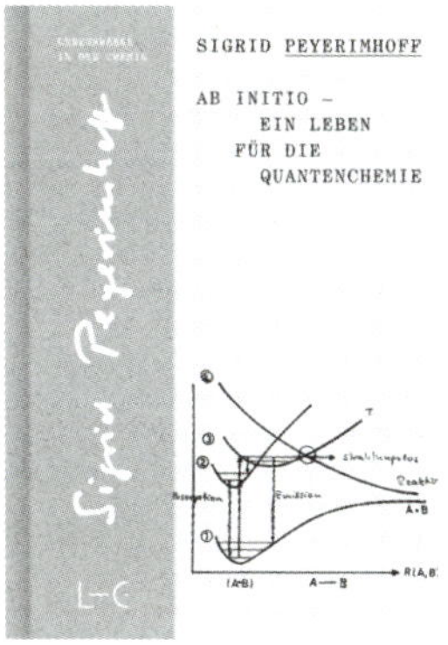

202 PP.
80 FIGS.
39,80 €
L-I-C.ORG/1138

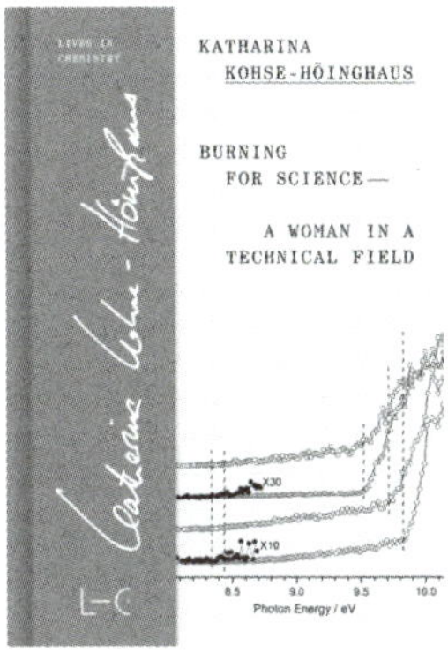

314 PP.
227 FIGS.
39,80 €
L-I-C.ORG/1137

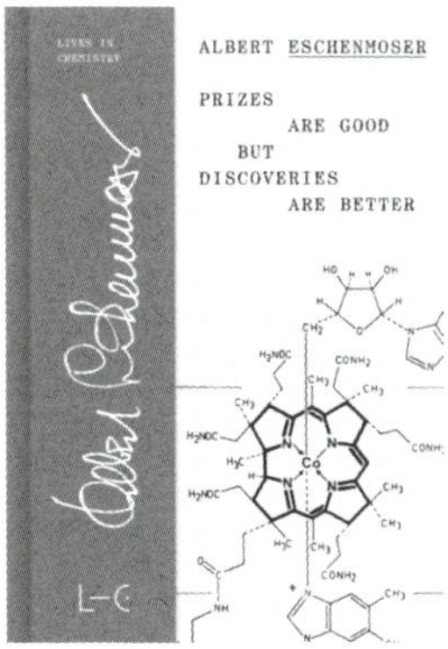

346 PP.
332 FIGS.
39,80 €
L-I-C.ORG/1136

314 PP.
230 FIGS.
39,80 €
L-I-C.ORG/1139

FOLLOW US
BSKY.APP/PROFILE/LIVESINCHEM.BSKY.SOCIAL
INSTAGRAM.COM/LIVESINCHEM
LINKEDIN.COM/IN/LIVESINCHEM
X.COM/LIVESINCHEM
YOUTUBE.COM/@LIVESINCHEM